CUSTOMIZED COMPLETE STATISTICAL PHYSICS

BERKELEY PHYSICS COURSE
VOLUME 5

F. REIF
Professor of Physics
University of California, Berkeley

The McGraw-Hill Companies, Inc
Primis Custom Publishing

New York St. Louis San Francisco Auckland Bogota
Caracas Lisbon London Madrid Mexico Milan Montreal
New Delhi Paris San Juan Singapore Sydney Tokyo Toronto

McGraw-Hill

A Division of The ***McGraw-Hill*** *Companies*

McGraw-Hill's College Custom Series consists of products that are produced from camera-ready copy. Peer review, class testing, and accuracy are primarily the responsibility of the author(s).

2 3 4 5 6 7 8 9 QSR QSR 9 8 7 6 5 4 3 2

ISBN 0-07-038662-5

Preface to the Berkeley Physics Course

This is a two-year elementary college physics course for students majoring in science and engineering. The intention of the writers has been to present elementary physics as far as possible in the way in which it is used by physicists working on the forefront of their field. We have sought to make a course which would vigorously emphasize the foundations of physics. Our specific objectives were to introduce coherently into an elementary curriculum the ideas of special relativity, of quantum physics, and of statistical physics.

This course is intended for any student who has had a physics course in high school. A mathematics course including the calculus should be taken at the same time as this course.

There are several new college physics courses under development in the United States at this time. The idea of making a new course has come to many physicists, affected by the needs both of the advancement of science and engineering and of the increasing emphasis on science in elementary schools and in high schools. Our own course was conceived in a conversation between Philip Morrison of Cornell University and C. Kittel late in 1961. We were encouraged by John Mays and his colleagues of the National Science Foundation and by Walter C. Michels, then the Chairman of the Commission on College Physics. An informal committee was formed to guide the course through the initial stages. The committee consisted originally of Luis Alvarez, William B. Fretter, Charles Kittel, Walter D. Knight, Philip Morrison, Edward M. Purcell, Malvin A. Ruderman, and Jerrold R. Zacharias. The committee met first in May 1962, in Berkeley; at that time it drew up a provisional outline of an entirely new physics course. Because of heavy obligations of several of the original members, the committee was partially reconstituted in January 1964, and now consists of the undersigned. Contributions of others are acknowledged in the prefaces to the individual volumes.

The provisional outline and its associated spirit were a powerful influence on the course material finally produced. The outline covered in detail the topics and attitudes which we believed should and could be taught to beginning college students of science and engineering. It was never our intention to develop a course limited to honors students or to students with advanced standing. We have sought to present the principles of physics from fresh and unified viewpoints, and parts of the course may therefore seem almost as new to the instructor as to the students.

The five volumes of the course as planned will include:

I. Mechanics (Kittel, Knight, Ruderman)
II. Electricity and Magnetism (Purcell)
III. Waves (Crawford)
IV. Quantum Physics (Wichmann)
V. Statistical Physics (Reif)

The authors of each volume have been free to choose that style and method of presentation which seemed to them appropriate to their subject.

The initial course activity led Alan M. Portis to devise a new elementary physics laboratory, now known as the Berkeley Physics Laboratory. Because the course emphasizes the principles of physics, some teachers may feel that it does not deal sufficiently with experimental physics. The laboratory is rich in important experiments, and is designed to balance the course.

The financial support of the course development was provided by the National Science Foundation, with considerable indirect support by the University of California. The funds were administered by Educational Services Incorporated, a nonprofit organization established to administer curriculum improvement programs. We are particularly indebted to Gilbert Oakley, James Aldrich, and William Jones, all of ESI, for their sympathetic and vigorous support. ESI established in Berkeley an office under the very competent direction of Mrs. Mary R. Maloney to assist the development of the course and the laboratory. The University of California has no official connection with our program, but it has aided us in important ways. For this help we thank in particular two successive Chairman of the Department of Physics, August C. Helmholz and Burton J. Moyer; the faculty and nonacademic staff of the Department; Donald Coney, and many others in the University. Abraham Olshen gave much help with the early organizational problems.

Your corrections and suggestions will always be welcome.

Eugene D. Commins
Frank S. Crawford, Jr.
Walter D. Knight
Philip Morrison
Alan M. Portis
Edward M. Purcell
Frederick Reif
Malvin A. Ruderman
Eyvind H. Wichmann
Charles Kittel, *Chairman*

January, 1965
Berkeley, California

Preface to Volume V

This last volume of the Berkeley Physics Course is devoted to the study of large-scale (i.e., *macro*scopic) systems consisting of many atoms or molecules; thus it provides an introduction to the subjects of statistical mechanics, kinetic theory, thermodynamics, and heat. The approach which I have followed is not patterned upon the historical development of these subjects and does not proceed along conventional lines. My aim has been rather to adopt a modern point of view and to show, in as systematic and simple a way as possible, how the basic notions of atomic theory lead to a coherent conceptual framework capable of describing and predicting the properties of macroscopic systems.

In writing this book I have tried to keep in mind a student who, unencumbered by any prior familiarity with the subject matter, is encountering it for the first time from the vantage point of his previous study of elementary physics and atomic properties. I have therefore chosen an order of presentation which might seem well-motivated to such a student if he attempted to discover by himself how to gain an understanding of macroscopic systems. In seeking to make the presentation coherent and unified, I have based the entire discussion upon the systematic elaboration of a single principle, the tendency of an isolated system to approach its situation of greatest randomness. Although I have restricted my attention to simple systems, I have treated them by methods which are widely applicable and easily generalized. Above all, I have tried throughout to stress physical insight, the ability to perceive significant relationships quickly and simply. Thus I have attempted to discuss physical ideas at length without getting lost in mathematical formalism, to provide simple examples to illustrate abstract general concepts, to make numerical estimates of significant quantities, and to relate the theory to the real world of observation and experiment.

The subject matter to be covered in this volume had to be selected with great care. My intention has been to emphasize the most fundamental concepts which would be useful to physicists as well as to students of chemistry, biology, or engineering. The Teaching and Study Notes summarize the organization and contents of the book and provide some guides for the prospective teacher and student. The unconventional order of presentation, designed to stress the relation between the macroscopic and atomic levels of description, does not necessarily sacrifice the virtues inherent in more traditional approaches. In particular, the following features may be worth mentioning:

(i) The student who completes Chap. 7 (even if he omits Chap. 6) will know the fundamental principles and basic applications of classical thermodynamics as well as if he had studied the subject along traditional lines. Of course, he will also have acquired much more insight into the meaning of entropy and a considerable knowledge of statistical physics.

(ii) I have been careful to emphasize that the statistical theory leads to certain results which are purely macroscopic in content and completely independent of whatever models one might assume about the atomic structure of the systems under consideration. The generality and model-independence of the classical thermodynamic laws is thus made explicitly apparent.

(iii) Although a historical approach rarely provides the most logical or illuminating introduction to a subject, an acquaintance with the evolution of scientific ideas is both interesting and instructive. I have, therefore, included in the text some pertinent remarks, references, and photographs of prominent scientists, all designed to give the student some perspective concerning the historical development of the subject.

The prerequisites needed for a study of this volume include, besides a rudimentary knowledge of classical mechanics and electromagnetism, only an acquaintance with the simplest atomic concepts and with the following quantum ideas in their most unsophisticated form: the meaning of quantum states and energy levels, Heisenberg's uncertainty principle, the de Broglie wavelength, the notion of spin, and the problem of a free particle in a box. The mathematical tools required do not go beyond simple derivatives and integrals, plus an acquaintance with Taylor's series. Any student familiar with the essential topics covered in the preceding volumes of the Berkeley Physics Course (particularly in Vol. IV) would, of course, be amply prepared for the present book. The book could be used equally well, however, for the last part of any other modern introductory physics course or for any comparable course at, or above, the level of second-year college students.

As I indicated at the beginning of this preface, my aim has been to penetrate the essence of a sophisticated subject sufficiently to make it seem simple, coherent, and easily accessible to beginning students. Although the goal is worth pursuing, it is difficult to attain. Indeed, the writing of this book was for me an arduous and lonely task that consumed an incredible amount of time and left me feeling exhausted. It would be some slight compensation to know that I had achieved my aim sufficiently well so that the book would be found useful.

F. Reif

Acknowledgments

I am grateful to Professor Allan N. Kaufman for reading the final manuscript critically and for always being willing to let me have the benefit of his opinions. Professors Charles Kittel and Edward M. Purcell made valuable comments concerning an earlier version of the first couple of chapters. Among graduate students, I should like to mention Richard Hess, who made many helpful observations about the preliminary edition of this volume, and Leonard Schlessinger, who worked out complete solutions for the problems and who provided the answers listed at the end of the book. I feel especially indebted to Jay Dratler, an undergraduate student who read both the preliminary edition and a substantial portion of the final manuscript. Starting out unfamiliar with the subject matter, he learned it himself from the book and exhibited in the process a fine talent for detecting obscurities and making constructive suggestions. He is probably the person who has contributed most significantly toward the improvement of the book.

The making of the computer-constructed pictures took an appreciable amount of time and effort. I wish, therefore, to express my warmest thanks to Dr. Berni J. Alder who helped me enormously in this task by personal cooperation uncontaminated by financial compensation. My ideas about these pictures could never have come to fruition if he had not put his computing experience at my disposal. We hope to continue our collaboration in the future by making available some computer-constructed movies which should help to illustrate the same ideas in more vivid form.

Mrs. Beverly Sykes, and later on Mrs. Patricia Cannady, were my loyal secretaries during the long period that I was occupied with this book. I feel very much indebted to them for their skill in deciphering and typing its successive handwritten versions. I also owe thanks to several other persons for their assistance in the production of the book. Among these are Mrs. Mary R. Maloney and Mrs. Lila Lowell, who were always willing to help with miscellaneous chores, and Mr. Felix Cooper, who is responsible for the execution of the artwork. Finally, I am grateful to Mr. William R. Jones, of Educational Services, Inc., for his efforts in handling relations with the National Science Foundation.

This volume owes an enormous debt to *Fundamentals of Statistical and Thermal Physics* (FSTP), my earlier book published by McGraw-Hill in 1965, which represented an attempt at educational innovation at the more advanced level of upper-division college students. My extensive experience derived from FSTP, and many details of presentation, have been incorporated in the present volume.† I wish, therefore, to express my gratitude to

† Some portions of the present volume are thus also subject to the copyright provisions of *Fundamentals of Statistical and Thermal Physics.*

those individuals who assisted me during the writing of FSTP as well as to those who have provided me with constructive criticisms since its publication. I should also like to thank the McGraw-Hill Book Company for disregarding the conflicting copyright provisions to give me unrestricted permission to include material from FSTP in the present volume. Although I am not dissatisfied with the general approach developed in FSTP, I have come to recognize that the exposition there could often have been simpler and more penetrating. I have, therefore, made use of these new insights to include in the present volume all the improvements in organization and wording intended for a second edition of FSTP. By virtue of its similar point of view, FSTP may well be a useful reference for students interested in pursuing topics beyond the level of the present book; such students should, however, be cautioned to watch out for certain changes of notation.

Although the present volume is part of the Berkeley Physics Course project, it should be emphasized that the responsibility for writing this volume has been mine alone. If the book has any flaws (and I myself am aware of some even while reading proof), the onus for them must, therefore, rest upon my own shoulders.

Teaching and Study Notes

Organization of the Book

The book is divided into three main parts which I shall describe in turn:

Part A: Preliminary Notions (Chapters 1 and 2)

Chapter 1: This chapter provides a qualitative introduction to the most fundamental physical concepts to be explored in this book. It is designed to make the student aware of the characteristic features of macroscopic systems and to orient his thinking along fruitful lines.

Chapter 2: This chapter is somewhat more mathematical in nature and is intended to familiarize the student with the basic notions of probability theory. No prior knowledge of probability ideas is assumed. The concept of ensembles is stressed throughout and all examples are designed to illuminate physically significant situations. Although this chapter is oriented toward subsequent applications in the remainder of the book, the probability concepts discussed are, of course, intended to be useful to the student in far wider contexts.

These chapters should not take too much time. Indeed, some students may well have sufficient background preparation to be familiar with some of the material in these chapters. Nevertheless, I would definitely recommend that such students *not* skip these two chapters, but that they consider them a useful review.

Part B: Basic Theory (Chapters 3, 4, and 5)

This part constitutes the heart of the book. The logical and quantitative development of the subject of this volume really starts with Chapter 3. (In this sense the first two chapters could have been omitted, but this would have been very unwise pedagogically.)

Chapter 3: This chapter discusses how a system consisting of many particles is described in statistical terms. It also introduces the basic postulates of the statistical theory. By the end of this chapter the student should have come to realize that the quantitative understanding of macroscopic systems hinges essentially on considerations involving the counting of the states accessible to the systems. He may not yet perceive, however, that this insight has much useful value.

Chapter 4: This chapter constitutes the real pay-off. It starts out, innocently enough, by investigating how two systems interact by heat transfer alone. This investigation leads very quickly, however, to the fundamental concepts of entropy, of absolute temperature, and of the canonical distribution (or Boltzmann factor). By the end of the chapter the student is in a position to deal with thoroughly practical problems—indeed, he has learned how to calculate from first principles the paramagnetic properties of a substance or the pressure of an ideal gas.

Chapter 5: This chapter brings the ideas of the theory completely down to earth. Thus it discusses how one relates atomic concepts to macroscopic measurements and how one determines experimentally such quantities as absolute temperature or entropy.

An instructor thoroughly pressed for time can stop at the end of these five chapters without too many pangs of conscience. At this point a student should have acquired a fairly good understanding of absolute temperature, entropy, and the Boltzmann factor—i.e., of the most fundamental concepts of statistical mechanics and thermodynamics. (Indeed, the only thermodynamic result still missing concerns the fact that the entropy remains unchanged in a quasi-static adiabatic process.) At this stage I would consider the minimum aims of the course to have been fulfilled.

Part C: Elaboration of the Theory (Chapters 6, 7, and 8)

This part consists of three chapters which are independent of each other in the sense that any one of them can be taken up without requiring the others as prerequisites. In addition, it is perfectly possible to cover only the first few sections of any one of these chapters before turning to another of these chapters. Any instructor can thus use this flexibility to suit his own predilections or the interests of his students. Of the three chapters, Ch. 7 is the one of greatest fundamental importance in rounding out the theory; since it completes the discussion of thermodynamic principles, it is also the one likely to be most useful to students of chemistry or biology.

Chapter 6: This chapter discusses some particularly important applications of the canonical distribution by introducing approximate classical notions into the statistical description. The Maxwell velocity distribution of molecules in a gas and the equipartition theorem are the main topics of this chapter. Illustrative applications include molecular beams, isotope separation, and the specific heat of solids.

Chapter 7: This chapter begins by showing that the entropy remains unchanged in a process which is adiabatic and quasi-static. This completes the discussion of the thermodynamic laws which are then summarized in their full generality. The chapter then proceeds to examine a few important applications: general equilibrium conditions, including properties of the Gibbs free energy; equilibrium between phases; and implications for heat engines and biological organisms.

Chapter 8: This last chapter is intended to illustrate the discussion of the nonequilibrium properties of a system. It treats the transport properties of a dilute gas by the simplest mean-free-path arguments and deals with viscosity, thermal conductivity, self-diffusion, and electrical conductivity.

This completes the description of the essential organization of the book. In the course as taught at Berkeley, the aim is to cover the major part of this book in about eight weeks of the last quarter of the introductory physics sequence.

The preceding outline should make clear that, although the presentation of the subject matter of the book is unconventional, it is characterized by a tight logical structure of its own. This logical development may well seem more natural and straightforward to the student, who approaches the topics without any preconceptions, than to the instructor whose mind is molded by conventional ways of teaching the subject. I would advise the instructor to think the subject through afresh himself. If sheer force of habit should lead him to inject traditional points of view injudiciously, he may disrupt the logical development of the book and thus confuse, rather than enlighten, the student.

Other Features of the Book

Appendix: The four sections of the appendix contain some peripheral material. In particular, the Gaussian and Poisson distributions are specifically discussed because they are important in so many diverse fields and because they are also relevant in the laboratory part of the Berkeley Physics Course.

Mathematical Notes: These notes constitute merely a collection of mathematical tidbits found useful somewhere in the text or in some of the problems.

Mathematical Symbols and Numerical Constants: These can be found listed at the end of the book and also on its inside front and back covers.

Summaries of Definitions: These are given at the ends of chapters for ease of reference and convenience of review.

Problems: The problems constitute a very important part of the book. I have included about 160 of them to provide an ample and thought-provoking collection to choose from. Although I would not expect a student to work through all of them, I would encourage him to solve an appreciable fraction of the problems at the end of any chapter he has studied; otherwise he is likely to derive little benefit from the book. Problems marked by stars are somewhat more difficult. The supplementary problems deal mostly with material discussed in the appendices.

Answers to Problems: Answers to most of the problems are listed at the end of the book. The availability of these answers should facilitate the use of the book for self-study. Furthermore, although I would recommend that a student try to solve each problem before looking at the answer given for it, I believe that it is pedagogically valuable if he can check his own answer immediately after he has worked out a solution. In this way he can become aware of his mistakes early and thus may be stimulated to do further thinking instead of being lulled into unjustified complacency. (Although I have tried to assure that the answers listed in the book are correct, I cannot guarantee it. I should appreciate being informed of any mistakes that might be uncovered.)

Subsidiary material: Material which consists of illustrations or various remarks is set in two-column format with smaller type in order to differentiate it from the main skeleton of logical development. Such subsidiary material should not be skipped in a first reading, but might be passed over in subsequent reviews.

Equation numbering: Equations are numbered consecutively within each chapter. A simple number, such as (8), refers to equation number 8 in the chapter under consideration. A double number is used to refer to equations in other chapters. Thus (3.8) refers to Equation (8) in Chap. 3, (A.8) to Equation (8) in the Appendix, (M.8) to Equation (8) in the Mathematical Notes.

Advice to the Student

Learning is an active process. Simply reading or memorizing accomplishes practically nothing. Treat the subject matter of the book as though you were trying to discover it yourself, using the text merely as a guide that you should leave behind. The task of science is to learn ways of thinking which are effective in describing and predicting the behavior of the observed world. The only method of learning new ways of thinking is to practice thinking. Try to strive for insight, to find new relationships and simplicity where before you saw none. Above all, do not simply memorize formulas; learn modes of reasoning. The *only* relations worth remembering deliberately are the few Important Relations listed explicitly at the end of each chapter. If these are not sufficient to allow you to reconstruct in your head any other significant formula in about twenty seconds or less, you have not understood the subject matter.

Finally, it is much more important to master a few fundamental concepts than to acquire a vast store of miscellaneous facts and formulas. If in the text I have seemed to belabor excessively some simple examples, such as the system of spins or the ideal gas, this has been deliberate. It is particularly true in the study of statistical physics and thermodynamics that some apparently innocent statements are found to lead to remarkable conclusions of unexpected generality. Conversely, it is also found that many problems can easily lead one into conceptual paradoxes or seemingly hopeless calculational tasks; here again, a consideration of simple examples can often resolve the conceptual difficulties and suggest new calculational procedures or approximations. Hence my last advice is that you try to understand simple basic ideas well and that you then proceed to work many problems, both those given in the book and those resulting from questions you may pose yourself. Only in this way will you test your understanding and learn how to become an independent thinker in your own right.

Contents

Chapter 5 ***Microscopic Theory and Macroscopic Measurements*** ***191***

Chapter 6 ***Canonical Distribution in the Classical Approximation*** ***223***

statistical physics

Chapter 1

Characteristic Features of Macroscopic Systems

Chapter 1 Characteristic Features of Macroscopic Systems

Dass ich erkenne, was die Welt
Im Innersten zusammenhält,
Schau' alle Wirkenskraft und Samen,
Und tu' nicht mehr in Worten kramen.
Goethe, *Faust*†

The entire world of which we are aware through our senses consists of objects that are *macro*scopic, i.e., large compared to atomic dimensions and thus consisting of very many atoms or molecules. This world is enormously varied and complex, encompassing gases, liquids, solids, and biological organisms of the most diverse forms and compositions. Accordingly, its study has formed the subject matter of physics, chemistry, biology, and several other disciplines. In this book we want to undertake the challenging task of gaining some insights into the fundamental properties of all macroscopic systems. In particular, we should like to investigate how the few unifying concepts of atomic theory can lead to an understanding of the observed behavior of macroscopic systems, how quantities describing the directly measurable properties of such systems are interrelated, and how these quantities can be deduced from a knowledge of atomic characteristics.

Scientific progress made in the first half of this century has led to very basic knowledge about the structure of matter on the *micro*scopic level, i.e., on the small scale of the order of atomic size (10^{-8} cm). Atomic theory has been developed in quantitative detail and has been buttressed by an overwhelming amount of experimental evidence. Thus we know that all matter consists of molecules built up of atoms which, in turn, consist of nuclei and electrons. We also know the quantum laws of microscopic physics governing the behavior of atomic particles. Hence we should be in a good position to exploit these insights in discussing the properties of macroscopic objects.

Indeed, let us justify this hope in greater detail. Any macroscopic system consists of very many atoms. The laws of quantum mechanics describing the dynamical behavior of atomic particles are well established. The electromagnetic forces responsible for the interactions between these atomic particles are also very well understood.

† From Faust's opening soliloquy in Goethe's play, Part I, Act I, Scene I, lines 382–385. Translation: "That I may recognize what holds the world together in its inmost essence, behold the driving force and source of everything, and rummage no more in empty words."

Ordinarily they are the only forces relevant because gravitational forces between atomic particles are usually negligibly small compared to electromagnetic forces. In addition, a knowledge of nuclear forces is usually not necessary since the atomic nuclei do not get disrupted in most ordinary macroscopic physical systems and in all chemical and biological systems.† Hence we can conclude that our knowledge of the laws of microscopic physics should be quite adequate to allow us, in principle, to deduce the properties of any macroscopic system from a knowledge of its microscopic constituents.

It would, however, be quite misleading to stop on this optimistic note. A typical macroscopic system of the type encountered in everyday life contains about 10^{25} interacting atoms. Our concrete scientific aim is that of understanding and predicting the properties of such a system on the basis of a minimum number of fundamental concepts. We may well know that the laws of quantum mechanics and electromagnetism describe completely all the atoms in the system, whether it be a solid, a liquid, or a human being. But this knowledge is utterly useless in achieving our scientific aim of prediction unless we have available methods for coping with the tremendous complexity inherent in such systems. The difficulties involved are not of a type which can be solved merely by the brute force application of bigger and better electronic computers. The problem of 10^{25} interacting particles dwarfs the capabilities of even the most fanciful of future computers; furthermore, unless one asks the right questions, reams of computer output tape are likely to provide no insight whatever into the essential features of a problem. It is also worth emphasizing that complexity involves much more than questions of quantitative detail. In many cases it can lead to remarkable qualitative features that may seem quite unexpected. For instance, consider a gas of identical simple atoms (e.g., helium atoms) which interact with each other through simple known forces. It is by no means evident from this microscopic information that such a gas can condense very abruptly so as to form a liquid. Yet this is precisely what happens. An even more striking illustration is provided by any biological organism. Starting solely from a knowledge of atomic structure, would one suspect that a few simple kinds of atoms, forming certain types of molecules, can give rise to systems capable of biological growth and self-reproduction?

The understanding of macroscopic systems consisting of very many particles thus requires primarily the formulation of new concepts capable of dealing with complexity. These concepts, based ultimately

† Gravitational and nuclear interactions may, however, become pertinent in some astrophysical problems.

upon the known fundamental laws of microscopic physics, should achieve the following aims: make apparent the parameters most useful in describing macroscopic systems; permit us to discern readily the essential characteristics and regularities exhibited by such systems; and finally, provide us with relatively simple methods capable of predicting quantitatively the properties of these systems.

The discovery of concepts sufficiently powerful to achieve these aims represents clearly a major intellectual challenge, even when the fundamental laws of microscopic physics are assumed known. It is thus not surprising that the study of complex systems consisting of many atoms occupies much attention in research at the forefront of physics. On the other hand, it is remarkable that quite simple reasoning is sufficient to lead to substantial progress in the understanding of macroscopic systems. As we shall see, the basic reason is that the very presence of a large number of particles allows one to use statistical methods with singular effectiveness.

It is by no means obvious how we should go about achieving our aim of understanding macroscopic systems. Indeed, their apparent complexity may seem forbidding. In setting out on our path of discovery we shall, therefore, follow good scientific procedure by first examining some simple examples. At this stage we shall not let our imagination be stifled by trying to be rigorous or overly critical. Our purpose in this chapter is rather to recognize the essential features characteristic of macroscopic systems, to see the main problems in qualitative outline, and to get some feeling for typical magnitudes. This preliminary investigation should serve to suggest to us appropriate methods for attacking the problems of macroscopic systems in a systematic and quantitative way.

1.1 Fluctuations in Equilibrium

A simple example of a system consisting of many particles is a gas of identical molecules, e.g., argon (Ar) or nitrogen (N_2) molecules. If the gas is *dilute* (i.e., if the number of molecules per unit volume is small), the average separation between the molecules is large and their mutual interaction is correspondingly small. The gas is said to be *ideal* if it is sufficiently dilute so that the interaction between its molecules is almost negligible.† An ideal gas is thus particularly simple. Each of its molecules spends most of its time moving like a free particle unin-

† The interaction is "almost" negligible if the total potential energy of interaction between the molecules is negligible compared to their total kinetic energy, but is sufficiently large so that the molecules can interact and thus exchange energy with each other.

fluenced by the presence of the other molecules or the container walls; only rarely does it come sufficiently close to the other molecules or the container walls so as to interact (or *collide*) with them. In addition, if the gas is sufficiently dilute, the average separation between its molecules is much larger than the average de Broglie wavelength of a molecule. In this case quantum-mechanical effects are of negligible importance and it is permissible to treat the molecules as distinguishable particles moving along classical trajectories.†

Consider then an ideal gas of N molecules confined within a container or box. In order to discuss the simplest possible situation, suppose that this whole system is *isolated* (i.e., that it does not interact with any other system) and that it has been left undisturbed for a very long time. We now imagine that we can observe the gas molecules, without affecting their motion, by using a suitable camera to take a motion picture of the gas. Successive frames of the film would show the positions of the molecules at regular intervals separated by some short time τ_0. We then could examine the frames individually, or alternatively, could run the movie through a projector.

In the latter case we would observe on the projection screen a picture showing the gas molecules in constant motion: Thus any given molecule moves along a straight line until it collides with some other molecule or with the walls of the box; it then continues moving along some other straight line until it collides again; and so on and so forth. Each molecule moves strictly in accordance with the laws of motion of mechanics. Nevertheless, N molecules moving throughout the box and colliding with each other represent a situation so complex that the picture on the screen appears rather chaotic (unless N is very small).

Let us now focus attention on the positions of the molecules and their distribution in space. To be precise, consider the box to be divided by some imaginary partition into two equal parts (see Fig. 1.1). Denote the number of molecules in the left half of the box by n and the number in the right half by n'. Of course

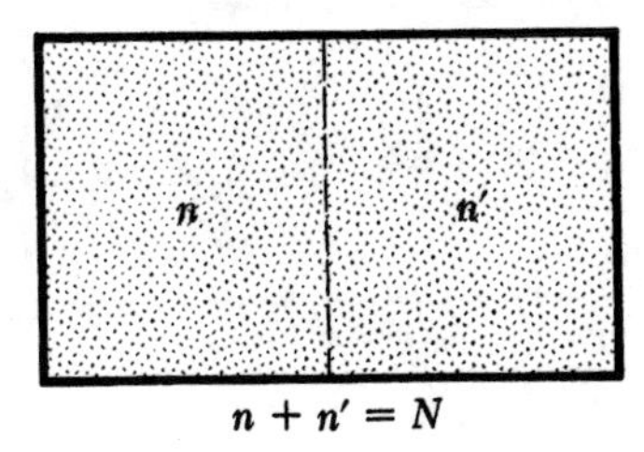

Fig. 1.1 A box containing an ideal gas consisting of N molecules. The box is shown subdivided into two equal parts by an imaginary partition. The number of molecules in the left half is denoted by n, the number of molecules in the right half by n'.

$$n + n' = N, \tag{1}$$

the total number of molecules in the box. If N is large, we would *ordinarily* find that $n \approx n'$, i.e., that roughly half of the molecules are in each half of the box. We emphasize, however, that this statement is only approximately true. Thus, as the molecules move throughout the box, colliding occasionally with each other or with the walls, some

† The validity of the classical approximation will be examined more extensively in Sec. 6.3.

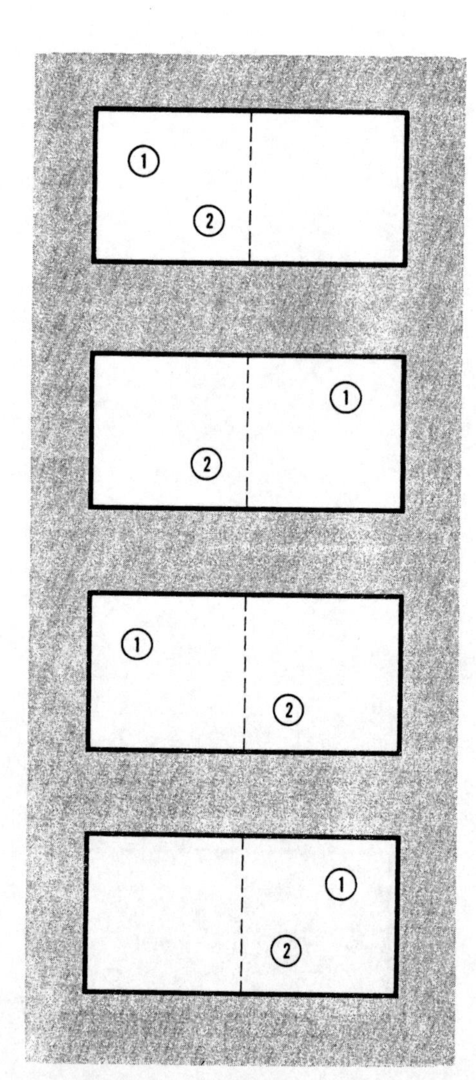

Fig. 1.2 Schematic diagram illustrating the 4 different ways in which 2 molecules can be distributed between the two halves of a box.

of them enter the left half of the box, while others leave it. Hence the number n of molecules actually located in the left half fluctuates constantly in time (see Figs. 1.3 through 1.6). *Ordinarily* these fluctuations are small enough so that n does not differ too much from $\frac{1}{2}N$. There is, however, nothing that prevents all molecules from being in the left half of the box (so that $n = N$, while $n' = 0$). Indeed this *might* happen. But how likely is it that it actually *does* happen?

To gain some insight into this question, let us ask in how many ways the molecules can be distributed between the two halves of the box. We shall call each distinct way in which the molecules can be distributed between these two halves a *configuration.* A single molecule can then be found in the box in two possible configurations, i.e., it can be either in the left half or in the right half. Since the two halves have equal volumes and are otherwise equivalent, the molecule is equally likely to be found in either half of the box.† If we consider 2 molecules, each one of them can be found in either of the 2 halves. Hence the total number of possible configurations (i.e., the total number of possible ways in which the 2 molecules can be distributed between the two halves) is equal to $2 \times 2 = 2^2 = 4$ since there are, for each possible configuration of the first molecule, 2 possible configurations of the other (see Fig. 1.2). If we consider 3 molecules, the total number of their possible configurations is equal to $2 \times 2 \times 2 = 2^3 = 8$ since there are, for each of the 2^2 possible configurations of the first 2 molecules, 2 possible configurations of the last one. Similarly, if we consider the general case of N molecules, the total number of possible configurations is $2 \times 2 \times \cdots \times 2 = 2^N$. These configurations are listed explicitly in Table 1.1 for the special case where $N = 4$.

Note that there is only one way of distributing the N molecules so that all N of them are in the left half of the box. It represents only one special configuration of the molecules compared to the 2^N possible configurations of these molecules. Hence we would expect that, among a very large number of frames of the film, on the average only one out of every 2^N frames would show all the molecules to be in the left half. If P_N denotes the fraction of frames showing all the N molecules located in the left half of the box, i.e., if P_N denotes the relative frequency, or *probability,* of finding all the N molecules in the left half, then

$$P_N = \frac{1}{2^N}. \tag{2}$$

† We assume that the likelihood of finding a particular molecule in any half of the box is unaffected by the presence there of any number of other molecules. This will be true if the total volume occupied by the molecules themselves is negligibly small compared to the volume of the box.

1	2	3	4	n	n'	$C(n)$
L	L	L	L	4	0	1
L	L	L	R	3	1	4
L	L	R	L	3	1	
L	R	L	L	3	1	
R	L	L	L	3	1	
L	L	R	R	2	2	6
L	R	L	R	2	2	
L	R	R	L	2	2	
R	L	L	R	2	2	
R	L	R	L	2	2	
R	R	L	L	2	2	
L	R	R	R	1	3	4
R	L	R	R	1	3	
R	R	L	R	1	3	
R	R	R	L	1	3	
R	R	R	R	0	4	1

Table 1.1 Enumeration of the 16 possible ways in which $N = 4$ molecules (denoted by 1, 2, 3, 4) can be distributed between two halves of a box. The letter L indicates that the particular molecule is in the left half of the box, the letter R that it is in the right half. The number of molecules in each of the halves is denoted by n and n', respectively. The symbol $C(n)$ denotes the number of possible configurations of the molecules when n of them are in the left half of the box.

Computer-constructed pictures

The following pages and several subsequent ones show figures constructed by means of a high-speed electronic digital computer. The situation investigated in every case is the classical motion of several particles in a box, the particles being represented by disks moving in two dimensions. The forces between any two particles, or between a particle and a wall, are assumed to be like those between "hard" objects (i.e., to vanish when they do not touch and to become infinite when they do touch). All resulting collisions are thus elastic. The computer is given some initial specified positions and velocities of the particles. It is then asked to solve numerically the equations of motion of these particles for all subsequent (or prior) times and to display pictorially on a cathode-ray oscilloscope the positions of the molecules at successive times $t = j\tau_0$ where τ_0 is some small fixed time interval and where $j = 0, 1, 2, 3, \ldots$. A movie camera photographing the oscilloscope screen then yields the successive picture frames reproduced in the figures. (The time interval τ_0 was chosen long enough so that several molecular collisions occur between the successive frames displayed in the figures.) The computer is thus used to simulate in detail a hypothetical experiment involving the dynamical interaction between many particles.

All the computer-made pictures were produced with the generous cooperation of Dr. B. J. Alder of the Lawrence Radiation Laboratory at Livermore.

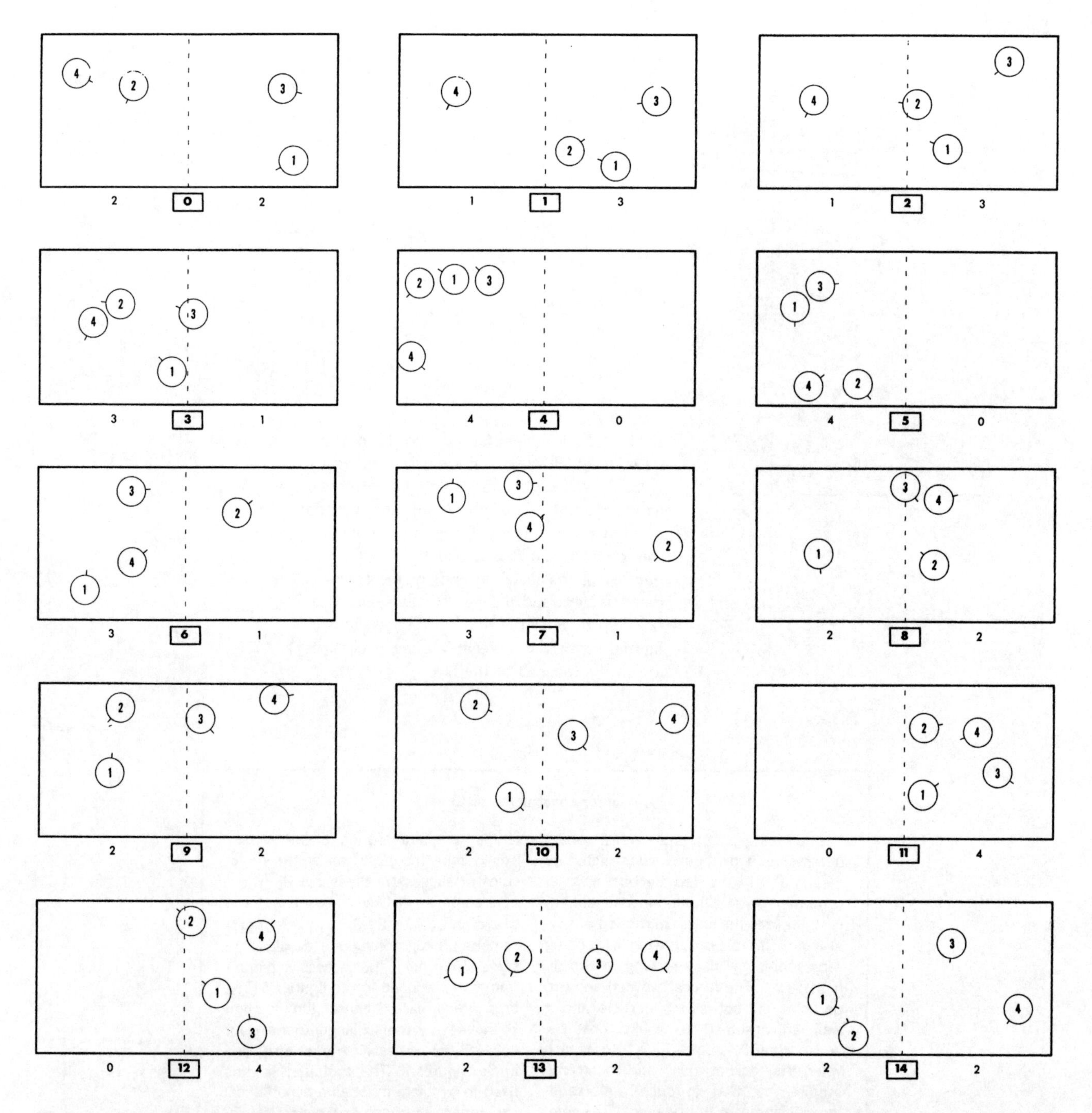

Fig. 1.3 Computer-made pictures showing 4 particles in a box. The fifteen successive frames (labeled by $j = 0, 1, 2, \ldots, 14$) are pictures taken a long time after the beginning of the computation with assumed initial conditions. The number of particles located in each half of the box is printed directly beneath that half. The short line segment emanating from each particle indicates the direction of the particle's velocity.

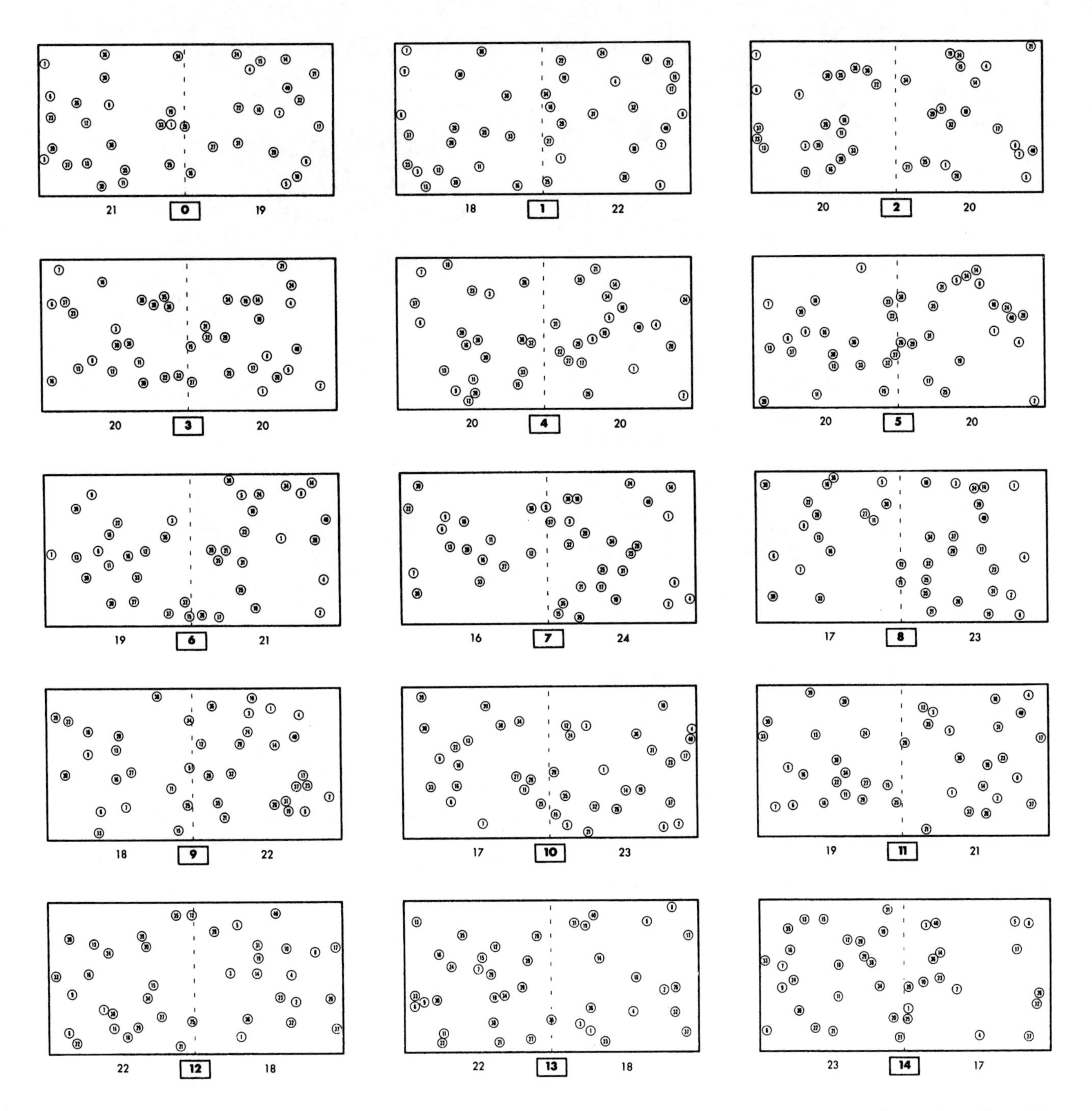

Fig. 1.4 Computer-made pictures showing 40 particles in a box. The fifteen successive frames (labeled by $j = 0, 1, 2, \ldots, 14$) are pictures taken a long time after the beginning of the computation with assumed initial conditions. The number of particles located in each half of the box is printed directly beneath that half. The velocities of the particles are not indicated.

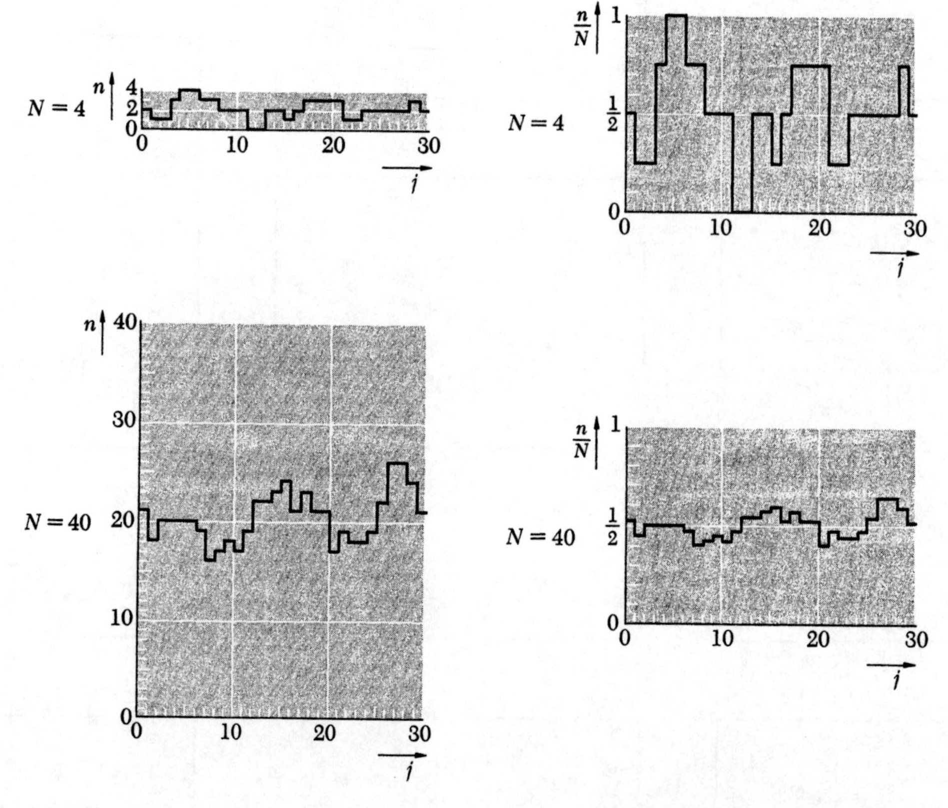

Fig. 1.5 The number n of particles in the left half of the box as a function of the frame index j or the elapsed time $t = j\tau_0$. The number n in the jth frame is indicated by a horizontal line extending from j to $j + 1$. The graphs describe Fig. 1.3 for $N = 4$ particles and Fig. 1.4 for $N = 40$ particles, but contain information about more frames than were shown there.

Fig. 1.6 The *relative* number n/N of particles in the left half of the box as a function of the frame index j or the elapsed time $t = j\tau_0$. The information presented is otherwise the same as that in Fig. 1.5.

Similarly, the case where no molecule at all is in the left half is also very special since there is again only one such configuration of the molecules out of 2^N possible configurations. Thus the probability P_0 of finding no molecule located in the left half should also be given by

$$P_0 = \frac{1}{2^N}. \tag{3}$$

More generally, consider a situation where n of the N molecules of the gas are located in the left half of the box and let us denote by $C(n)$ the number of possible configurations of the molecules in this case. [That is, $C(n)$ is the number of possible ways the molecules can be distributed in the box so that n of them are found in the left half of the box.] Since the total number of possible configurations of the molecules is 2^N, one would expect that, among a very large number of frames of the film, on the average $C(n)$ out of every 2^N such frames would show n molecules to be in the left half of the box. If P_n denotes the fraction of frames showing n molecules located in the left half, i.e., if P_n denotes the relative frequency, or probability, of finding n molecules in the left half, then

$$P_n = \frac{C(n)}{2^N}. \tag{4}$$

Example

Consider the special case where the gas consists of only four molecules. The number $C(n)$ of possible configurations of each kind is listed in Table 1.1. Suppose that a movie of this gas consists of a great many frames. Then we expect that the fraction P_n of these frames showing n molecules in the left half (and correspondingly $n' = N - n$ molecules in the right half) is given by:

$$\begin{aligned} P_4 &= P_0 = \tfrac{1}{16}, \\ P_3 &= P_1 = \tfrac{4}{16} = \tfrac{1}{4}, \\ P_2 &= \tfrac{6}{16} = \tfrac{3}{8}. \end{aligned} \tag{4a}$$

As we have seen, a situation where $n = N$ (or where $n = 0$) corresponds only to a single possible molecular configuration. More generally, if N is large, then $C(n) \ll 2^N$ if n is even moderately close to N (or even moderately close to 0). In other words, a situation where the distribution of molecules is so nonuniform that $n \gg \frac{1}{2}N$ (or that $n \ll \frac{1}{2}N$) corresponds to relatively few configurations. A situation of this kind, which can be obtained in relatively few ways, is rather special and is accordingly said to be relatively *nonrandom* or *orderly;*

according to (4), it occurs relatively infrequently. On the other hand, a situation where the distribution of the molecules is almost uniform, so that $n \approx n'$, corresponds to many possible configurations; indeed, as is illustrated in Table 1.1, $C(n)$ is maximum if $n = n' = \frac{1}{2}N$. A situation of this kind, which can be obtained in many different ways, is said to be *random* or *disordered;* according to (4) it occurs quite frequently. In short, more random (or uniform) distributions of the molecules in the gas occur more frequently than less random ones. The physical reason is clear: All molecules must move in a very special way if they are to concentrate themselves predominantly in one part of the box; similarly, if they are all located in one part of the box, they must move in a very special way if they are to remain concentrated there.

The preceding statements can be made quantitative by using Eq. (4) to calculate the actual probability of occurrence of a situation where any number n of molecules are in the left half of the box. We shall postpone until the next chapter the requisite computation of the number of molecular configurations $C(n)$ in the general case. It is, however, very easy and illuminating to consider an extreme case and ask how frequently one would expect all the molecules to be located in the left half of the box. Indeed, (2) asserts that a fluctuation of this kind would, on the average, be observed to occur only once in every 2^N frames of the film.

To gain an appreciation for magnitudes, we consider some specific examples. If the gas consisted of only 4 molecules, all of them would, on the average, be found in the left half of the box once in every 16 frames of the film. A fluctuation of this kind would thus occur with moderate frequency. On the other hand, if the gas consisted of 80 molecules, all of these would, on the average, be found in the left half of the box in only one out of $2^{80} \approx 10^{24}$ frames of the film. This means that, even if we took a million pictures every second, we would have to run the film for a time appreciably greater than the age of the universe before we would have a reasonable chance of obtaining one frame showing all the molecules in the left half of the box.† Finally, suppose that we consider as a realistic example a box having a volume of 1 cm^3 and containing air at atmospheric pressure and room temperature. Such a box contains about 2.5×10^{19} molecules. [See Eq. (27) later in this chapter.] A fluctuation where all of these are located in one half of the box would, on the average, appear in only one out of

---

† There are about 3.15×10^7 seconds in a year and the estimated age of the universe is of the order of 10^{10} years.

$$2^{2.5\times10^{19}} \approx 10^{7.5\times10^{18}}$$

frames of the film. (This number of frames is so fantastically large that it could not be accumulated even though our film ran for times incredibly larger than the age of the universe.) Fluctuations where not all, but a predominant majority of the molecules are found in one half of the box, would occur somewhat more frequently; but this frequency of occurrence would still be exceedingly small. Hence we arrive at the following general conclusion: *If the total number of particles is large, fluctuations corresponding to an appreciably nonuniform distribution of these molecules occur almost never.*

Let us now conclude by summarizing our discussion of the isolated ideal gas which has been left undisturbed for a long time. The number n of molecules in one half of the box fluctuates in time about the constant value $\frac{1}{2}N$ which occurs most frequently. The frequency of occurrence of a particular value of n decreases rapidly the more n differs from $\frac{1}{2}N$, i.e., the greater the difference $|\Delta n|$ where

$$\Delta n \equiv n - \tfrac{1}{2}N. \tag{5}$$

Indeed, if N is large, only values of n with $|\Delta n| \ll N$ occur with significant frequency. Positive and negative values of Δn occur equally often. The time dependence of n has thus the appearance indicated schematically in Fig. 1.7.

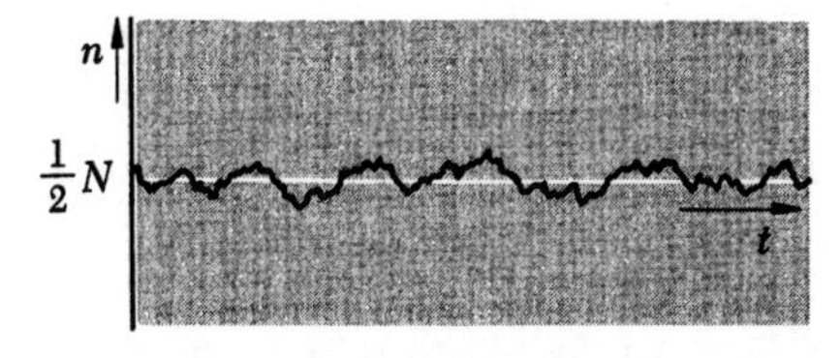

Fig. 1.7 Schematic illustration showing how the number n of molecules in one half of a box fluctuates as a function of the time t. The total number of molecules is N.

The gas can be described in greatest detail by specifying its *microscopic state*, or *microstate*, at any time, i.e., by specifying the maximum possible information about the gas molecules at this time (e.g., the position and velocity of each molecule). From this microscopic point of view, a hypothetical film of the gas appears very complex since the locations of the individual molecules are different in every frame of the film. As the individual molecules move about, the microscopic state of the gas changes thus in a most complicated way. From a large-scale or *macro*scopic point of view, however, one is not interested in the behavior of each and every molecule, but in a much less detailed description of the gas. Thus the *macro*scopic state, or *macrostate*, of the gas at any time might be quite adequately described by specifying merely the *number* of molecules located in any part of the box at this time.† From this macroscopic point of view, the isolated gas which

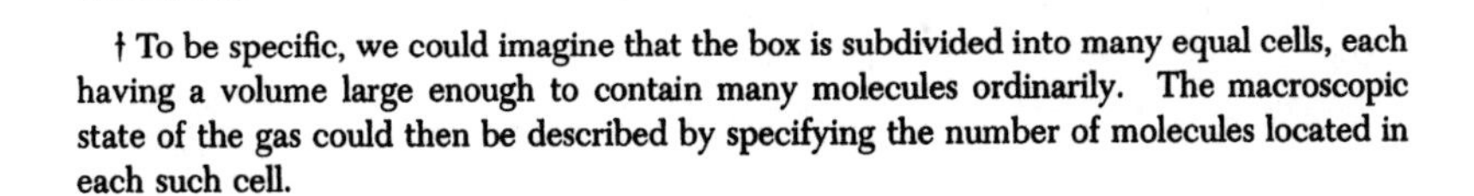
† To be specific, we could imagine that the box is subdivided into many equal cells, each having a volume large enough to contain many molecules ordinarily. The macroscopic state of the gas could then be described by specifying the number of molecules located in each such cell.

has been left undisturbed for a long time represents a very simple situation since its macroscopic state does not tend to change in time. Indeed, suppose that, starting at some time t_1, we observed the gas over some moderately long period of time τ by taking a movie of it. Alternatively, suppose that, starting at some other time t_2, we again observed the gas over the same time τ by taking a movie of it. From a macroscopic point of view these two movies would ordinarily look indistinguishable. In each case the number n of particles in the left half of the box would ordinarily fluctuate about the same value $\frac{1}{2}N$, and the magnitude of the observed fluctuations would ordinarily also look alike. Disregarding very exceptional occurrences (to be discussed in the next section), the observed macrostate of the gas is thus independent of the starting time of our observations; i.e., we can say that the macrostate of our gas does not tend to change in time. In particular, the value about which n fluctuates (or, more precisely, its *average* value) does not tend to change in time. A system of many particles (such as our gas) whose macroscopic state does not tend to change in time is said to be in *equilibrium*.

Remark

To define the concept of a time average in precise terms, let us denote by $n(t)$ the number of molecules in the left half of the box at any time t. The time average value of n at any time t, taken over the time interval τ, can then be denoted by $[\bar{n}(t)]_\tau$ and is defined by

$$[\bar{n}(t)]_\tau \equiv \frac{1}{\tau}\int_t^{t+\tau} n(t')\, dt'. \tag{6}$$

Equivalently, if a movie strip beginning at the time t extends over a time τ and contains $g = \tau/\tau_0$ frames occurring at successive times $t_1 = t$, $t_2 = t + \tau_0$, $t_3 = t + 2\tau_0, \ldots, t_g = t + (g - 1)\tau_0$, the definition (6) becomes

$$[\bar{n}(t)]_\tau = \frac{1}{g}[n(t_1) + n(t_2) + \cdots + n(t_g)].$$

If we omit explicit indication of the time interval τ to be considered, $\bar{n}(t)$ implies an average over some appropriately chosen time interval τ of appreciable length. In the equilibrium situation of our gas, $\bar{n}$ tends to be constant and equal to $\frac{1}{2}N$.

1.2 *Irreversibility and the Approach to Equilibrium*

Consider an isolated gas consisting of a large number N of molecules. If the fluctuations occurring in this gas in equilibrium are such that n is ordinarily very close to its most probable value $\frac{1}{2}N$, under what conditions can we expect to find situations where n differs very appreciably from $\frac{1}{2}N$? Such situations can occur in two different ways which we shall discuss in turn.

Rarely occurring large fluctuations in equilibrium

Although in the gas in equilibrium n is ordinarily always close to $\frac{1}{2}N$, values of n far from $\frac{1}{2}N$ *can* occur, but they do so quite rarely. If we observe the gas long enough, we may thus succeed in observing, at some particular instant of time t, a value of n appreciably different from $\frac{1}{2}N$.

Suppose that such a large spontaneous fluctuation of $|\Delta n|$ has occurred, e.g., that n at some particular time t_1 assumes a value n_1 which is much greater than $\frac{1}{2}N$. What can we then say about the probable behavior of n as time goes on? To the extent that $|n_1 - \frac{1}{2}N|$ is large, the particular value n_1 corresponds to a highly nonuniform distribution of the molecules and occurs very rarely in equilibrium. It is then most likely that the value n_1 occurs as a result of a fluctuation which is represented by a peak whose maximum is near n_1 (as indicated by the peak marked X in Fig. 1.8). The reason is the following: It *might* be possible for a value as large as n_1 to occur also as a result of a fluctuation represented by a peak whose maximum is larger than n_1 (such as the peak marked Y in Fig. 1.8); but the occurrence of such a large fluctuation is much less likely still than the already rare occurrence of a smaller fluctuation such as X. Thus we may conclude that it is indeed most likely that the time t_1, where $n = n_1$, corresponds to a peak (such as X) where n is maximum. The general behavior of n as a function of time is then, however, apparent from Fig. 1.8. As time proceeds, n must tend to decrease (with accompanying small fluctuations) until it reverts to the usual equilibrium situation where it no longer tends to change, but fluctuates merely about the constant average value $\frac{1}{2}N$. The approximate time required for the large fluctuation (where $n = n_1$) to decay back to the equilibrium situation (where $n \approx \frac{1}{2}N$) is called the *relaxation time* for decay of this fluctuation. Note that, among many movie strips of duration τ, a movie strip showing the gas near the time t_1 where a large fluctuation occurs is highly exceptional. Not only does such a movie strip occur very rarely but, if it does occur, it

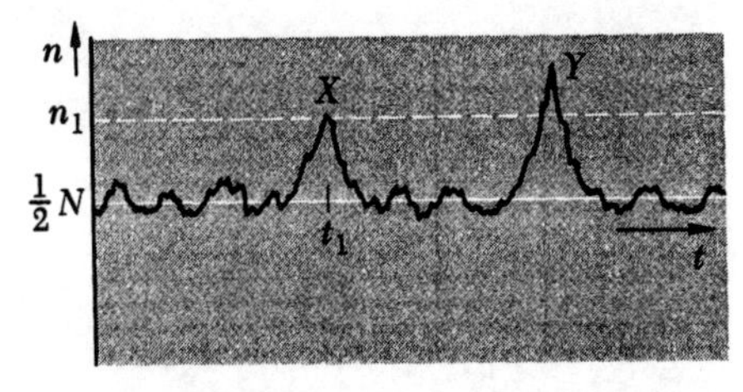

Fig. 1.8 Schematic illustration showing rare instances where the number n of molecules in one half of a box exhibits large fluctuations about its equilibrium value $\frac{1}{2}N$.

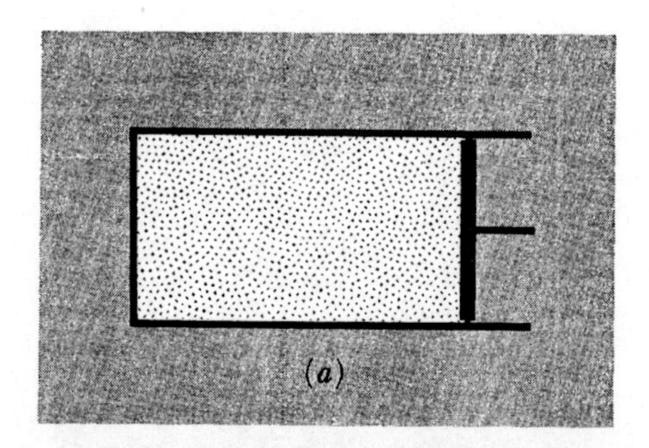

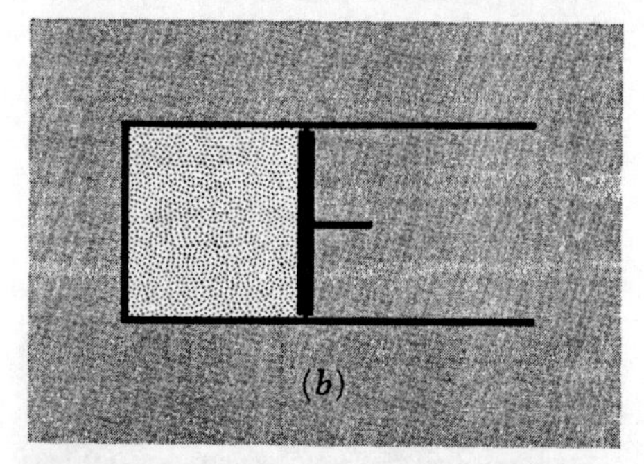

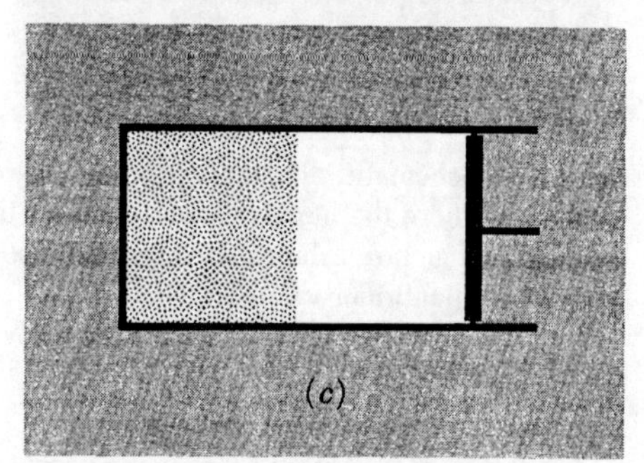

Fig. 1.9 The piston in (*a*) is moved to the position (*b*) so as to compress the gas into the left half of the box. When the piston is suddenly restored to its initial position, as shown in (*c*), the molecules immediately afterward are all located in the left half of the box, while the right half is empty.

is distinguishable from other movie strips since it shows a situation which tends to change in time.†

Our comments can thus be summarized in the following way: If n is known to assume a value n_1 which is appreciably different from its equilibrium average value $\frac{1}{2}N$, then n will almost‡ always change in such a direction as to approach the equilibrium value $\frac{1}{2}N$. In more physical terms, the value n_1 corresponds to a very nonuniform distribution of the molecules and the molecules would have to move in a very special way in order to preserve such nonuniformity. The incessant motion of the molecules thus almost always results in mixing them up so thoroughly that they become distributed over the entire box in the most random (or uniform) possible way. (See Figs. 1.15 through 1.20 at the end of this section.)

Remarks

Note that the statements of the preceding paragraph are equally applicable whether the large fluctuation $(n_1 - \frac{1}{2}N)$ is positive or negative. If it is positive, the value n_1 will almost always correspond to the maximum of a fluctuation in n (such as that indicated by the peak X in Fig. 1.8). If it is negative, it will almost always correspond to the minimum of a fluctuation in n. The argument leading to the conclusion of the paragraph remains, however, essentially identical.

Note also that the statements of the paragraph remain equally valid irrespective of whether the change in time is in the forward or backward direction (i.e., irrespective of whether a movie of the gas is played forward or backward through a projector). If n_1 corresponds to a maximum such as that indicated by the peak X at time t_1, then n must decrease both for $t > t_1$ and for $t < t_1$.

† This does not contradict the statement that the gas is in equilibrium in the long run, i.e., when observed over a very long time which may contain several occurrences of fluctuations as large as n_1.

‡ We use the qualifying word "almost" since, instead of corresponding to a maximum of a peak such as X, the value n_1 may very rarely lie on the rising side of a peak such as Y. In this case n would initially increase, i.e., change *away* from its equilibrium value $\frac{1}{2}N$.

Specially prepared initial situations

Although a nonrandom situation, where n is appreciably different from $\frac{1}{2}N$, may occur as a result of a spontaneous fluctuation of the gas in equilibrium, such a large fluctuation occurs so rarely that it would almost never be observed in practice. [Remember the numerical estimates based on Eq. (2) or (3).] Most macroscopic systems with which we deal, however, have not remained isolated and undisturbed for very long periods of time and, accordingly, are not in equilibrium. Nonrandom situations thus occur quite commonly, *not* as a result of spontaneous fluctuations of a system in equilibrium, but as a result of interactions which affected the system at some not too distant time in the past. Indeed, it is quite easy to bring about a nonrandom situation of a system by means of external intervention.

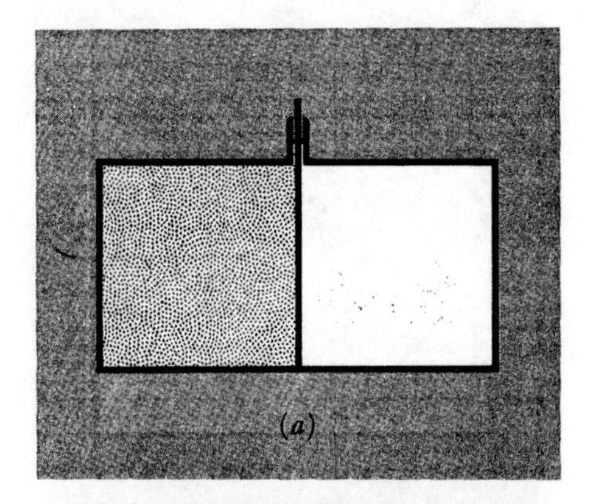

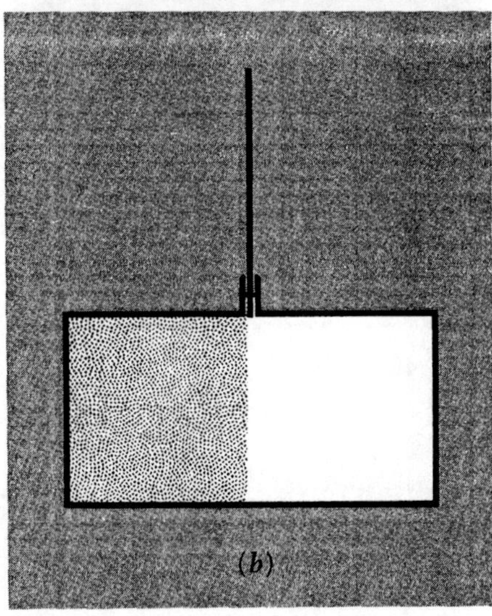

Fig. 1.10 When the partition in (*a*) is suddenly removed, all the molecules immediately afterward are located in the left half of the box, as shown in (*b*).

Examples

When a wall of a box is made movable, it becomes a *piston*. One can use such a piston (as shown in Fig. 1.9) to compress a gas into the left half of a box. When the piston is suddenly restored to its initial position, all the molecules immediately afterward are still in the left half of the box. Thus one has produced an extremely nonuniform distribution of the molecules in the box.

Equivalently, consider a box divided into two equal parts by a partition (see Fig. 1.10). Its left half is filled with N molecules of a gas, while its right half is empty. If the gas is in equilibrium under these circumstances, the distribution of its molecules is essentially uniform throughout the left half of the box. Imagine that the partition is now suddenly removed. The molecules immediately afterward are then still uniformly distributed throughout the left half of the box. This distribution, however, is highly nonuniform under the new conditions which leave the molecules free to move throughout the entire box.

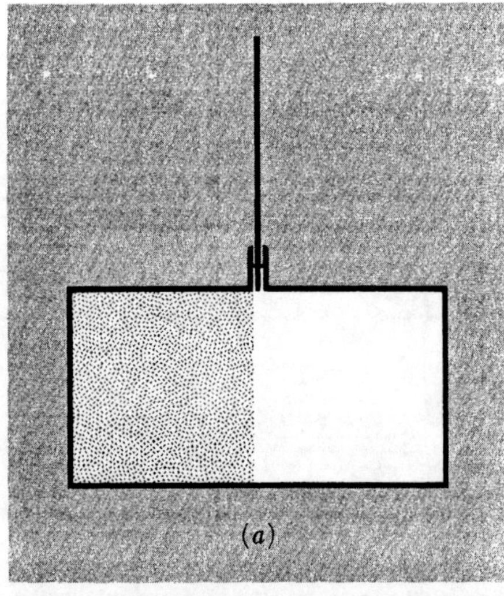

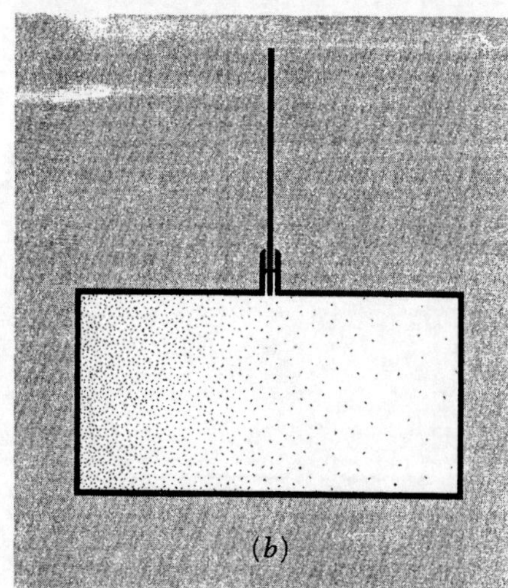

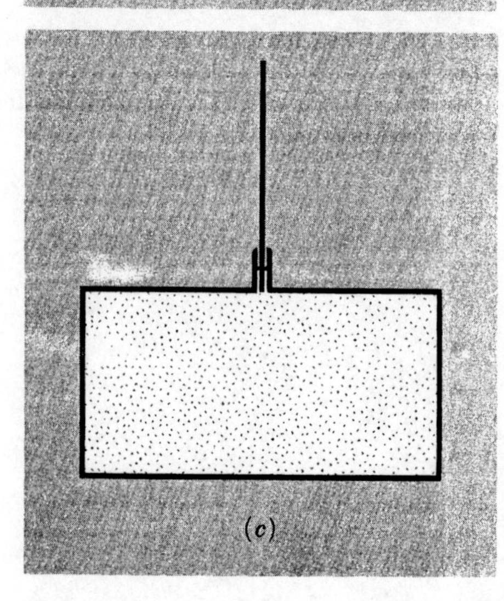

Suppose that an isolated system is known to be in a highly nonrandom situation, e.g., suppose that all the molecules in a gas are known to be predominantly in the left half of the box so that n is appreciably different from $\frac{1}{2}N$. Then it is essentially irrelevant whether the system got into this condition by virtue of a very rare spontaneous fluctuation in equilibrium, or whether it got there by virtue of some form of prior external intervention. Irrespective of its past history, the subsequent behavior of the system in time will thus be similar to that discussed previously when we considered the decay of a large fluctuation in equilibrium. In short, since nearly all possible ways in which the molecules of the system can move will result in a more random distribution of these molecules, the situation of the system will almost always tend to change in time so as to become as random as possible. After the most random condition has been attained, it exhibits then no further tendency to change, i.e., it represents the ultimate equilibrium condition attained by the system. For example, Fig. 1.11 indicates schematically what happens after the partition of Fig. 1.10 is suddenly removed. The number n of molecules in the left half tends to change from its initial value $n = N$ (corresponding to a highly nonuniform distribution of the molecules in the box) until the ultimate equilibrium situation is attained where $n \approx \frac{1}{2}N$ (corresponding to an essentially uniform distribution of molecules in the box). (See Figs. 1.12 and 1.18.)

The important conclusion which we have reached in this section can thus be summarized as follows:

> If an isolated system is in an appreciably nonrandom situation, it will (except for fluctuations which are unlikely to be large) change in time so as to approach ultimately its most random situation where it is in equilibrium. (7)

Note that the preceding statement does *not* make any assertions about the *relaxation time*, i.e., the approximate time required for the system to reach its final equilibrium situation. The actual magnitude of this time depends sensitively on details of the system under consideration; it might be of the order of microseconds or of the order of centuries.

Fig. 1.11 The box of Fig. 1.10 is here shown (*a*) immediately after the partition has been removed, (*b*) a short time thereafter, and (*c*) a long time thereafter. A movie played backward would show the pictures in the reverse order (*c*), (*b*), (*a*).

Example

Referring to Fig. 1.10, consider again the box divided into two equal parts by a partition. The left half of the box contains N molecules of gas, while its right half is empty. Imagine now that the partition is suddenly removed—but only partly (as shown in Fig. 1.13) rather than fully (as in the previous experiment of Fig. 1.10). In *both* experiments the nonrandom situation immediately after the partition is removed (when the number n of molecules in the left half is equal to N) tends to change in time until the molecules become essentially uniformly distributed throughout the entire box (so that $n \approx \frac{1}{2}N$). But the time required until the final equilibrium condition is reached will be longer in the experiment of Fig. 1.13 than in that of Fig. 1.10.

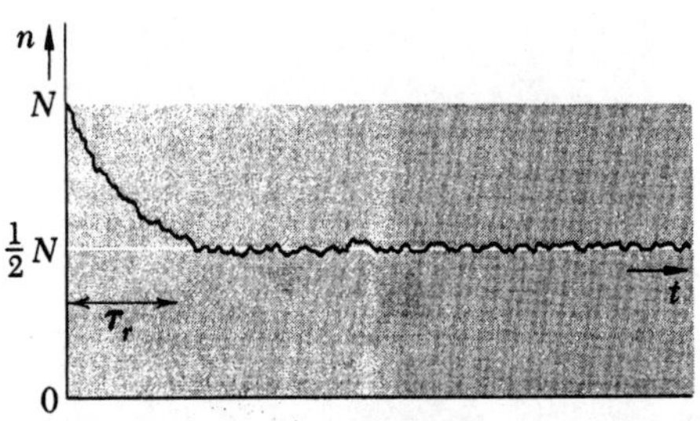

Fig. 1.12 Schematic illustration showing how the number n of molecules in the left half of the box in Fig. 1.11 varies as a function of the time t, beginning with the instant immediately after the partition is removed. The relaxation time is indicated by τ_r.

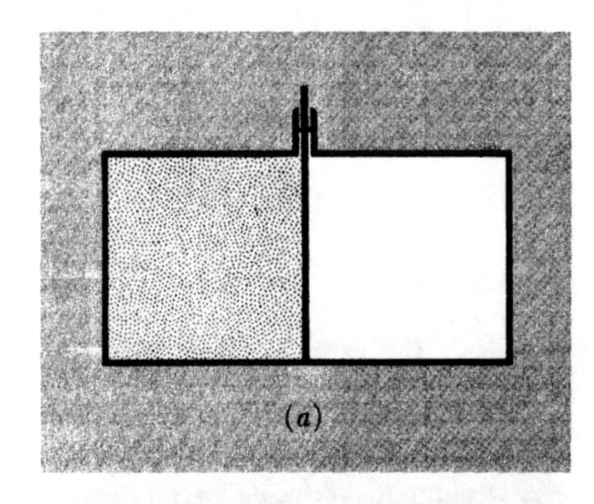

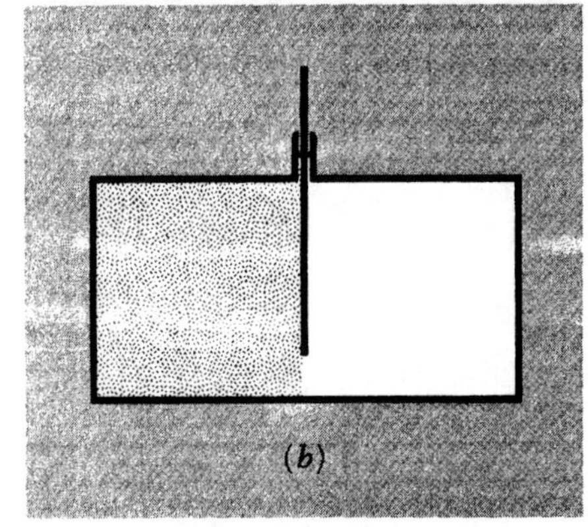

Fig. 1.13 The partition in (*a*) is suddenly removed partway only, as shown in (*b*).

Irreversibility

The statement (7) asserts that, when an isolated macroscopic system changes in time, it tends to do so in a very definite direction—namely, from a less random to a more random situation. We could observe the process of change by taking a movie of the system. Suppose now that we played the movie backward through a projector (i.e., that we ran the film through the projector in a direction opposite to that used in the original movie camera photographing the process). We would then observe on the screen the *time-reversed process,* i.e., the process that would occur if one imagined the direction of time to be reversed. The movie on the screen would look very peculiar indeed since it would portray a process in which the system changes from a more random to a much less random situation—something which one would almost never observe in actuality. Just by watching the movie on the screen, we could conclude with almost complete certainty that the movie is being played backward through the projector.

Example

For example, suppose one filmed the process that takes place after the partition in Fig. 1.10 is suddenly removed. The movie played forward through the projector would show the gas spreading out (as indicated in Fig. 1.11) until it becomes essentially uniformly distributed throughout the entire box. Such a process is quite familiar. On the other hand, the movie played backward would show the gas, initially distributed uniformly throughout the box, concentrating itself spontaneously in the left half of the box so as to leave the right half empty. Such a process is virtually never observed in actuality. This does not mean that this process is impossible, only that it is exceedingly improbable. What the backward movie shows *could* happen in actuality if all the molecules moved in an extremely special way.† But it is *exceedingly* improbable that all the molecules would ever move in this special way; indeed, it is as improbable as the occurrence of a fluctuation where $n = N$ (in a situation where the gas is in equilibrium throughout the entire box).

Fig. 1.14 This cartoon is humorous because it portrays the reverse of an irreversible process. The indicated sequence of events *could* happen, but it is exceedingly unlikely that it ever would. *(Reprinted by special permission of The Saturday Evening Post and James Frankfort, © 1965 The Curtis Publishing Company.)*

A process is said to be ***irreversible*** if the time-reversed process (the one which would be observed in a movie played backward) is such that it would almost never occur in actuality. But all macroscopic systems not in equilibrium tend to approach equilibrium, i.e., situations of greatest randomness. Hence we see that all such systems exhibit irreversible behavior. Since in everyday life we are constantly surrounded by systems which are not in equilibrium, it becomes clear why time seems to have an unambiguous direction which allows us to distinguish clearly the past from the future. Thus we expect people to be born, grow up, and die. We never see the time-reversed process (in principle possible, but fantastically unlikely) where someone rises from his grave, grows progressively younger, and disappears into his mother's womb.

Note that there is nothing intrinsic in the laws of motion of the particles of a system which gives time a preferred direction. Indeed, suppose that one took a movie of the isolated gas in equilibrium, as shown in Fig. 1.4 (or considered the time dependence of the number n of molecules in one half of the box, as indicated in Fig. 1.5). Looking at this movie projected on a screen, we would have no way of

† Indeed, consider the molecules at some time t_1 after they have become uniformly distributed throughout the box. Now suppose that, at some subsequent time t_2, each molecule would again be in exactly the same position as at the time t_1 and would have a velocity of equal magnitude but precisely opposite direction. Then each molecule would retrace its path in time. The gas would thus reconcentrate itself in the left half of the box.

telling whether it was running through the projector forward or backward. The preferred direction of time arises only when one deals with an isolated macroscopic system which is somehow *known* to be in a very special nonrandom situation at a specified time t_1. If the system has been left undisturbed for a very long time and got into this situation as a result of a very rare spontaneous fluctuation in equilibrium, there is indeed nothing special about the direction of time. As already pointed out in connection with the peak X in Fig. 1.8, the system then tends to change toward the most random situation as time goes either forward or backward (i.e., both when the movie is played forward, and when it is played backward starting at the time t_1). The only other way the system can have got into the special nonrandom situation at the time t_1 is by virtue of interaction with some other system at some prior time. But in this case a specific direction of time is singled out by the knowledge that, before being left undisturbed, the system was made to interact with some other system at some time *before* the time t_1.

Finally it is worth pointing out that the irreversibility of spontaneously occurring processes is a matter of degree. The irreversibility becomes more pronounced to the extent that the system contains many particles since the occurrence of an orderly situation then becomes increasingly unlikely compared to the occurrence of a random one.

Example

Consider a box which contains only a single molecule moving around and colliding elastically with the walls. If one took a movie of this system and then watched its projection on a screen, there would never be any way of telling whether the movie is being played forward or backward through the projector.

Consider now a box which contains N molecules of an ideal gas. Suppose that a movie of this gas is projected on a screen and portrays a process in which the molecules of the gas, originally distributed uniformlv throughout the box, all become concentrated in the left half of the box. What could we conclude? If $N = 4$, this kind of process could, in actuality, occur relatively frequently as a result of a spontaneous fluctuation. (On the average, 1 out of every 16 frames of the film would show all the molecules in the left half of the box.) Hence we cannot really tell with appreciable certainty whether the film is being played forward or backward. (See Fig. 1.15.) But if $N = 40$, this kind of process would very rarely occur in actuality as a result of a spontaneous fluctuation. (On the average, only 1 out of $2^N = 2^{40} \approx 10^{12}$ frames of the film would show all the molecules in the left half of the box.) It is much more likely that the film is being played backward and portrays the result of a prior intervention, e.g., the removal of a partition which had previously confined the molecules in the left half of the box. (See Fig. 1.17.) For an ordinary gas where $N \sim 10^{20}$, spontaneous fluctuations of the kind seen on the screen would almost never occur in actuality. One could then be almost completely certain that the film is being played backward.

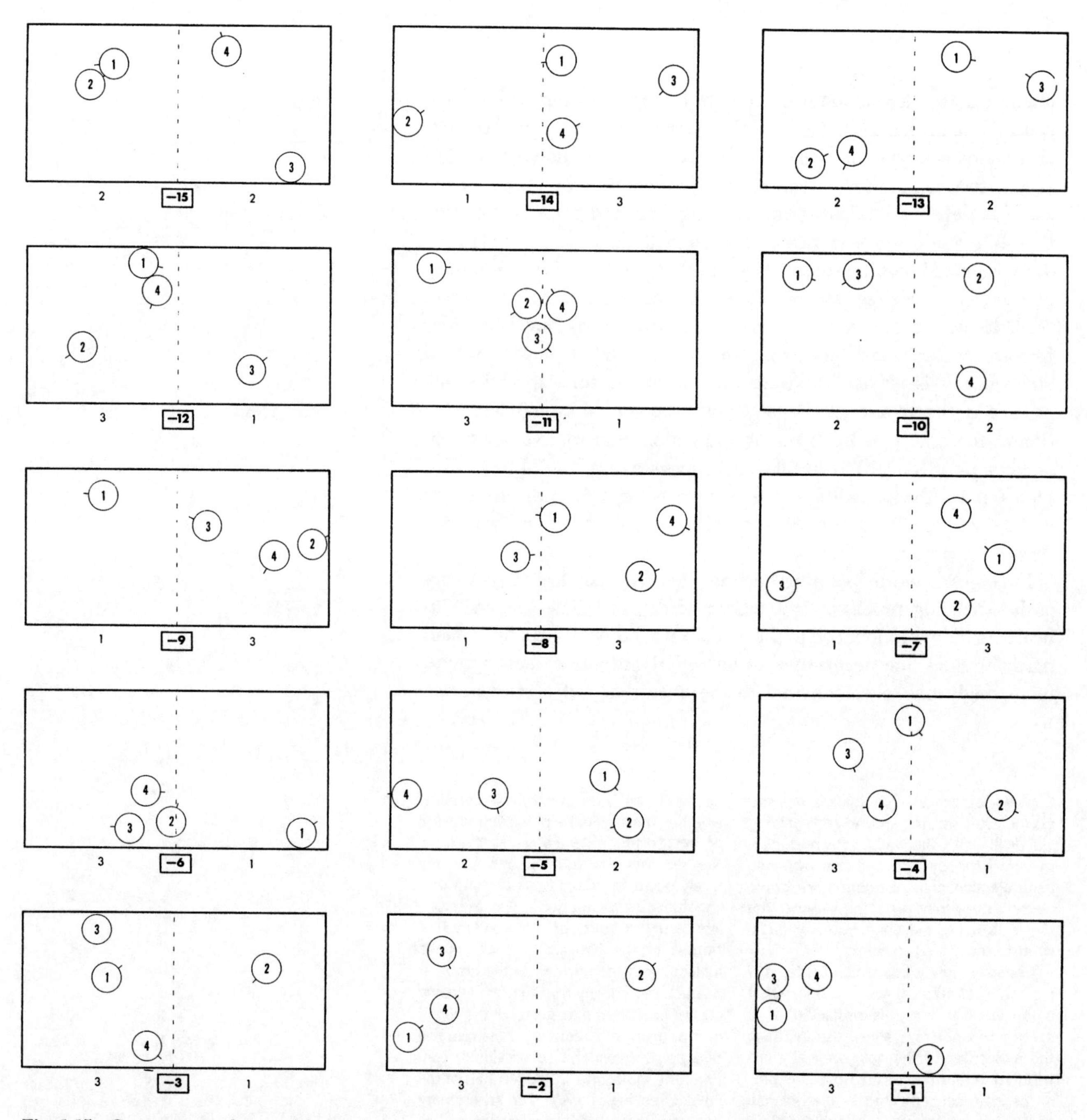

Fig. 1.15 Construction of a possible past history for Fig. 1.16. The pictures on this left-hand page were computed by starting with all the particles in the left half of the box in the positions shown in the frame $j = 0$ on the right-hand page and *reversing* the directions of all the assumed particle velocities in this frame. The resulting evolution of the system in time is then shown by the sequence of frames read in the order $j = 0, -1, -2, \ldots, -15$. The short line segment emanating from each particle indicates the direction of its velocity in this case.

If the velocity of every particle on this left page is now imagined to be reversed in direction, then the sequence of frames in the order $j = -15, -14, \ldots, -1, 0, 1, \ldots, 14$ (extending over both left and right pages) represents a possible motion of the particles in time. This motion, starting from the very special initial situation prevailing in the frame $j = -15$ by virtue of the way it was constructed, then results in a fluctuation where all the particles are found in the left half of the box in the frame $j = 0$.

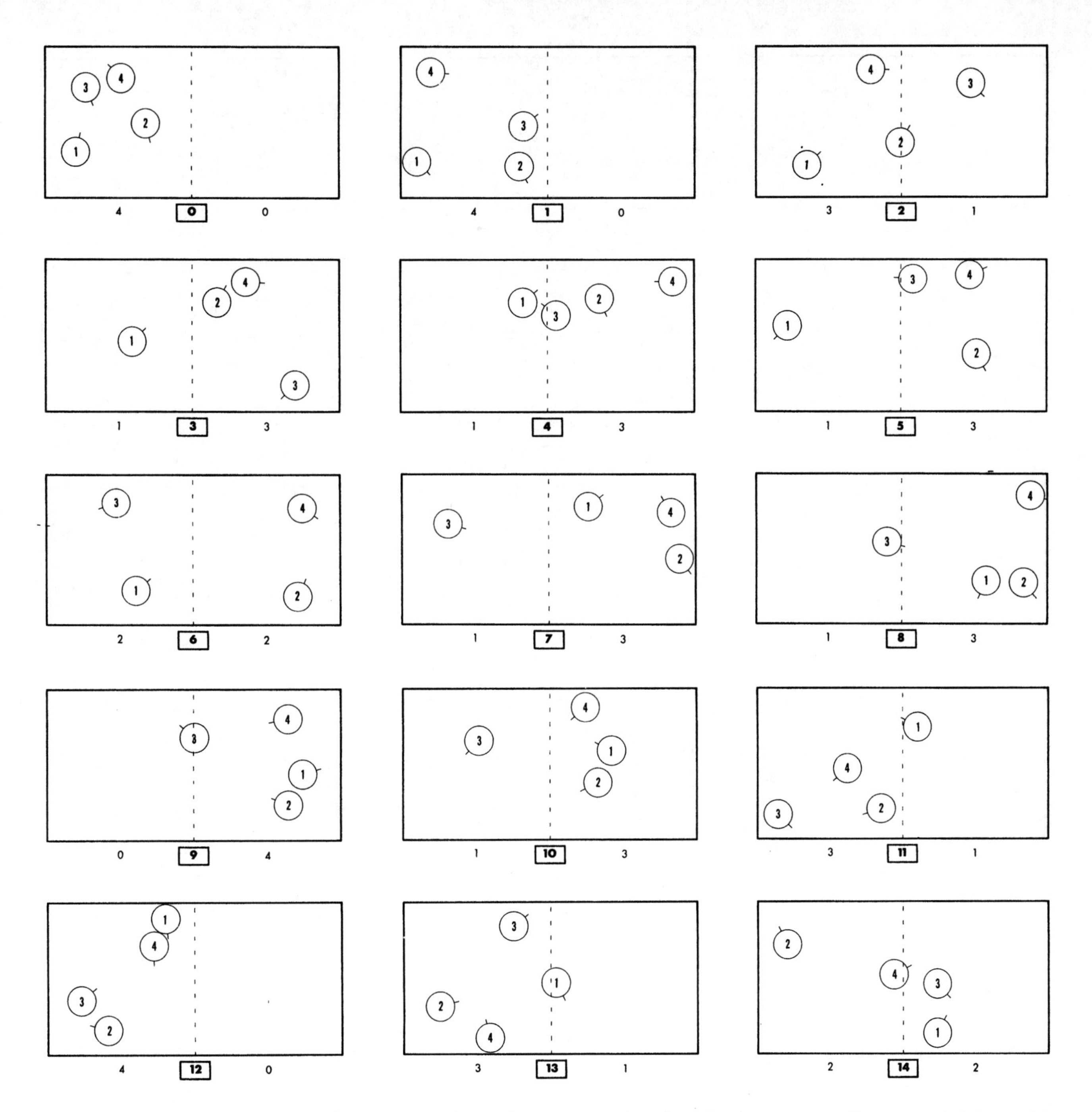

Fig. 1.16 Computer-made pictures showing 4 particles in a box. The pictures were constructed by starting with the special situation where all the particles are in the left half of the box in the positions shown in the frame $j = 0$ and are given some arbitrary assumed velocities. The resulting evolution of the system in time is then shown by the sequence of frames $j = 0, 1, 2, \ldots, 14$. The number of particles located in each half of the box is printed directly beneath that half. The short line segment emanating from each particle indicates the direction of the particle's velocity.

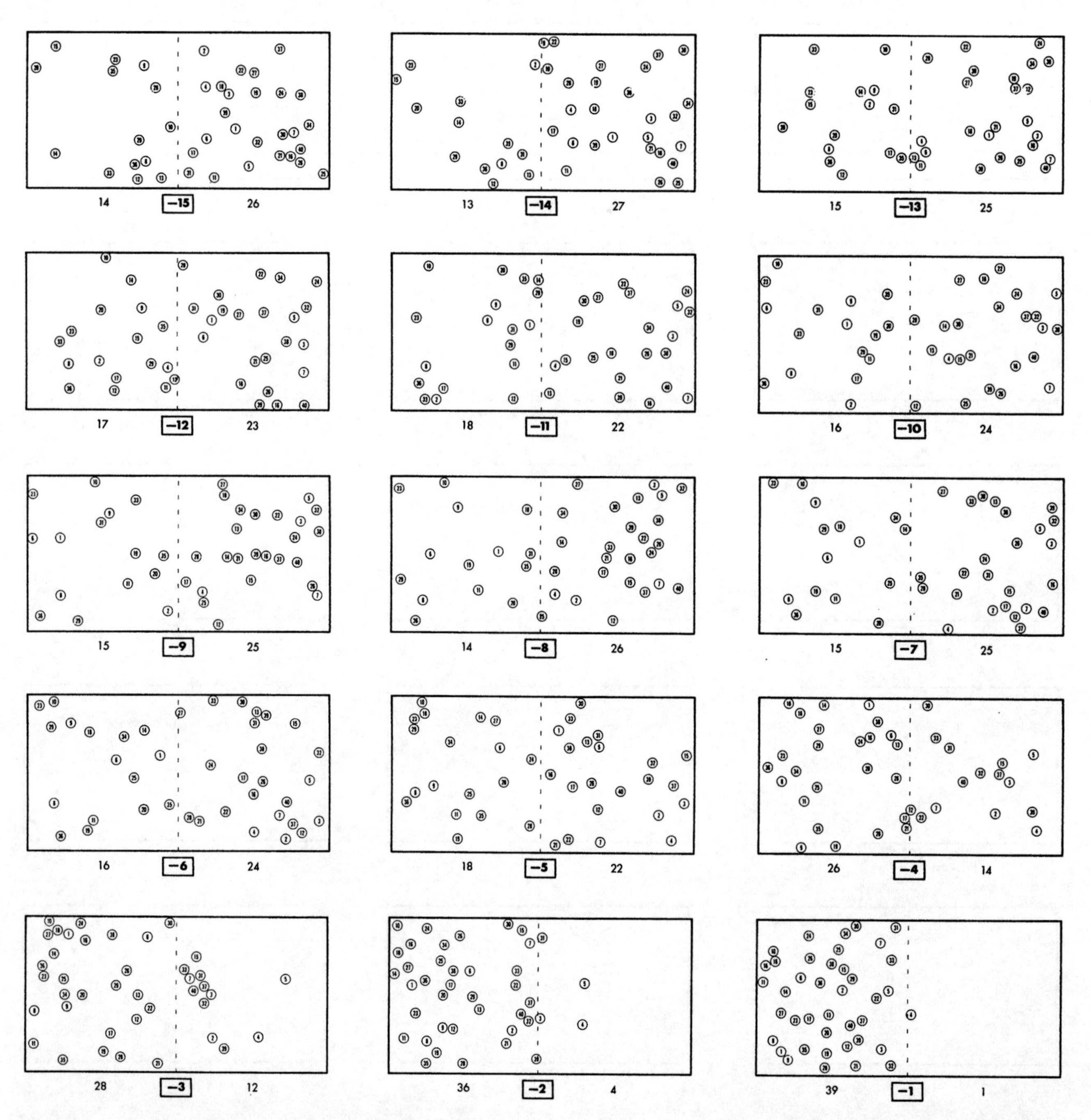

Fig. 1.17 Construction of a possible past history for Fig. 1.18. The pictures on this left-hand page were computed by starting with all the particles in the left half of the box in the positions shown in the frame $j = 0$ on the right-hand page and *reversing* the directions of all the assumed particle velocities in this frame. The resulting evolution of the system in time is then shown by the sequence of frames read in the order $j = 0, -1, -2, \ldots, -15$. No velocities are indicated.

If the velocity of every particle on this left page is now imagined to be reversed in direction, then the sequence of frames in the order $j = -15, -14, \ldots, -1, 0, 1, \ldots, 14$ (extending over both left and right pages) represents a possible motion of the particles in time. This motion, starting from the very special initial situation prevailing in the frame $j = -15$ by virtue of the way it was constructed, then results in a fluctuation where all the particles are found in the left half of the box in the frame $j = 0$.

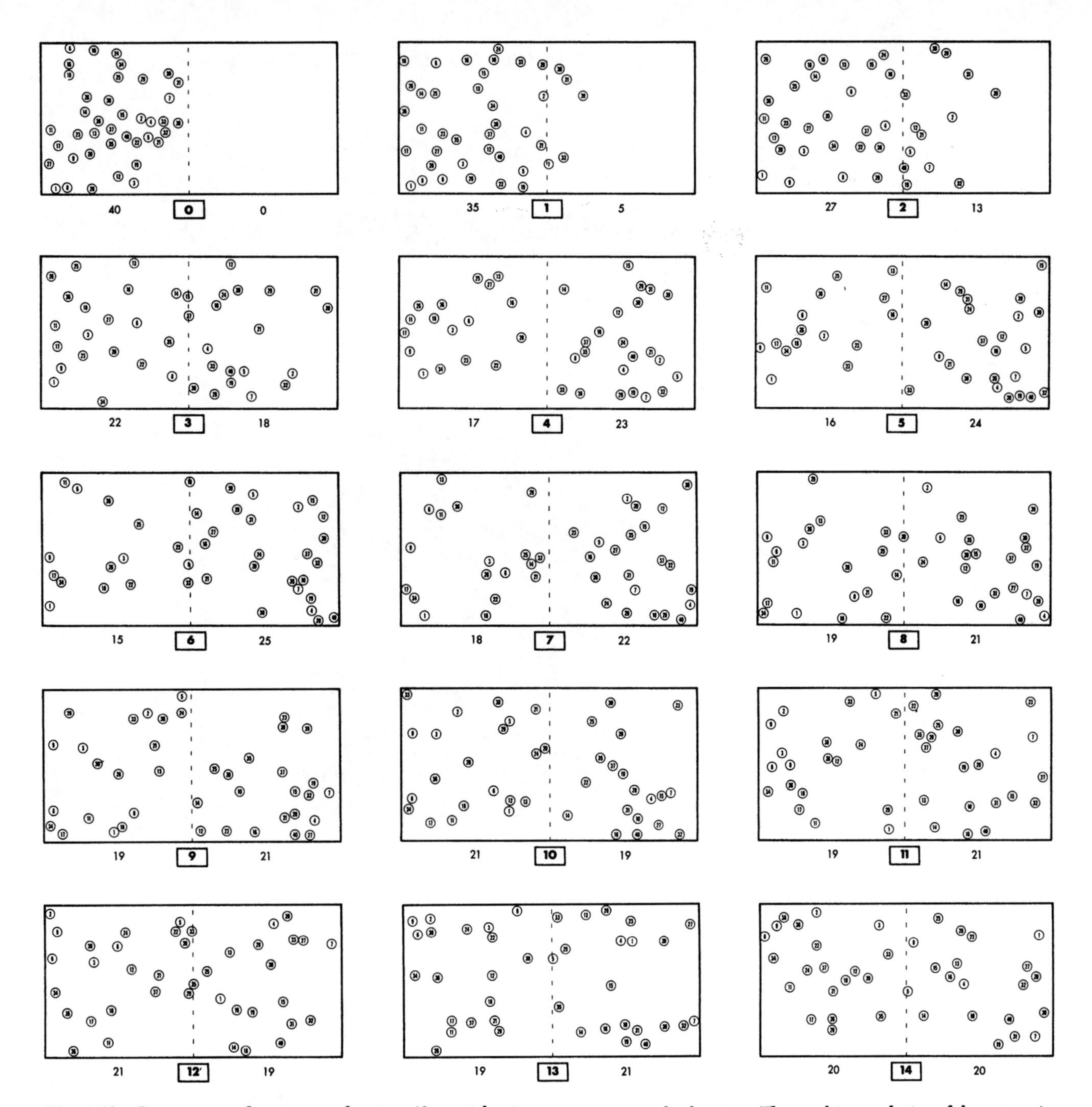

Fig. 1.18 Computer-made pictures showing 40 particles in a box. The pictures were constructed by starting with the special situation where all the particles are in the left half of the box in the positions shown in the frame $j = 0$ and are given some arbitrary assumed velocities. The resulting evolution of the system in time is then shown by the sequence of frames $j = 0, 1, 2, \ldots, 14$. The number of particles located in each half of the box is printed directly beneath that half. No velocities are indicated.

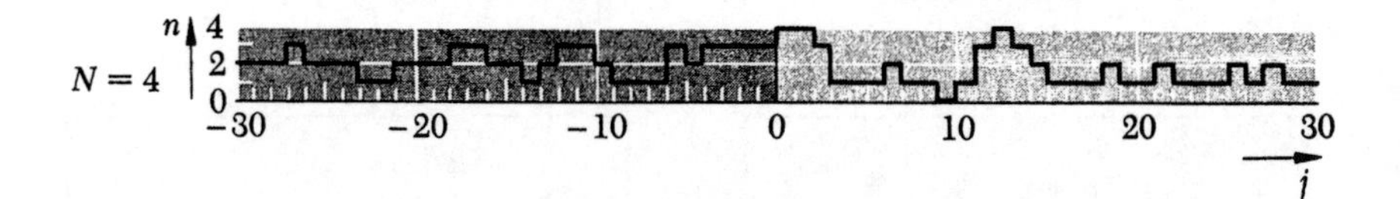

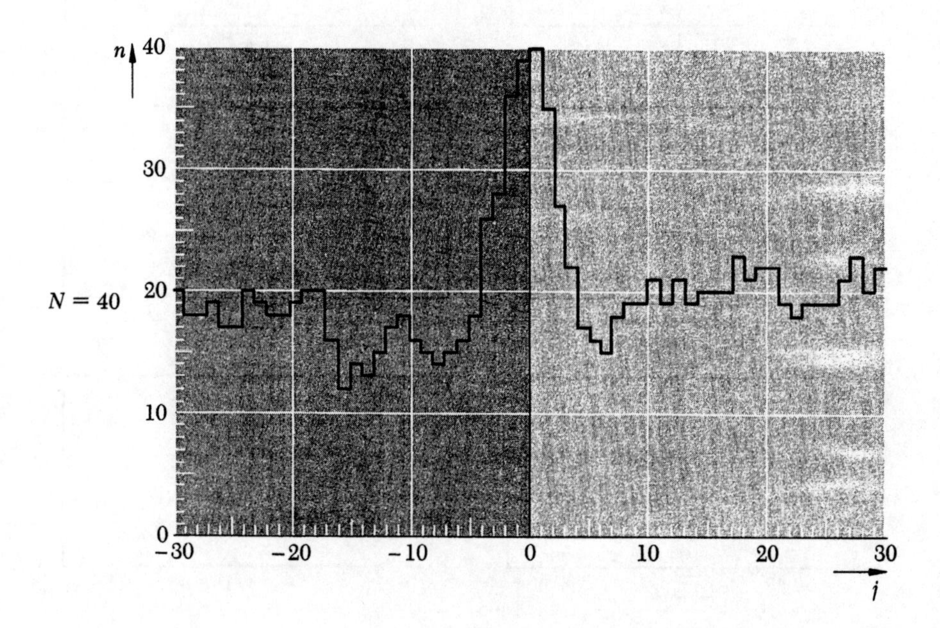

Fig. 1.19 The number n of particles in the left half of the box as a function of the frame index j or the time $t = j\tau_0$. The number n in the jth frame is indicated by a horizontal line extending from j to $j + 1$. The graphs describe Figs. 1.15 and 1.16 for $N = 4$ particles, and Figs. 1.17 and 1.18 for $N = 40$ particles, but contain information about more frames than were shown there. The right half of each graph shows the approach of the system to equilibrium. The entire domain of each graph shows the occurrence of a rare fluctuation which might occur in an equilibrium situation.

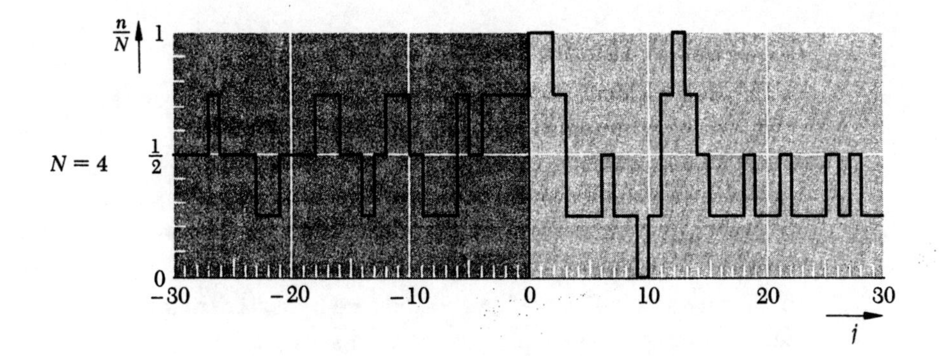

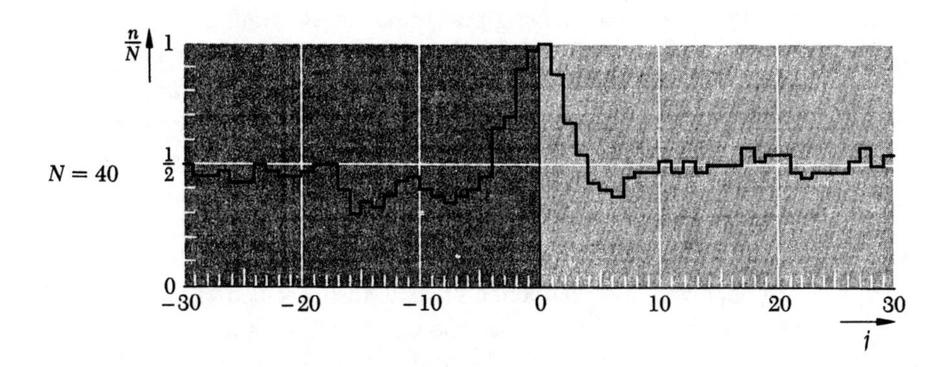

Fig. 1.20 The *relative* number n/N of particles in the left half of the box as a function of the frame index j or the time $t = j\tau_0$. The information presented is otherwise the same as that in Fig. 1.19.

1.3 *Further Illustrations*

By thinking in detail about the simple case of the ideal gas of N molecules, we have grappled with all the essential problems involved in understanding systems consisting of very many particles. Indeed, most of the remainder of this book will consist merely of the systematic elaboration and refinement of the ideas which we have already discussed. To begin, let us illustrate the universal applicability of the basic concepts which we have introduced by considering briefly a few further examples of simple macroscopic systems.

Ideal system of N spins

Consider a system of N particles each of which has a "spin $\frac{1}{2}$" and an associated magnetic moment of magnitude μ_0. The particles might be electrons, atoms having one unpaired electron, or nuclei such as protons. The concept of spin must be described in terms of quantum ideas. The statement that a particle has spin $\frac{1}{2}$ thus implies that a measurement of the component (along a specified direction) of the spin angular momentum of the particle can have only two discrete possible outcomes: The measured component can be either $+\frac{1}{2}\hbar$ or $-\frac{1}{2}\hbar$ (where $\hbar$ denotes Planck's constant divided by 2π), i.e., the spin can be said to point either parallel or antiparallel to the specified direction. Correspondingly, the component (along the specified direction) of the magnetic moment of the particle can be either $+\mu_0$ or $-\mu_0$, i.e., the magnetic moment can also be said to point either parallel or antiparallel to the specified direction. For the sake of simplicity, we shall designate these two possible orientations as "up" or "down," respectively.†

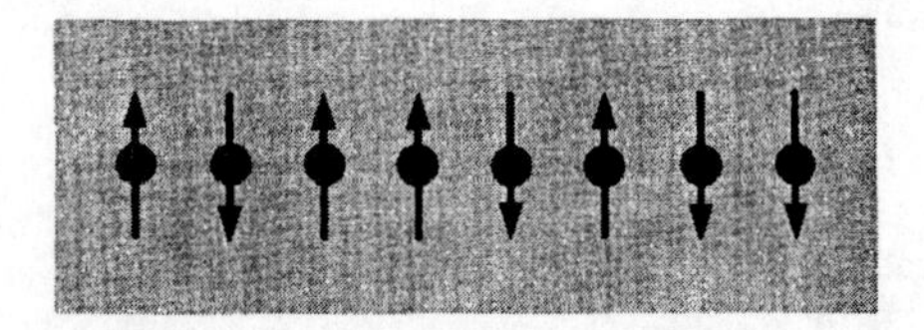

Fig. 1.21 A simple system of particles, each having spin $\frac{1}{2}$. Each spin can point either up or down.

The system of N particles with spin $\frac{1}{2}$ is thus quite similar to a collection of N bar magnets, each having a magnetic moment μ_0 which can point either up or down. For the sake of simplicity, we may regard the particles as essentially fixed in position, as they would be if they were atoms located at the lattice sites of a solid.‡ We shall call the system of spins *ideal* if the interaction between the spins is almost negligible. (This is the case if the average distance between the particles with spin is so large that the magnetic field produced by one moment at the position of another moment is small enough to be nearly negligible.)

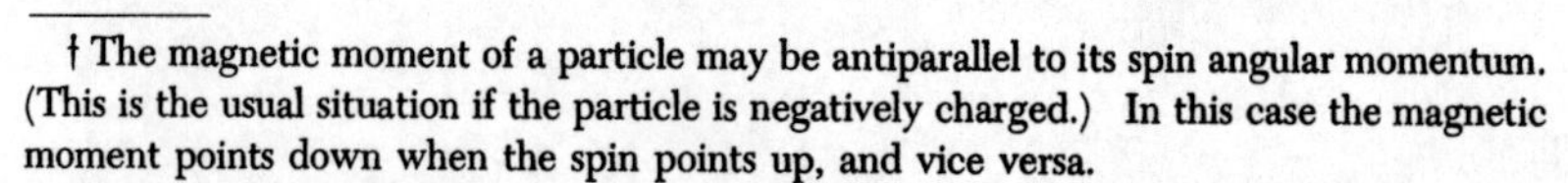

† The magnetic moment of a particle may be antiparallel to its spin angular momentum. (This is the usual situation if the particle is negatively charged.) In this case the magnetic moment points down when the spin points up, and vice versa.

‡ If the particles are free to move in space, their translational motion can usually be treated separately from the orientation of their spins.

The ideal system of N spins has been described entirely in terms of quantum mechanics, but is otherwise completely analogous to the ideal gas of N molecules. In the case of the gas, each molecule moves about and collides occasionally with other molecules; hence it is found sometimes in the left and sometimes in the right half of the box. In the case of the system of spins, each magnetic moment interacts slightly with the other magnetic moments, so that its orientation occasionally changes; hence each magnetic moment is found to point sometimes up and sometimes down. In the case of the isolated ideal gas in equilibrium, each molecule is equally likely to be found in either the left or right half of the box. Similarly, in the case of the isolated system of spins in equilibrium in the absence of any externally applied magnetic field, each magnetic moment is equally likely to be found pointing either up or down. We can denote by n the number of spins pointing up, and by n' the number of spins pointing down. In equilibrium the most random situation where $n \approx n' \approx \frac{1}{2}N$ thus occurs most frequently, while fluctuations where n differs appreciably from $\frac{1}{2}N$ occur very rarely. Indeed, when N is large, nonrandom situations where n differs appreciably from $\frac{1}{2}N$ occur almost always as a result of prior interaction of the isolated system of spins with some other system.

Distribution of energy in an ideal gas

Consider again the isolated ideal gas of N molecules. We reached the general conclusion that the time-independent equilibrium situation attained by the system after a sufficiently long time corresponds to the most random distribution of the molecules. In our previous discussion we focused attention solely on the positions of the molecules. We then saw that the equilibrium of the gas corresponds to the most random distribution of the molecules in space, i.e., to the essentially uniform distribution of the molecules throughout the fixed volume of the box. But what can we say about the velocities of the molecules? Here it is useful to recall the fundamental mechanical principle that the total energy E of the gas must remain constant since the gas is an isolated system. This total energy E is equal to the sum of the energies of the individual gas molecules, since the potential energy of interaction between molecules is negligible. Hence the basic question becomes: How is the fixed total energy of the gas distributed over the individual molecules? (If a molecule is monatomic, its energy ϵ is, of course, merely its kinetic energy $\epsilon = \frac{1}{2}m\mathbf{v}^2$, where m is its mass and $\mathbf{v}$ its velocity.) It is possible that one group of molecules might have very high energies while another group might have very low energies. But

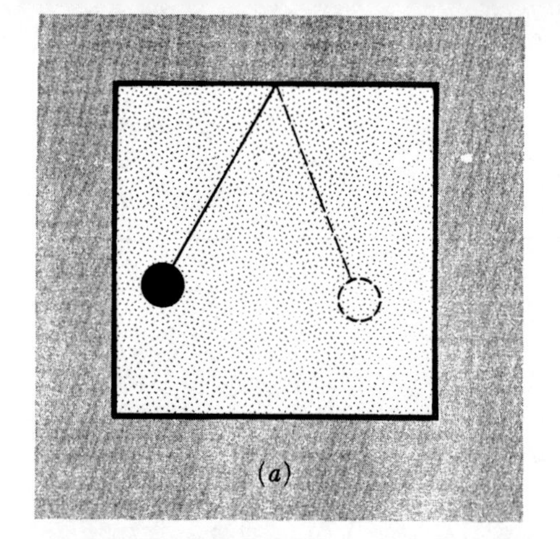

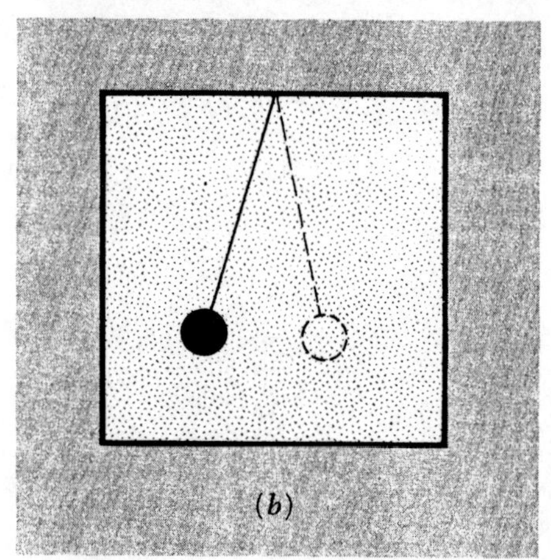

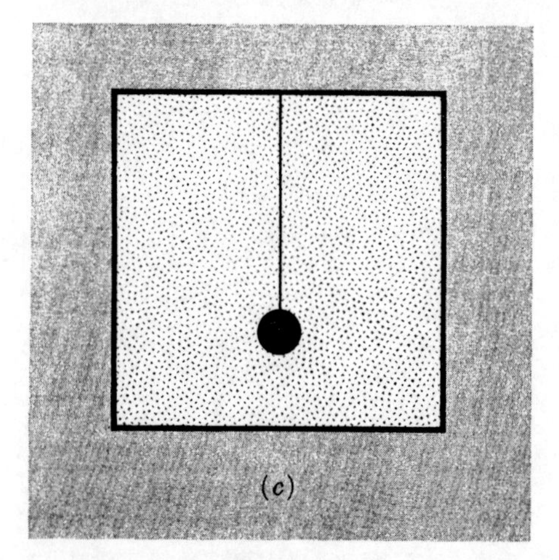

Fig. 1.22 Oscillation of a pendulum immersed in a gas. The pendulum is shown successively (*a*) soon after it is set into oscillation, (*b*) a short time thereafter, and (*c*) after a very long time. A movie played backward would show the pictures in the reverse order (*c*), (*b*), (*a*).

this kind of situation is quite special and would not persist as the molecules collide with each other and thus exchange energy. The time-independent equilibrium situation which is ultimately attained corresponds therefore to the most random distribution of the total energy of the gas over all the molecules. Each molecule has then, on the average, the same energy and thus also the same speed.† In addition, since there is no preferred direction in space, the most random situation of the gas is that where the velocity of each molecule is equally likely to point in any direction.

Pendulum oscillating in a gas

Consider a pendulum which is set oscillating in a box containing an ideal gas. If the gas were not present, the pendulum would continue oscillating indefinitely with no change in amplitude. (We neglect frictional effects that might arise at the point of support of the pendulum.) But in the presence of the gas, the situation is quite different. The molecules of the gas collide constantly with the pendulum bob. In each such collision energy is transferred from the pendulum bob to a molecule, or vice versa. What is the net effect of these collisions? Again this question can be answered by our familiar general argument without the need to consider the collisions in detail.‡ The energy E_b (kinetic plus potential) of the pendulum bob plus the total energy E_g of all the gas molecules must remain constant since the total system is isolated (if one includes the earth which provides the gravitational attraction). If the energy of the bob were transferred to the gas molecules, it could be distributed over these many molecules in many different ways instead of remaining entirely associated with the bob. A much more random situation of the system would thus result. Since an isolated system tends to approach its most random situation, the pendulum gradually transfers practically all its energy to the gas molecules and thus oscillates with ever-decreasing amplitude. This is again a typical irreversible process. After the final equilibrium situation has been reached, the pendulum hangs vertically, except for very small oscillations about this position.

---

† This does *not* mean that each molecule has the same energy at any one time. The energy of any one molecule fluctuates quite appreciably in the course of time as a result of its collisions with other molecules. But when each molecule is observed over a sufficiently long time interval τ, its average energy over that time interval is the same as that of any other molecule.

‡ A detailed analysis would argue that the pendulum bob suffers more collisions per unit time with the molecules located on the side toward which the bob moves than with the molecules located on the other side. As a result, collisions in which the bob loses energy to a molecule are more frequent than collisions in which it gains energy from a molecule.

Note one further point of interest. In the initial nonrandom situation, where the pendulum bob has a large amount of energy associated with it, this energy can be exploited to do useful work on a macroscopic scale. For example, the pendulum bob could be made to hit a nail so as to drive it some distance into a piece of wood. After the final equilibrium has been reached, the energy of the pendulum bob has not been lost; it has merely become redistributed over the many molecules of the gas. But there is now no easy way to use this energy to do the work necessary to drive the nail into the wood. Indeed, this would necessitate some method for concentrating energy, randomly distributed over the many gas molecules moving in many directions, so that it can exert a net force in only *one* particular direction over an appreciable distance.

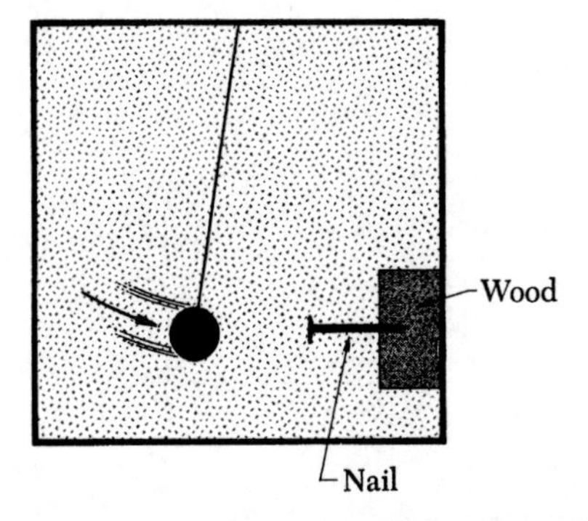

Fig. 1.23 Arrangement in which the pendulum bob can be made to strike a nail and thus do work on it by driving it into a piece of wood.

1.4 *Properties of the Equilibrium Situation*

Simplicity of the equilibrium situation

The discussion of the preceding sections shows that the equilibrium situation of a macroscopic system is particularly simple. The reasons are the following:

(i) The macrostate of a system in equilibrium is time-independent, except for ever-present fluctuations. Quite generally, the macrostate of a system can be described by certain *macroscopic parameters,* i.e., parameters which characterize the properties of the system on a large scale. (For example, the number n of molecules located in the left half of a box of gas is such a macroscopic parameter.) When the system is in equilibrium, the average values of all its macroscopic parameters remain constant in time, although the parameters themselves may exhibit fluctuations (ordinarily quite small) about their average values. The equilibrium situation of a system is, therefore, simpler to treat than the more general nonequilibrium case where some macroscopic parameters of the system tend to change in time.

(ii) The macrostate of a system in equilibrium is, except for fluctuations, the most random macrostate of the system under the specified conditions. The system in equilibrium is thus characterized in a unique way. In particular, this has the following implications:

(*a*) The equilibrium macrostate of a system is independent of its past history. For example, consider an isolated gas of N molecules in a box. These molecules may originally have been confined by a partition to one half of the box or to one quarter of the box (the total energy of the molecules being assumed the same in each case). But after the partition is removed and equilibrium has been attained, the macrostate of the gas is the same in both cases; it corresponds merely to the uniform distribution of all the molecules through the entire box.

(*b*) The equilibrium macrostate of a system can be completely specified by very few macroscopic parameters. For example, consider again the isolated gas of N identical molecules in a box. Suppose that the volume of the box is V, while the constant total energy of all the molecules is E. If the gas is in equilibrium and thus known to be in its most random situation, then the molecules must be uniformly distributed throughout the volume V and must, on the average, share equally the total energy E available to them. A knowledge of the macroscopic parameters V and E is, therefore, sufficient to conclude that the average number $\bar{n}_s$ of molecules in any subvolume V_s of the box is $\bar{n}_s = N(V_s/V)$, and that the average energy $\bar{\epsilon}$ per molecule is $\bar{\epsilon} = E/N$. If the gas were not in equilibrium, the situation would of course be much more complicated. The distribution of molecules would ordinarily be highly nonuniform and a mere knowledge of the total number N of molecules in the box would thus be completely insufficient for determining the average number $\bar{n}_s$ of molecules in any given subvolume V_s of the box.

Observability of fluctuations

Consider a macroscopic parameter describing a system consisting of many particles. If the number of particles in the system is large, the relative magnitude of the fluctuations exhibited by the parameter is ordinarily very small. Indeed, it is often so small as to be utterly negligible compared to the average value of the parameter. As a result, we remain usually unaware of the existence of fluctuations when we are dealing with large macroscopic systems. On the other hand, the ever-present fluctuations may be readily observed and may become of great practical importance if the macroscopic system under consideration is fairly small or if our methods of observation are quite sensitive. Several examples will serve to illustrate these comments.

Density fluctuations in a gas

Consider an ideal gas which is in equilibrium and consists of a large number N of molecules confined within a box of volume V. Focus attention on the number n_s of molecules located within some specified subvolume V_s inside the box. This number n_s fluctuates in time about an average value

$$\bar{n}_s = \frac{V_s}{V}N,$$

the magnitude of its fluctuation at any time being given by the difference

$$\Delta n_s = n_s - \bar{n}_s.$$

If we consider the left half of the box as the part of interest, $V_s = \frac{1}{2}V$ and $\bar{n}_s = \frac{1}{2}N$. When V_s is large, the average number $\bar{n}_s$ of molecules is also large. In accordance with our discussion of Sec. 1.1, the only fluctuations which occur with appreciable frequency are then those sufficiently small so that $|\Delta n_s| \ll \bar{n}_s$.

On the other hand, suppose that we wanted to investigate the scattering of light by a substance. Then we would be interested in knowing what happens in a volume element V_s having linear dimensions of the order of the wavelength of light. [Since the wavelength of visible light (about 5×10^{-5} cm) is much larger than atomic dimensions, such a volume element is still macroscopic in size, although it is small.] If the number of molecules in every such volume element were the same (as is very nearly the case in a solid such as glass), then the substance under consideration would be spatially uniform and would merely refract a beam of light without scattering it. But in the case of our ideal gas, the average number $\bar{n}_s$ of molecules in a volume as small as V_s is quite small and fluctuations Δn_s in the number n_s of molecules within V_s are no longer negligible compared to $\bar{n}_s$. The gas can, therefore, be expected to scatter light to an appreciable extent. Indeed, the fact that the sky does not look black is due to the fact that light from the sun is scattered by the gas molecules of the atmosphere. The blue color of the sky thus provides visible evidence for the importance of fluctuations.

Fluctuations of a torsion pendulum

Consider a thin fiber stretched between two supports (or suspended from one support under the influence of gravity) and carrying an attached mirror. When the mirror turns through a small angle, the twisted fiber provides a restoring torque. The mirror is thus capable of performing small angular oscillations and constitutes accordingly a torsion pendulum. Since the restoring torque of a thin fiber can be made very small and since a beam of light reflected from the mirror provides a very effective way of detecting small angular deflections of the mirror, a torsion fiber is commonly used for very sensitive measurements of small torques. For example, it may be recalled that a torsion pendulum was used by Cavendish to measure the universal constant of gravitation and by Coulomb to measure the electrostatic force between charged bodies.

When a sensitive torsion pendulum is in equilibrium, its mirror is not perfectly still, but can be seen to perform erratic angular oscillations about its average equilibrium orientation. (The situation is analogous to that of the ordinary pendulum, discussed in Sec. 1.3, which exhibits in equilibrium small fluctuations about its vertical position.) These fluctuations are ordinarily caused by the random impacts of the surrounding air molecules on the mirror.

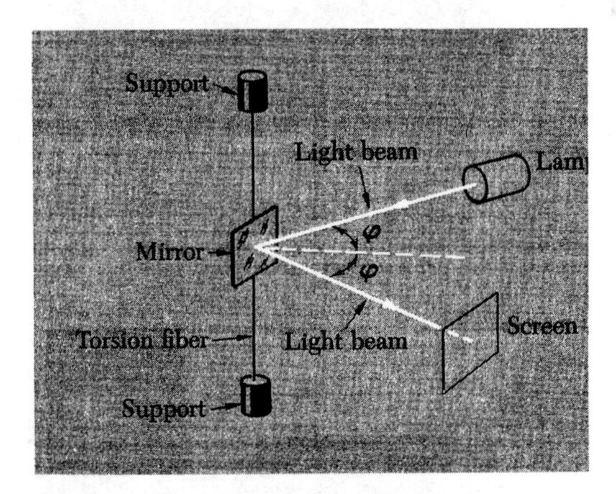

Fig. 1.24 Torsion pendulum formed by a mirror mounted on a thin fiber. A beam of light reflected from the mirror indicates the angle of rotation φ of the mirror.

[The fluctuations of the mirror would be modified, but would not disappear, even if all the molecules of the surrounding gas were removed. In that case the total energy of the torsion pendulum would still consist of two parts, the energy E_ω due to the angular velocity of the mirror moving as a whole, plus the energy E_i due to the internal motion of all the atoms of the mirror and fiber. (The atoms are free to perform small vibrations about their sites in the solids which constitute the mirror and fiber.) Although the total energy $E_\omega + E_i$ of the torsion pendulum is constant, fluctuations do occur in the way this energy is shared between E_ω and E_i. Any fluctuation in which E_ω gains energy at the expense of the internal motion of the atoms thus results in an increased angular velocity of the mirror, and vice versa.]

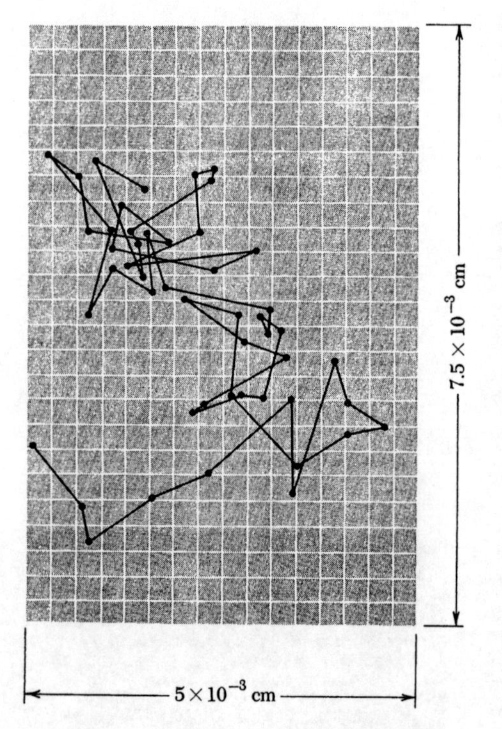

Brownian motion of a particle

Small solid particles, about 10^{-4} cm in size, can be introduced into a drop of liquid and then observed under a microscope. Such particles are not at rest, but are seen to be constantly moving about in a highly irregular way. This phenomenon is called *Brownian motion* because it was first observed in the last century by an English botanist named Brown. He did not understand the origin of the phenomenon. It remained for Einstein in 1905 to give the correct explanation in terms of random fluctuations to be expected in equilibrium. A solid particle is subject to a fluctuating net force due to the many random collisions of the particle with the molecules of the liquid. Since the particle is small, the number of molecules with which it collides per unit time is relatively small and accordingly fluctuates appreciably. In addition, the mass of the particle is so small that any collision has a noticeable effect on the particle. The resulting random motion of the particle thus becomes large enough to be observable.

Fig. 1.25 Brownian motion of a solid particle, 10^{-4} cm in diameter, suspended in water and observed through a microscope. The three-dimensional motion of such a particle, as seen projected in the horizontal plane of the field of view of the microscope, is shown by this diagram where the lines join consecutive positions of the particle as observed at 30-second intervals. [The data are from J. Perrin, *Atoms,* p. 115 (D. Van Nostrand Company, Inc., Princeton, N.J., 1916).]

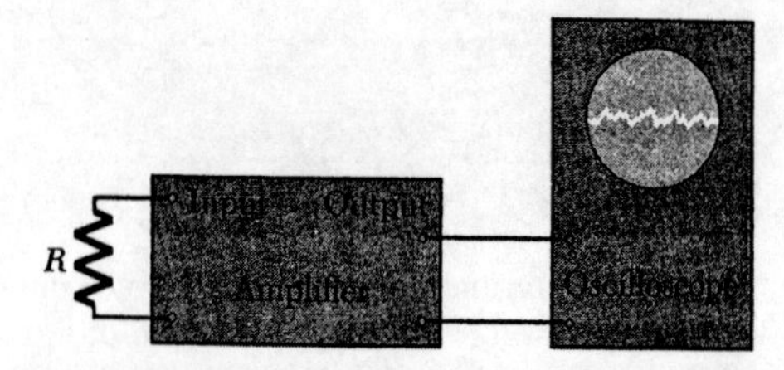

Fig. 1.26 A resistor R connected to the input terminals of a sensitive amplifier whose output is displayed on an oscilloscope.

Voltage fluctuations across a resistor

If an electrical resistor is connected across the input terminals of a sensitive electronic amplifier, the output of the amplifier is observed to exhibit random voltage fluctuations. Neglecting noise originating in the amplifier itself, the basic reason is the random Brownian motion of the electrons in the resistor. Suppose, for example, that this random motion leads to a fluctuation where the number of electrons is greater in one half of the resistor than in the other. The resulting charge difference then leads to an electric field in the resistor, and hence to a potential difference across its ends. Variations in this potential difference thus give rise to the voltage fluctuations amplified by the electronic instrument.

The existence of fluctuations can have important practical consequences. This is particularly true whenever one is interested in measuring small effects or signals, since these may be obscured by the intrinsic fluctuations ever-present in the measuring instrument. (These fluctuations are then said to constitute *noise* since their presence makes measurements difficult.) For example, it is difficult to use a torsion fiber to measure a torque which is so small that the angular deflection produced by it is less than the magnitude of the intrinsic fluctuations in angular position displayed by the mirror. Similarly, in the case of the resistor connected to the amplifier, it is difficult to measure an applied voltage across this resistor if this voltage is smaller than the magnitude of the intrinsic voltage fluctuations always present across the resistor.†

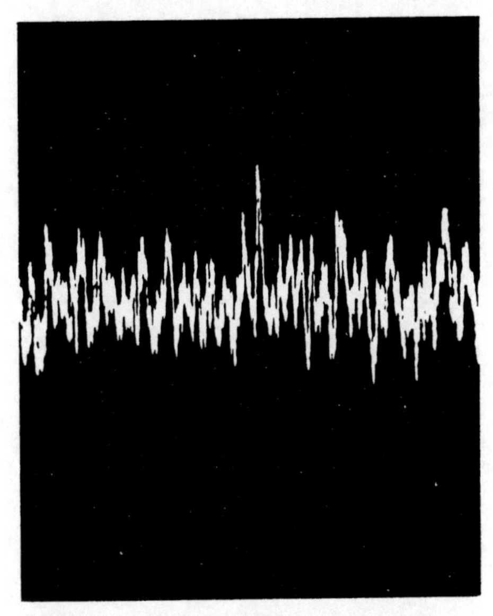

Fig. 1.27 Actual photograph of the noise output voltage displayed on an oscilloscope in the experimental arrangement of Fig. 1.26. (*Photograph by courtesy of Dr. F. W. Wright, Jr., University of California, Berkeley.*)

1.5 *Heat and Temperature*

Macroscopic systems which are not isolated can interact and thus exchange energy. One obvious way in which this can happen is if one system does macroscopically recognizable work upon some other system. For example, in Fig. 1.28 the compressed spring A' exerts a net force on the piston confining the gas A. Similarly, in Fig. 1.29, the compressed gas A' exerts a net force on the piston confining the gas A. When the piston is released and moves through some macroscopic distance, the force exerted by A' performs some work‡ on the system A.

It is, however, quite possible that two macroscopic systems can interact under circumstances where no macroscopic work is done. This type of interaction, which we shall call *thermal interaction*, occurs because energy can be transferred from one system to the other system on an atomic scale. The energy thus transferred is called *heat*. Imagine that the piston in Fig. 1.29 is clamped in position so that it cannot move. In this case one system cannot do macroscopic work upon the other, irrespective of the net force exerted on the piston. On the other hand, the atoms of system A do interact (or collide) with each other almost constantly and thus exchange energy among themselves.§

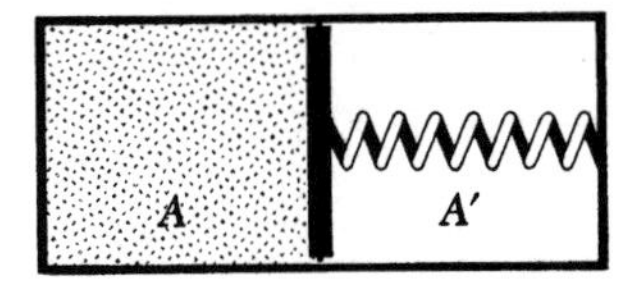

Fig. 1.28 The compressed spring A' does work on the gas A when the piston moves through some net macroscopic distance.

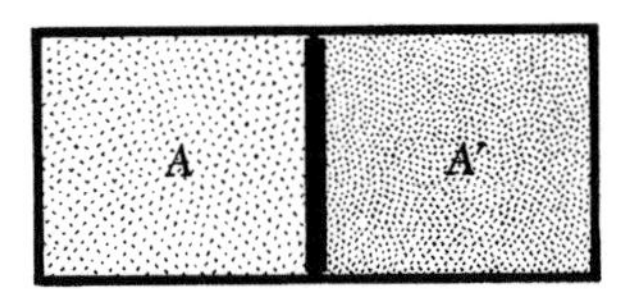

Fig. 1.29 The compressed gas A' does work on the gas A when the piston moves through some net macroscopic distance.

† Typically, voltages of the order of one microvolt or less may become difficult to measure. By averaging the measurement over a sufficiently long time, one may, however, discriminate against the random fluctuations in favor of the applied signal which does not fluctuate in time.

‡ The term *work* is used here in its usual sense familiar from mechanics and is thus defined basically as a force multiplied by the distance through which it acts.

§ If each molecule of the gas consists of more than one atom, different molecules can exchange energy by colliding with each other; furthermore, the energy of a single molecule can also be exchanged between its constituent atoms as a result of interaction between them.

Similarly, the atoms of system A' exchange energy among themselves. At the piston, which forms the boundary between the gases A and A', the atoms of A interact with the atoms of the piston, which then interact among themselves and then in turn interact with the atoms of A'. Hence energy can be transferred from A to A' (or from A' to A) as a result of many successive interactions between the atoms of these systems.

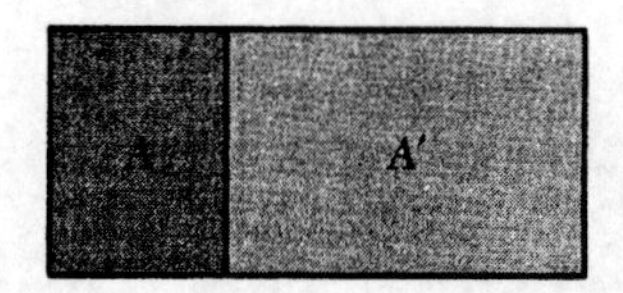

Fig. 1.30 Schematic diagram showing two general systems A and A' in thermal contact with each other.

Consider then any two systems A and A' in thermal interaction with each other. For example, the two systems might be the two gases A and A' just discussed; or A might be a copper block immersed in a system A' consisting of a container filled with water. Let us denote by E the energy of system A (i.e., the total energy, kinetic plus potential, of all the atoms in A); similarly let us denote by E' the energy of system A'. Since the combined system $A^* \equiv A + A'$ consisting of both A and A' is supposed to be insulated, the total energy

$$E + E' = \text{constant.}\dagger \tag{8}$$

The question arises, however, how this total energy is actually distributed between the systems A and A'. In particular, suppose that the systems A and A' are in equilibrium with each other, i.e., that the combined system A^* is in equilibrium. Except for small fluctuations, this equilibrium situation of A^* must then correspond to the most random distribution of energy in this system.

Let us first discuss the simple situation where the systems A and A' are two ideal gases consisting of molecules of the same kind. [For example, both A and A' might consist of nitrogen (N_2) molecules.] In this case the most random situation of the combined system A^* is clearly that where its total energy $E + E'$ is shared indiscriminately among all the identical molecules of A^*. Each molecule of A and A' should then have the same average energy. In particular, the average energy $\bar{\epsilon}$ of a molecule of gas A should be the same as the average energy $\bar{\epsilon}'$ of a molecule of gas A', i.e.,

$$\text{in equilibrium,} \qquad \bar{\epsilon} = \bar{\epsilon}'. \tag{9}$$

Of course, if there are N molecules in gas A and N' molecules in gas A', then

$$\bar{\epsilon} = \frac{E}{N} \quad \text{and} \quad \bar{\epsilon}' = \frac{E'}{N'}. \tag{10}$$

† In the case of the system of Fig. 1.29, we assume for the sake of simplicity that the container walls and the piston are so thin that their energies are negligibly small compared to the energies of the gases.

Hence the condition (9) can also be written in the form

$$\frac{E}{N} = \frac{E'}{N'}.$$

Suppose that the gases A and A' are initially separated from each other and separately in equilibrium. We denote their energies under these circumstances by E_i and E_i', respectively. Imagine now that the systems A and A' are placed in contact with each other so that they are free to exchange energy by thermal interaction. There are two cases which may arise:

(i) Ordinarily, the initial energies E_i and E_i' of the systems are such that the average initial energy $\bar{\epsilon}_i = E_i/N$ of a molecule of A is not the same as the average initial energy $\bar{\epsilon}_i' = E_i'/N'$ of a molecule of A'; i.e.,

ordinarily, $$\bar{\epsilon}_i \neq \bar{\epsilon}_i'. \tag{11}$$

In this case the initial distribution of energy in the combined system A^* is quite nonrandom and will not persist in time. Instead, the systems A and A' will exchange energy until they ultimately attain the equilibrium situation corresponding to the most random distribution of energy, that where the average energy per molecule is the same in both systems. The energies E_f and E_f' of the systems A and A' in the final equilibrium situation must then be such that

$$\bar{\epsilon}_f = \bar{\epsilon}_f' \quad \text{or} \quad \frac{E_f}{N} = \frac{E_f'}{N'}. \tag{12}$$

In the interaction process leading to the final equilibrium situation the system with the smaller average initial energy per molecule thus gains energy, while the system with the larger average initial energy per molecule loses energy. Of course, the total energy of the isolated combined system A^* remains constant so that

$$E_f' + E_f = E_i' + E_i.$$

Thus $$\Delta E + \Delta E' = 0, \tag{13}$$

or $$Q + Q' = 0, \tag{14}$$

where we have written

and $$\begin{aligned} Q &\equiv \Delta E \equiv E_f - E_i, \\ Q' &\equiv \Delta E' \equiv E_f' - E_i'. \end{aligned} \tag{15}$$

The quantity Q is called *the heat absorbed by* A in the interaction process and is defined as the increase in energy of A resulting from the thermal interaction process. A similar definition holds for the heat Q' absorbed by A'.

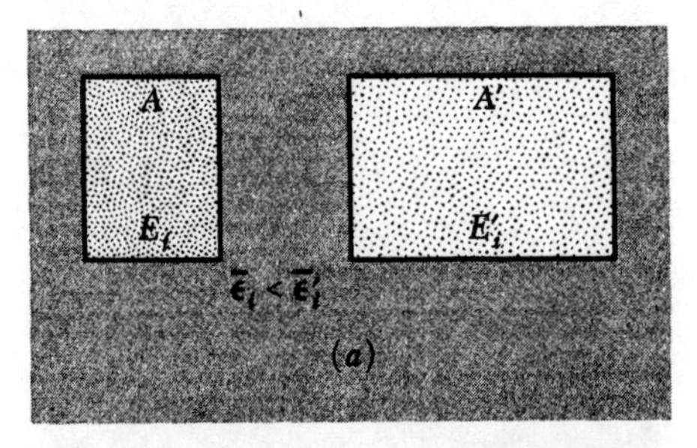

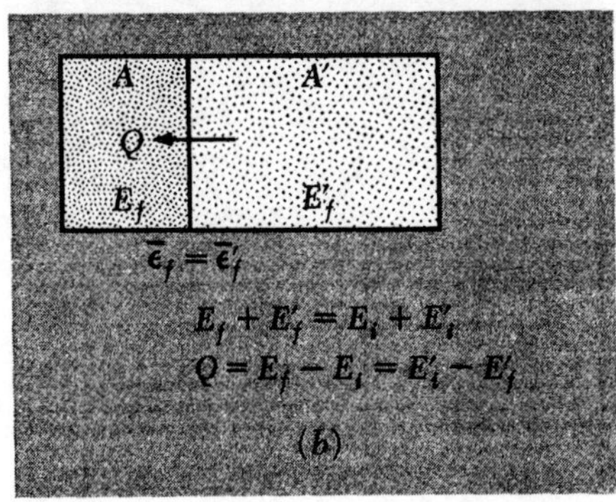

Fig. 1.31 Two gases A and A', consisting of molecules of the same kind, are initially separated from each other in (*a*). They are then brought into thermal contact with each other in (*b*) and are allowed to exchange heat until they reach equilibrium. The energy of a gas is denoted by E, the average energy per molecule by $\bar{\epsilon}$.

Note that the heat $Q = \Delta E$ absorbed by a system may be either positive or negative. Indeed, in the thermal interaction between two systems, one loses energy while the other gains energy; i.e., in (14) either Q is positive and Q' is negative, or vice versa. By definition, the system which gains energy by absorbing a positive amount of heat is said to be the *colder* system; on the other hand, the system which loses energy by absorbing a negative amount of heat (i.e., by "giving off" a positive amount of heat) is said to be the *warmer* or *hotter* system.

(ii) A special case may arise where the initial energies E_i and E_i' of the systems A and A' are such that the average energy of a molecule of A happens to be the same as that of a molecule of A'; i.e., it may happen that

$$\bar{\epsilon}_i = \bar{\epsilon}_i'. \tag{16}$$

In this case, when the systems A and A' are brought into thermal contact with each other, they automatically give rise to the most random situation of the combined system A^*. Thus the systems remain in equilibrium and there occurs no net transfer of energy (or heat) between the systems.

Temperature

Consider now the general case of thermal interaction between two systems A and A'. These systems might be different gases whose molecules have different masses or consist of different kinds of atoms; one or both of the systems might also be liquids or solids. Although the conservation of energy (8) is still valid, it now becomes more difficult to characterize the equilibrium situation corresponding to the most random distribution of the energy over all the atoms of the combined system $A + A'$. Qualitatively, most of our preceding comments concerning the case of two similar gases ought, however, still to be applicable. What we should then expect (and shall explicitly show later on) is the following: Each system, such as A, is characterized by a parameter T (conventionally called its "absolute temperature") which is related to the average energy per atom in the system. In the equilibrium situation corresponding to the most random distribution of energy we then expect, analogously to (9), a condition of the form

$$T = T'. \tag{17}$$

It is not possible to define the concept of absolute temperature more precisely until we have given a more precise definition of the concept of randomness (applicable to the distribution of energy over unlike atoms). It is, however, easy to introduce the concept of a temperature (*not* the "absolute temperature") measured with a particular thermometer. By a *thermometer* we mean any small macroscopic system M arranged so that only one of its macroscopic parameters changes when the system M absorbs or gives off heat. This parameter is called the *thermometric parameter* of the thermometer and will be denoted by the Greek letter θ. For example, the familiar mercury or alcohol thermometer is a special example of a thermometer. Here the length L of the column of liquid in the glass capillary tube of the thermometer changes when the energy of the liquid changes as a result of a transfer of heat. The thermometric parameter θ of this kind of thermometer is then the length L. If the thermometer M is placed in thermal contact with some other system A and allowed to come to equilibrium, its thermometric parameter θ assumes some value θ_A. This value θ_A is called the *temperature of the system A with respect to the particular thermometer M.*†

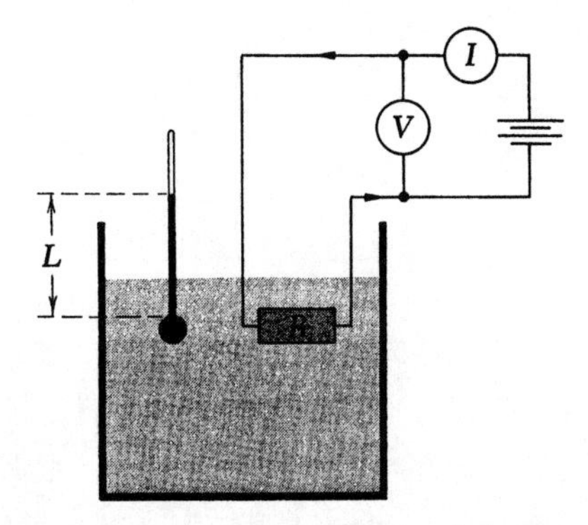

Fig. 1.32 Two different kinds of thermometers in thermal contact with a system consisting of a container filled with a liquid. One thermometer is a mercury thermometer; its thermometric parameter is the length L of the mercury column in the glass capillary tube. The other thermometer is an electrical resistor R made, for example, either of platinum wire or carbon; its thermometric parameter is its electrical resistance R (determined by passing a small current I through the resistor and measuring the voltage V across it).

The usefulness of a thermometer is made apparent by the following considerations. Suppose that the thermometer M is placed in thermal contact first with a system A and then with a system B. In each case the thermometer is allowed to come to equilibrium; it then indicates the respective temperatures θ_A and θ_B. Two cases may arise: either $\theta_A \neq \theta_B$ or $\theta_A = \theta_B$. It is a familiar experimental fact (which will later be justified theoretically) that if $\theta_A \neq \theta_B$, the systems A and B will exchange heat when brought into thermal contact with each other; but, if $\theta_A = \theta_B$, the systems will not exchange heat on thermal contact. The temperature θ of a system measured with respect to the particular thermometer is thus a very useful parameter characterizing the system, since a knowledge of the temperature allows the following predictive statement: If two systems are brought into thermal contact with each other, they will exchange heat if their temperatures are unequal and will not exchange heat if their temperatures are equal.

1.6 Typical Magnitudes

The preceding sections have shown us in broad outline how the behavior of macroscopic systems can be understood in terms of their constituent molecules or atoms. Our considerations have, however, been quite qualitative. To complete our preliminary orientation, we should also like to acquire some perspective about representative orders of

† The system A is supposed to be much larger than the thermometer M so that it is disturbed negligibly by the small amount of energy gained or lost to the thermometer. Note also that, according to our definition, the temperature measured by a mercury thermometer is a length and is thus to be measured in units of centimeters.

magnitude. For example, it would be worth knowing how fast a typical molecule moves or how frequently it collides with other molecules. To answer such questions, we may again turn to the simple case of an ideal gas.

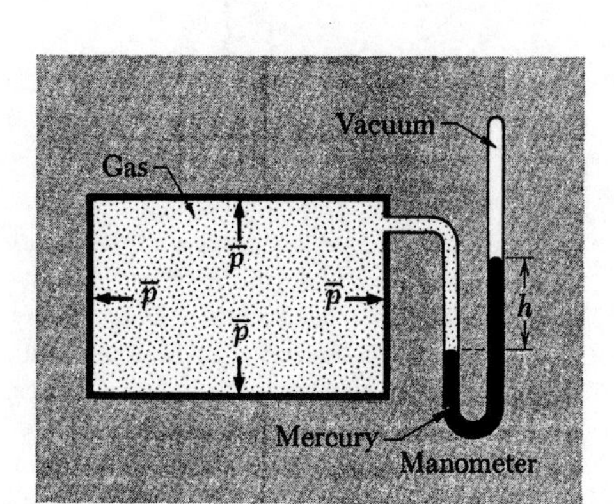

Fig. 1.33 A gas whose average pressure $\bar{p}$ is measured by means of a manometer consisting of a U-tube filled with mercury. In order to be in mechanical equilibrium, the mercury levels must adjust themselves so that the mercury column of height h has a weight, per unit cross-sectional area, equal to the pressure exerted by the gas.

Pressure of an ideal gas

When a gas is confined within a container, the many collisions of the gas molecules with the walls of the container give rise to a net force on each element of area of the walls. The force per unit area is called the *pressure* p of the gas. The average pressure $\bar{p}$ exerted by the gas is readily measured, for example by means of a manometer. The gas pressure should be calculable in terms of molecular quantities; conversely, it should then be possible to use the measured gas pressure to deduce the magnitudes of molecular quantities. We shall begin, therefore, by giving a simple approximate calculation of the pressure exerted by an ideal gas.

To be specific, we shall consider an ideal gas of N molecules, each having a mass m. We suppose that the gas is in equilibrium and that it is confined within a box in the shape of a rectangular parallelepiped having a volume V. The number of molecules per unit volume is conveniently denoted† by $n \equiv N/V$. The edges of the box can be assumed to be parallel to a set of cartesian x, y, z axes, as shown in Fig. 1.34.

We focus attention on a wall of the box, e.g., the right wall which is perpendicular to the x axis. Let us first ask how many molecules collide with an area A of this wall during some short time interval t. All the molecules do not have the same velocity at any one time. But since we are satisfied with approximate results, we can simplify our considerations by assuming that each molecule moves with the same speed, equal to its average speed $\bar{v}$. The molecules move, however, in random directions so that, on the average, one-third of them (or $\frac{1}{3}n$ molecules per unit volume) move predominantly along the x axis, one-third along the y axis, and one-third along the z axis. Of the $\frac{1}{3}n$ molecules per unit volume moving predominantly along the x axis, one-half (or $\frac{1}{6}n$ molecules per unit volume) move in the $+x$ direction toward the area A, while the remaining half move in the $-x$ direction away from the area A. Any molecule having a velocity directed predominantly in the $+x$ direction travels during the short time t a distance $\bar{v}t$ in the $+x$ direction. If such a molecule is located within a distance

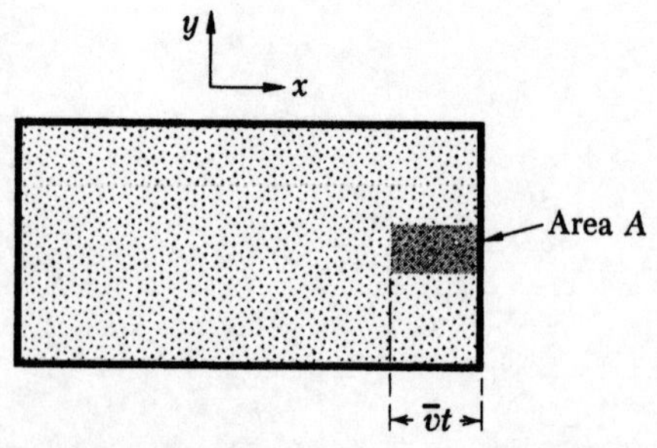

Fig. 1.34 Diagram illustrating collisions of gas molecules with an area A of a container wall. (The z axis points out of the paper.)

† The fact that the symbol n was previously used in a different context to denote the number of molecules in one half of a box of gas should not lead to confusion.

$\bar{v}t$ from the area A of the wall, it will thus strike this area in the time t; but if it lies further than a distance $\bar{v}t$ from the area A, it will not reach this area and thus will not collide with it.† Hence the average number of molecules which collide with the area A during the time t is simply equal to the average number of molecules which have a velocity predominantly in the $+x$ direction and which are contained within the cylinder of area A and length $\bar{v}t$. It is, therefore, obtained by multiplying $\frac{1}{6}n$ (the average number of molecules per unit volume which have their velocity predominantly in the $+x$ direction) by the volume $A\bar{v}t$ of the cylinder and is thus given by

$$(\tfrac{1}{6}n)(A\bar{v}t).$$

If we divide this result by the area A and the time t, we obtain an approximate expression for $\mathcal{J}_0$, the average number of molecules striking unit area of a wall per unit time (or the molecular *flux density*). Thus

$$\boxed{\mathcal{J}_0 \approx \tfrac{1}{6}n\bar{v}.} \tag{18}$$

Let us now calculate the average force exerted by the colliding molecules on a unit area of the wall. When such a molecule, moving predominantly in the $+x$ direction, strikes the wall, its kinetic energy $\frac{1}{2}m\bar{v}^2$ remains unchanged. (This must be true, at least on the average, since the gas is in equilibrium.) The *magnitude* of the momentum of the molecule must then, on the average, also remain unchanged; i.e., the molecule, approaching the right wall with momentum $m\bar{v}$ in the $+x$ direction, must have a momentum $-m\bar{v}$ in this direction after it rebounds from the wall. The $+x$ component of momentum of the molecule changes then by an amount $-m\bar{v} - m\bar{v} = -2m\bar{v}$ as a result of the collision with the wall. Correspondingly it follows, by the principle of conservation of momentum, that the wall gains in the collision an amount of momentum $+2m\bar{v}$ in the $+x$ direction. But the average force exerted on the wall by the gas molecules is, by Newton's second law, equal to the average rate of change of momentum experienced by the wall as a result of the molecular collisions. The average force exerted on unit area of the wall (i.e., the average pressure $\bar{p}$ on the wall) is then simply obtained by multiplication, i.e.,

† Since the time interval t can be considered arbitrarily short, much shorter than the average time between collisions of the molecules among themselves, collisions of the given molecule with other molecules are very unlikely to occur during the time t and can be disregarded.

$$\bar{p} = \begin{bmatrix}\text{the average momentum} \\ 2m\bar{v} \text{ gained by the wall} \\ \text{in one molecular collision}\end{bmatrix} \times \begin{bmatrix}\text{the average number of colli-} \\ \text{sions experienced per unit time} \\ \text{by a unit area of the wall}\end{bmatrix}.$$

Thus
$$\bar{p} \approx (2m\bar{v})\mathcal{J}_0 = (2m\bar{v})(\tfrac{1}{6}n\bar{v})$$

or
$$\boxed{\bar{p} \approx \tfrac{1}{3}nm\bar{v}^2.} \tag{19}$$

As would be expected, the pressure $\bar{p}$ is seen to be increased (i) if n is made large so that there are more molecules colliding with the walls, and (ii) if $\bar{v}$ is increased so that the molecules collide with the walls more frequently and give up more momentum per collision.

Since the average kinetic energy $\overline{\epsilon^{(k)}}$ of a molecule is approximately given by†

$$\overline{\epsilon^{(k)}} \approx \tfrac{1}{2}m\bar{v}^2, \tag{20}$$

the relation (19) can also be written as

$$\bar{p} \approx \tfrac{2}{3}n\overline{\epsilon^{(k)}}. \tag{21}$$

Note that (19) and (21) depend only on the number of molecules per unit volume, but make no reference to the nature of these molecules. These relations are thus equally valid, irrespective of whether the gas consists of molecules of He, Ne, O_2, N_2, or CH_4. The average pressure of any ideal gas kept in a container of fixed volume thus provides a very direct measure of the average kinetic energy of a molecule of the gas.

Numerical estimates

Before making some numerical estimates, it will be useful to recall some important definitions. The mass m of an atom or molecule is conveniently expressed in terms of some standard mass unit m_0. In accordance with present international convention (adopted in 1960 and called the *unified scale of atomic weights*) this mass unit m_0 is defined in terms of the mass m_C of an atom of the particular carbon isotope ^{12}C by‡

$$m_0 \equiv \frac{m_C}{12}. \tag{22}$$

---

† We neglect here the distinction between $\overline{v^2}$, the average of a square, and $\bar{v}^2$, the square of an average.

‡ We recall that a particular isotope is defined as an atom X with a uniquely specified nucleus; the symbol ^nX indicates that the atom has n nucleons (protons + neutrons) in its nucleus. Atoms having nuclei with different numbers of neutrons but the same number of protons are chemically alike since they have the same number of extranuclear electrons.

The mass of a ^{12}C atom is thus exactly equal to 12 mass units. The mass of a H atom is then *approximately* equal to one mass unit.

The ratio of the mass m of an atom (or molecule) to the mass unit m_0 is called the *atomic weight* of the atom (or *molecular weight* of the molecule) and will be denoted by μ. Thus

$$\mu \equiv \frac{m}{m_0}. \tag{23}$$

The atomic weight of ^{12}C is thus, by definition, equal to 12.

A convenient macroscopic number of atoms (or molecules) is the number N_a of atoms of mass m_0 which would have a total mass of 1 gram; i.e., the number N_a is defined by

$$N_a \equiv \frac{1}{m_0}. \tag{24}$$

Alternatively, this defining relation can be written in the form

$$N_a = \frac{m}{mm_0} = \frac{\mu}{m} \tag{25}$$

where we have used (23). This means that N_a is also equal to the number of molecules, of molecular weight μ, which have a total mass of μ grams. The number N_a is called *Avogadro's number.*

One *mole* of a certain kind of molecule (or atom) is defined as a quantity consisting of N_a molecules (or atoms) of this kind. A mole of molecules of molecular weight μ has, therefore, a total mass of μ grams.

The numerical value of Avogadro's number is experimentally found to be

$$N_a = (6.02252 \pm 0.00009) \times 10^{23} \text{ molecules/mole.} \tag{26}$$

(See the table of numerical constants at the end of the book.)

Let us now use the relations (19) or (21) for the pressure of a gas to estimate molecular quantities for nitrogen (N_2) gas, the main constituent of air. At room temperature and atmospheric pressure (10^6 dynes/cm^2), the mass of N_2 gas contained within a vessel having a volume of one liter (10^3 cm^3) is experimentally found to be about 1.15 grams. Since the atomic weight of a N atom is about 14, the molecular weight of a N_2 molecule is $2 \times 14 = 28$. It follows that 28 gm of N_2 gas contains Avogadro's number $N_a = 6.02 \times 10^{23}$ of N_2 molecules. The total number N of molecules in the experimental vessel is thus

$$N = (6.02 \times 10^{23}) \frac{1.15}{28} = 2.47 \times 10^{22} \text{ molecules}$$

so that

$$n \equiv \frac{N}{V} = \frac{2.47 \times 10^{22}}{10^3} \approx 2.5 \times 10^{19} \text{ molecules/cm}^3. \tag{27}$$

Using (21), it then follows that the average kinetic energy of a N_2 molecule is

$$\overline{\epsilon^{(k)}} \approx \frac{3}{2}\frac{\bar{p}}{n} = \frac{3}{2}\left(\frac{10^6}{2.5 \times 10^{19}}\right)$$

$$\approx 6.0 \times 10^{-14} \text{ erg.} \quad (28)$$

Since N_a nitrogen molecules (where N_a is Avogadro's number) have a mass of 28 gm, the mass m of a single N_2 molecule is

$$m = \frac{28}{6.02 \times 10^{23}} = 4.65 \times 10^{-23} \text{ gm.} \quad (29)$$

Hence (20) implies that

$$\bar{v}^2 \approx \frac{2\overline{\epsilon^{(k)}}}{m} \approx \frac{2(6.0 \times 10^{-14})}{4.65 \times 10^{-23}} = 2.6 \times 10^9$$

or

$$\bar{v} \approx 5.1 \times 10^4 \text{ cm/sec.} \quad (30)$$

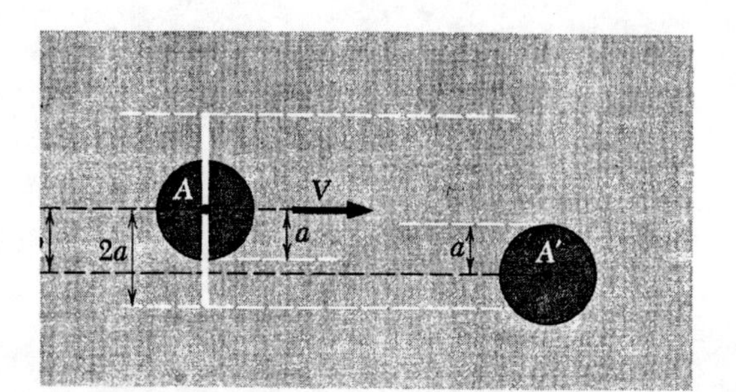

Fig. 1.35 Diagram illustrating an encounter between two hard spheres of radius a.

Mean free path

Focusing attention on a molecule in a gas at any instant of time, let us estimate the average distance l which this molecule travels before it collides with some other molecule in the gas. This distance l is called the *mean free path* of the molecule. To simplify matters, we can imagine that each molecule is spherical in shape and that the forces between any two molecules are similar to those between two hard spheres of radius a. This means that the molecules exert no force on each other as long as the separation R between their centers is greater than $2a$, but that they exert extremely large forces on each other (i.e., they *collide*) if $R < 2a$. Figure 1.35 illustrates an encounter between two such molecules. Here molecule A' can be considered stationary as molecule A approaches it with some relative velocity $\mathbf{V}$ in such a way that the centers of the molecules would approach within a distance b of each other if they remained undeflected. It is then apparent that the molecules will never collide if $b > 2a$, but that they will collide if $b < 2a$. Another way of expressing this geometrical relationship is to imagine that molecule A carries with it a disk of radius $2a$ (this disk being centered about the center of the molecule and oriented perpendicular to the velocity $\mathbf{V}$). A collision between the two molecules will then occur only if the center of molecule A' lies within the volume swept out by the disk carried by molecule A.

The area σ of the imaginary disk carried by a molecule is

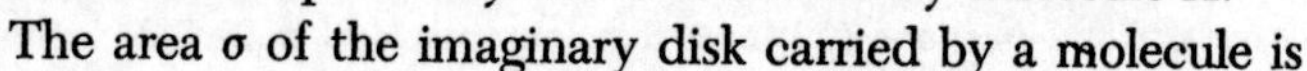

$$\sigma = \pi(2a)^2 = 4\pi a^2 \quad (31)$$

and is called the *total scattering cross section* for molecule-molecule collisions. The volume swept out by this disk as the molecule travels a distance l is equal to σl. Suppose that this volume is such that it contains, on the average, one other molecule, i.e., that

$$(\sigma l)n \approx 1,$$

where n is the number of molecules per unit volume. Then the distance l is the average distance traveled by the molecule before suffer-

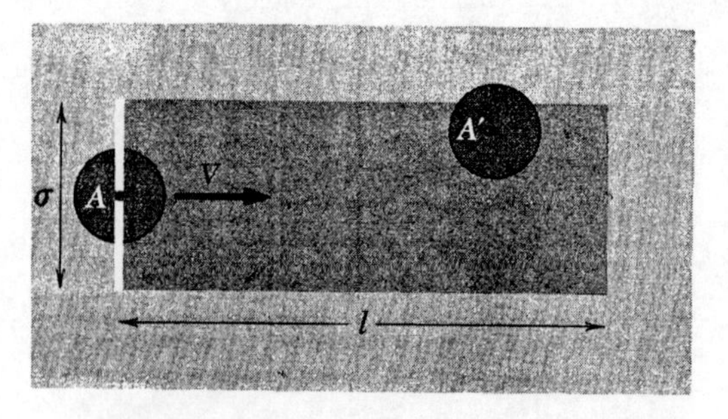

Fig. 1.36 Diagram illustrating the collisions suffered by a particular molecule A when it encounters another molecule A' whose center is located within the volume swept out by the area σ of the imaginary disk carried by A.

ing a collision with another molecule, i.e., it is the desired mean free path. Thus we get

$$\boxed{l \approx \frac{1}{n\sigma}.} \tag{32}$$

As one would expect, the mean free path becomes long (i) if n is small, so that there are only few molecules with which a given molecule can collide, and (ii) if the molecular radius is small, so that molecules must approach each other quite closely before they collide.

To estimate orders of magnitude, we return to the previously discussed case of N_2 gas at room temperature and atmospheric pressure. The radius of a molecule is of the order of 10^{-8} cm, i.e.,

$$a \sim 10^{-8} \text{ cm}.$$

Hence (31) gives for the cross section

$$\sigma = 4\pi a^2 \sim 12 \times 10^{-16} \text{ cm}^2.$$

Using the value (27) of n, (32) then yields the estimate

$$l \approx \frac{1}{n\sigma} \sim \frac{1}{(2.5 \times 10^{19})(12 \times 10^{-16})}$$

or

$$l \sim 3 \times 10^{-5} \text{ cm}. \tag{33}$$

Note that $l \gg a$, i.e., the mean free path is much larger than the radius of a molecule. The molecules thus interact with each other infrequently enough so that it is a good approximation to treat the gas as ideal. On the other hand, the mean free path is very small compared to the linear dimensions of the one-liter vessel containing the gas.

1.7 *Important Problems of Macroscopic Physics*

Although the discussion of this chapter has been mostly qualitative, it has revealed the most significant features of macroscopic systems. We have, therefore, gained sufficient perspective to recognize some of the questions which we should ultimately like to explore and understand.

Fundamental concepts

Our first task must clearly be that of transforming our qualitative insights into precisely formulated theoretical concepts capable of making quantitative predictions. For example, we have recognized that certain situations of a macroscopic system are more probable (or more random) than others. But precisely how does one assign a probability to a given macrostate of a system and how does one measure its degree of randomness? This question is of all-pervasive importance. We also concluded that the time-independent equilibrium situation is characterized as being the most random situation of an isolated system. The problem of defining randomness in a precise general way thus occurs again. Indeed, this problem proved troublesome to us when we tried to discuss the case of two arbitrary systems in thermal contact with each other. We suspected that the equilibrium condition of maximum randomness should imply that some parameter T (measuring roughly the mean energy per atom in a system) ought to be the same for both systems. But since we did not know how to define randomness in the general case, we were unable to arrive at an unambiguous definition of this parameter T (which we called the "absolute temperature"). Thus we can pose the following basic question: How can we use probability ideas to describe macroscopic systems in a systematic way so as to define concepts such as randomness or absolute temperature?

In discussing the example of the pendulum in Sec. 1.3, we saw that it is not readily apparent how to transform energy randomly distributed over many molecules to a less random form where it can do work by exerting a net macroscopic force over a macroscopic distance. This example illustrates questions of the most profound importance. Indeed, to what extent is it possible to take energy distributed randomly over many molecules of a substance (such as coal or gasoline) and to transform it into a less random form where it can be used to move pistons against opposing forces? In other words, to what extent is it possible to build the steam engines or gasoline engines responsible for our industrial revolution? Similarly, to what extent is it possible to take the energy distributed randomly over many molecules of certain chemical compounds and to transform it into a less random form where it can be used to produce muscular contraction or the synthesis of highly ordered polymer molecules such as proteins? In other words, to what extent can chemical energy be used to make biological processes possible? A good understanding of the concept of randomness should allow us to make significant statements about all these questions.

Properties of systems in equilibrium

Since macroscopic systems in equilibrium are particularly simple, their properties ought to be most readily amenable to quantitative discussion. There are, indeed, many equilibrium situations which are of great interest and importance. Let us mention some of the questions which are worthy of investigation.

A homogeneous substance is one of the simplest systems whose equilibrium properties one may hope to calculate. For example, suppose that some particular fluid (gas or liquid) is in equilibrium at a given temperature. How does the magnitude of the pressure which it exerts depend on its temperature and on its volume? Or suppose that some substance contains a certain concentration of iron atoms, each of which has a definite magnetic moment. If this substance is at a given temperature and is located in a given magnetic field, what is the value of its magnetization, i.e., of its net magnetic moment per unit volume? How does this magnetization depend on the temperature and on the magnetic field? Or suppose that a small amount of heat is added to some particular substance (which might be a liquid, a solid, or a gas). By what amount does its temperature increase?

Not only can we ask questions about macroscopic parameters of a system in equilibrium; we can also inquire into the behavior of the atoms of which it consists. For example, consider a container of gas maintained at some given temperature. All the molecules of the gas do not have the same speed and we can ask what fraction of the molecules have a speed in any specified range. If we make a very small hole in the container, some of the molecules will escape through the hole into the surrounding vacuum where their speed can be directly measured and the theory compared with the experimental results. Or consider an empty container whose walls are maintained at some elevated temperature. Since the atoms in the walls emit electromagnetic radiation, the container itself is then filled with radiation (or with photons) in equilibrium with the walls. What amount of energy of this electromagnetic radiation is concentrated in any given frequency range? If one makes a very small hole in the container through which some of the radiation escapes, then one can readily use a spectrometer to measure the amount of radiation emerging in any narrow frequency band and thus can compare the predictions with experimental results. This last problem is actually very important for understanding the radiation emitted by any hot body, whether it be the sun or the filament of a light bulb.

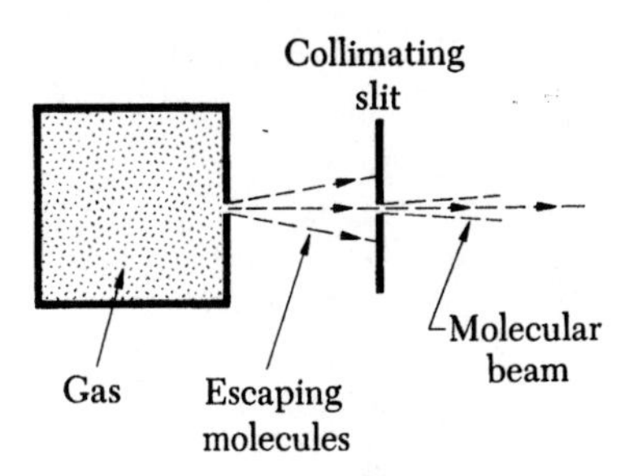

Fig. 1.37 A container filled with gas has a very small hole through which some molecules can escape into a surrounding vacuum. Collimation of these emerging molecules by one or more slits yields a well-defined molecular beam. The distribution of velocities of the molecules in the beam is closely related to that of the molecules in the container. Molecular beams of this sort represent a very powerful method for studying nearly isolated atoms or molecules and have been used to perform some of the most fundamental experiments of modern physics.

Another type of situation of great interest is that where chemical reactions can occur between various kinds of molecules. As a specific example, consider a container of volume V which is filled with carbon dioxide (CO_2) gas. It is possible to transform CO_2 molecules into carbon monoxide (CO) and oxygen (O_2) molecules, or vice versa, according to the chemical reaction

$$2\,CO_2 \rightleftarrows 2\,CO + O_2. \tag{34}$$

As the temperature of the container is raised, some of the CO_2 molecules will dissociate into CO and O_2 molecules. The container will then enclose a mixture of CO_2, CO, and O_2 gases in equilibrium with each other. One would like to know how to calculate, from first principles, the relative number of CO_2, CO, and O_2 molecules thus present in equilibrium at any specified temperature.

But even a simple substance consisting only of a single kind of molecule presents some intriguing problems. Such a substance can typically exist in several different forms, or *phases,* such as gas, liquid, or solid. A specific example might be water which can exist in the form of water vapor, liquid water, or ice. Each such phase consists of the same kind of molecule (H_2O, in the case of water), but the molecules are arranged differently. In the gas phase, the molecules are far apart from each other and thus move about almost independently of each other in random fashion. In the solid phase, on the other hand, the molecules are arranged in a very orderly manner. They are located in a regular crystal lattice near specific sites and are free merely to perform small oscillations about these sites. In the liquid phase the situation is intermediate, being neither as orderly as in the solid nor as random as in the gas. The molecules here are close to each other and always under each other's strong influence, yet they are free to move past each other over long distances. The evidence for these molecular arrangements in the various phases comes largely from studies of x-ray scattering.

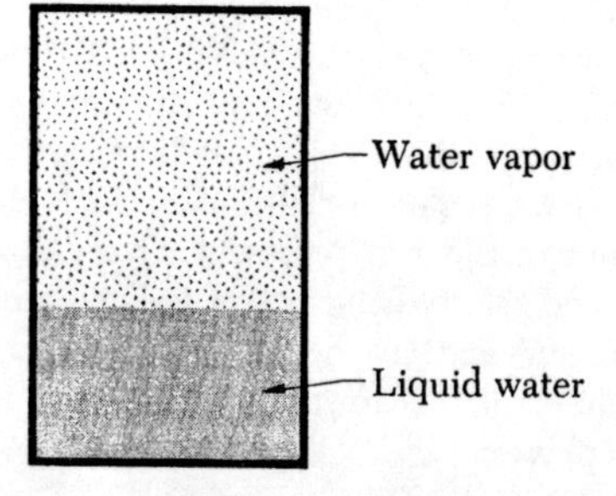

Fig. 1.38 Liquid water and its gas form, water vapor, are here shown coexisting in equilibrium at some specified temperature. The pressure exerted by the vapor then has a unique value which depends only on the temperature.

It is well known that a substance changes from one phase to another phase at a very well-defined temperature (absorbing or giving off heat in the process). For example, water changes from the solid ice form to the liquid water form at 0° centigrade and changes in turn (at a pressure of 1 atmosphere) from the liquid water form to the gaseous steam form at 100° centigrade. Under certain circumstances two phases can, therefore, coexist in equilibrium with each other (e.g., ice and liquid water at 0° centigrade). A good theory of systems in equilibrium should allow us to make statements about the conditions of temperature and pressure under which two phases are thus able to coexist in equilibrium. It should also allow us to predict at what temperature a particular solid substance melts so as to become a liquid, and at what temperature the liquid vaporizes to become a gas.

These are actually very difficult and fascinating problems. The concept of degree of randomness, or order, is again crucial. As the absolute temperature (or mean energy per atom) of a substance is increased, it first changes from the most orderly (or least random) solid form to the liquid form which is of an intermediate degree of order; as the temperature is increased further, it changes from this liquid form to the gas form which is the most nearly random or disordered. But it is very striking that these changes from one degree of order to the next take place very abruptly at extremely well-defined temperatures. The essential reason is a kind of critical instability involving all the molecules of the substance. For example, suppose that the absolute temperature of a solid is sufficiently high so that its molecules, as a result of their relatively large mean energy, can oscillate about their regular lattice positions with displacements large enough to be comparable with their intermolecular separation. Suppose then that there occurs a fluctuation in which a few adjacent molecules move simultaneously out of their regular lattice positions; this makes it easier for neighboring molecules slightly farther away also to move out of their regular positions; and so forth. The net result is then somewhat analogous to the collapse of a pile of dominoes; i.e., the high degree of order of the solid suddenly begins to disintegrate all at once and the solid turns into a liquid. This instability, which results in the melting of the solid, involves jointly *all* the molecules of the solid; it is, therefore, said to be a *cooperative phenomenon.* Since it does require the analysis of all the molecules in simultaneous interaction, the theoretical problem of treating any cooperative phenomenon such as melting or vaporization from a detailed microscopic point of view is very difficult and challenging.

Systems which are not in equilibrium

The discussion of systems which are not in equilibrium is ordinarily much more difficult than that of systems which are in equilibrium. Here it is necessary to deal with processes which change in time and the interest is focused on how fast or slowly they change. The discussion of such questions requires a detailed analysis of how effectively molecules interact with each other. Except in relatively simple cases such as dilute gases, this kind of analysis can become fairly complicated.

Let us mention again a few typical problems of interest. Consider, for example, the chemical reaction (34). Suppose that some CO_2 gas is introduced into the container at some given elevated temperature. How long would it take to achieve the equilibrium concentration of CO? Thus one would like to calculate the *rate* at which the chemical reaction (34) proceeds from left to right.

As another illustration, consider two large bodies which are connected by a rod and which are at different temperatures T_1 and T_2. Since this is not an equilibrium situation, heat will flow from one body to the other through the connecting rod. But how effectively is the energy transported down the rod, i.e., how long does it take for a given amount of heat to flow from one body to the other? This depends on an intrinsic property of the rod, its "thermal conductivity." For example, a rod of copper allows heat flow to proceed much more readily than a rod made of stainless steel; i.e., copper has a much higher thermal conductivity than stainless steel. The task of the theory is to define "thermal conductivity" precisely and to calculate this parameter from first principles.

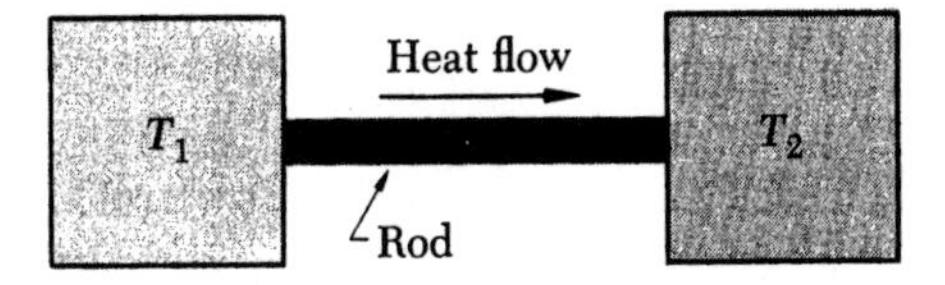

Fig. 1.39 Two bodies at different temperatures connected by a rod which conducts heat from one to the other.

Concluding Remarks

The preceding survey of problems indicates to some extent the range of macroscopic natural phenomena which we might like to treat quantitatively on the basis of fundamental microscopic considerations. We shall not learn how to discuss all such problems in the remainder of this book. Indeed, some of the questions which we have raised (e.g., calculations of phase transformations such as melting or vaporization), give rise to problems which are still not adequately solved and which are thus active areas of research. On the other hand, we are now well prepared to turn the qualitative considerations of this chapter into a more systematic quantitative discussion of the properties of macroscopic systems. This discussion will go a long way toward answering the most basic questions which we have raised.

Summary of Definitions

isolated system A system which does not interact with any other system.

ideal gas A gas of molecules whose mutual interaction is almost negligible (i.e., the interaction is large enough to ensure that the molecules can exchange energy among themselves, but is otherwise negligible).

ideal spin system A system of spins whose mutual interaction is almost negligible (i.e., the interaction is large enough to ensure that the spins can exchange energy among themselves, but is otherwise negligible).

microscopic Small, of the order of atomic dimensions or less.

macroscopic Very large compared to atomic dimensions.

microscopic state (or microstate) State of a system described in microscopic detail by the most complete specification, according to the laws of mechanics, of all the atoms in the system.

macroscopic state (or macrostate) State of a system described, without attention to microscopic details, by specifying only quantities which can be determined by macroscopic measurements.

macroscopic parameter A parameter which can be determined by large-scale measurements and which describes the macroscopic state of a system.

equilibrium A macroscopic state which does not tend to change in time, except for random fluctuations.

relaxation time Approximate time required for a system, starting from a situation far removed from equilibrium, to attain equilibrium.

irreversible process A process such that the time-reversed process (which would be observed in a movie played backward) would almost never occur in actuality.

thermal interaction Interaction which does not involve the performance of work on a macroscopic scale.

heat Energy transfer which is not associated with the performance of macroscopic work, but occurs on the atomic scale.

thermometer A small macroscopic system arranged so that only one of its macroscopic parameters changes when the system absorbs or gives off heat.

thermometric parameter The one variable macroscopic parameter of a thermometer.

temperature of a system with respect to a given thermometer The value of the thermometric parameter of the thermometer when the latter is placed in thermal contact with the system and allowed to come to equilibrium with it.

mean free path The average distance which a molecule travels in a gas before it collides with another molecule.

Suggestions for Supplementary Reading

F. J. Dyson, "What is Heat?", *Sci. American* **191,** 58 (Sept. 1954).

R. Furth, "The Limits of Measurement," *Sci. American* **183,** 48, (July 1950). A discussion of Brownian motion and other fluctuation phenomena.

B. J. Alder and T. E. Wainwright, "Molecular Motions," *Sci. American* **201,** 113 (Oct. 1959). This article discusses the application of modern high-speed computers to the study of molecular motions in various macroscopic systems.

Problems

1.1 *Fluctuations in a spin system*

Consider an ideal system of 5 spins in the absence of an external magnetic field. Suppose that one took a movie of this spin system in equilibrium. What fraction of the movie frames would show n spins pointing up? Consider all the possibilities $n = 0, 1, 2, 3, 4$, and 5.

1.2 *Diffusion of a liquid*

Suppose that a drop of a dye (having the same density as water) is introduced into a glass of water. The whole system is kept at a constant temperature and is left mechanically undisturbed. Suppose that one took a movie of the process occurring after the drop of dye had been put into the water. What would one see on a screen on which the movie is projected? What would one see if the movie is run backward through the projector? Is the process reversible or irreversible? Describe the process in terms of the motion of the dye molecules.

1.3 *Microscopic explanation of friction*

A wooden block, which has originally been given a push, is sliding on the floor and gradually comes to rest. Is this process reversible or irreversible? Describe the process as it would appear on a movie played backward. Discuss what happens during this process on the microscopic scale of atoms and molecules.

1.4 *The approach to thermal equilibrium*

Consider two gases A and A' in separate containers. Initially the average energy of a molecule of gas A is quite different from the average energy of a molecule of gas A'. The two containers are then placed in contact with each other so that energy in the form of heat can be transferred from gas A to the molecules of the container walls and thence to gas A'. Is the ensuing process reversible or irreversible? Describe in microscopic detail the process that would appear if one filmed the situation and ran this film backward through a projector.

1.5 *Variation of gas pressure with volume*

A container is divided into two parts by a partition. One of these parts has a volume V_i and is filled with a dilute gas; the other part is empty. Remove the partition and wait until the final equilibrium condition is attained where the molecules of the gas are uniformly distributed throughout the entire container of volume V_f.

(*a*) Has the total energy of the gas been changed? Use this result to compare the average energy per molecule and the average speed of a molecule in the equilibrium situations before and after the removal of the partition.

(*b*) What is the ratio of the pressure exerted by the gas in the final situation to that of the pressure exerted by it in the initial situation?

1.6 *Number of gas molecules incident on an area*

Consider nitrogen (N_2) gas at room temperature and atmospheric pressure. Using the numerical values given in the text, find the average number of N_2 molecules striking a 1-cm^2 area of the container walls per second.

1.7 *Leak rate*

A 1-liter glass bulb contains N_2 gas at room temperature and atmospheric pressure. The glass bulb, which is to be used in conjunction with some other experiment, is itself enclosed in a large evacuated chamber. Unfortunately the glass bulb has, unbeknown to the experimenter, a small pinhole about 10^{-5} cm in radius. To assess the importance of this hole, estimate the time required for 1 percent of the N_2 molecules to escape from the bulb into the surrounding vacuum.

1.8 *Average time between molecular collisions*

Consider nitrogen gas at room temperature and atmospheric pressure. Using the numerical values given in the text, find the average time a N_2 molecule travels before colliding with another molecule.

1.9 *Equilibrium between atoms of different masses

Consider a collision between two different atoms having masses m_1 and m_2. Denote the velocities of these atoms before the collision by $\mathbf{v}_1$ and $\mathbf{v}_2$, respectively; denote their velocities after the collision by $\mathbf{v}_1'$ and $\mathbf{v}_2'$, respectively. It is of interest to investigate the energy transferred from one atom to the other as a result of the collision.

(*a*) Introduce the relative velocity $\mathbf{V} \equiv \mathbf{v}_1 - \mathbf{v}_2$ and the center-of-mass velocity $\mathbf{c} \equiv (m_1\mathbf{v}_1 + m_2\mathbf{v}_2)/(m_1 + m_2)$. The relative velocity after the collision is then $\mathbf{V}' = \mathbf{v}_1' - \mathbf{v}_2'$. In a collision $\mathbf{c}$ remains unchanged by virtue of conservation of momentum, while $|\mathbf{V}'| = |\mathbf{V}|$ by virtue of conservation of energy. Show that the energy gain $\Delta\epsilon_1$ of atom 1 in a collision is given by

$$\Delta\epsilon_1 = \tfrac{1}{2}m_1(v_1'^2 - v_1^2) = m_1m_2(m_1 + m_2)^{-1}\mathbf{c}\cdot(\mathbf{V}' - \mathbf{V}). \qquad \text{(i)}$$

(*b*) Denote by θ the angle between $\mathbf{V}'$ and $\mathbf{V}$, by φ the angle between the plane containing $\mathbf{V}'$ and $\mathbf{V}$ and the plane containing $\mathbf{c}$ and $\mathbf{V}$, and by ψ the angle between $\mathbf{c}$ and $\mathbf{V}$. Show that (i) then becomes

$$\Delta\epsilon_1 = m_1m_2(m_1 + m_2)^{-1}cV[(\cos\theta - 1)\cos\psi + \sin\theta\sin\psi\cos\varphi] \qquad \text{(ii)}$$

where $cV\cos\psi = \mathbf{c}\cdot\mathbf{V} = (m_1 + m_2)^{-1}[m_1v_1^2 + (m_2 - m_1)\mathbf{v}_1\cdot\mathbf{v}_2 - m_2v_2^2]$.

(*c*) Consider two atoms of this kind in a gas where many collisions occur. On the average, the azimuthal angle φ is then as often positive as negative so that $\overline{\cos\varphi} = 0$; also, since $\mathbf{v}_1$ and $\mathbf{v}_2$ have random directions, the cosine of the angle between them is as often positive as negative so that $\overline{\mathbf{v}_1\cdot\mathbf{v}_2} = 0$. Show that (ii) therefore becomes, on the average,

$$\overline{\Delta\epsilon_1} = \frac{2m_1m_2}{(m_1 + m_2)^2}\,(\overline{1 - \cos\theta})(\bar{\epsilon}_2 - \bar{\epsilon}_1) \qquad \text{(iii)}$$

where $\epsilon_1 = \frac{1}{2}m_1v_1^2$ and $\epsilon_2 = \frac{1}{2}m_2v_2^2$.

In the equilibrium situation in particular the energy of an atom must, on the average, remain unchanged so that $\overline{\Delta\epsilon_1} = 0$. Show that (iii) then implies that in equilibrium,

$$\bar{\epsilon}_2 = \bar{\epsilon}_1. \qquad \text{(iv)}$$

Thus the average energies of interacting atoms in equilibrium are equal even if the masses of the atoms are different.

1.10 *Comparison of molecular speeds in a gas mixture*

Consider a gas mixture enclosed in some container and consisting of monatomic molecules of two different masses m_1 and m_2.

(*a*) Suppose that this gas mixture is in equilibrium. Use the result of the preceding problem to find the approximate ratio of the average speed $\bar{v}_1$ of a molecule of mass m_1 to the average speed $\bar{v}_2$ of a molecule of mass m_2.

(*b*) Suppose that the two kinds of molecules are He (helium) and Ar (argon) which have atomic weights equal to 4 and 40, respectively. What is the ratio of the average speed of a He atom to that of an Ar atom?

1.11 *Pressure of a gas mixture*

Consider an ideal gas which consists of two kinds of atoms. To be specific, suppose that there are, per unit volume, n_1 atoms of mass m_1 and n_2 atoms of mass m_2. The gas is assumed to be in equilibrium so that the average energy $\bar{\epsilon}$ per atom is the same for both kinds of atoms. Find an approximate relation for the average pressure $\bar{p}$ exerted by the gas mixture. Express your result in terms of $\bar{\epsilon}$.

1.12 *Mixing of two gases*

Consider a container divided into two equal parts by a partition. One of these parts contains 1 mole of helium (He) gas, the other 1 mole of argon (Ar) gas. Energy in the form of heat can pass through the partition from one gas to the other. After a sufficiently long time, the two gases will, therefore, come to equilibrium with each other. The average pressure of the helium gas is then $\bar{p}_1$, that of the argon gas is $\bar{p}_2$.

(*a*) Compare the pressures $\bar{p}_1$ and $\bar{p}_2$ of the two gases.

(*b*) What happens when the partition is removed? Describe the process as it would appear on a movie played backward. Is it reversible or irreversible?

(*c*) What is the average pressure exerted by the gas in the final equilibrium situation?

1.13 *Effect of a semipermeable partition ("osmosis")*

A glass bulb contains argon (Ar) gas at room temperature and at a pressure of 1 atmosphere. It is placed in a large chamber containing helium (He) gas also at room temperature and at a pressure of 1 atmosphere. The bulb is made of a glass which is permeable to the small He atoms, but is impermeable to the larger Ar atoms.

(*a*) Describe the process which ensues.

(*b*) What is the most random distribution of the molecules which is attained in the ultimate equilibrium situation?

(*c*) What is the average pressure of the gas inside the bulb when this ultimate equilibrium situation has been attained?

1.14 *Thermal vibrations of the atoms in a solid*

Consider nitrogen (N_2) gas in equilibrium within a box at room temperature. In accordance with the result of Prob. 1.9, it is then reasonable to assume that the average kinetic energy of a gas molecule is roughly equal to the average kinetic energy of an atom in the solid wall of the container. Each atom in such a solid is localized near a fixed site. It is, however, free to oscillate about this site and should, to good approximation, perform simple harmonic motion about this position. Its potential energy is then, on the average, equal to its kinetic energy.

Suppose that the walls consist of copper which has a density of 8.9 gm/cm^3 and an atomic weight of 63.5.

(*a*) Estimate the average speed with which a copper atom vibrates about its equilibrium position.

(*b*) Estimate roughly the mean spacing between the copper atoms in the solid. (You may assume them to be located at the corners of a regular cubic lattice.)

(*c*) When a force F is applied to a copper bar of cross-sectional area A and length L, the increase in length ΔL of the bar is given by the relation

$$\frac{F}{A} = Y\frac{\Delta L}{L}$$

where the proportionality constant Y is called *Young's modulus*. Its measured value for copper is $Y = 1.28 \times 10^{12}$ dynes/cm^2. Use this information to estimate the restoring force acting on a copper atom when it is displaced by some small amount x from its equilibrium position in the solid.

(*d*) What is the potential energy of an atom when it is displaced by an amount x from its equilibrium position? Use this result to estimate the average magnitude $|x|$ of the amplitude of vibration of a copper atom about its equilibrium position. Compare $|x|$ with the separation between the copper atoms in the solid.

Chapter 2

Basic Probability Concepts

Chapter 2 Basic Probability Concepts

The considerations of the preceding chapter have indicated that probability arguments are of fundamental significance for the understanding of macroscopic systems consisting of very many particles. It will, therefore, be useful to review the most basic notions of probability and to examine how they can be applied to some simple, but important, problems. Indeed, this discussion is likely to be valuable in contexts far wider than those of immediate interest to us. For example, probability concepts are indispensable in all games of chance, in the insurance business (with its need to assess the probable occurrence of death or disease among its clients), and in sampling procedures such as public opinion polls. In biology, they are profoundly important in the study of genetics. In physics, they are needed in treating the occurrence of radioactive decay, the arrival of cosmic rays at the surface of the earth, or the random emission of electrons from the hot filament of a vacuum tube; furthermore, they play a fundamental role in the quantum-mechanical description of atoms and molecules. Most important to us, they will form the basis of our entire discussion of macroscopic systems.

2.1 Statistical Ensembles

Consider a system A on which one can perform experiments or observations.† In many cases the particular outcome resulting from the performance of a single experiment cannot be predicted with certainty, either because this is intrinsically impossible‡ or because the information available about the system is insufficient to permit such a unique prediction. Although it is not possible to make statements about a single experiment, it may still be possible to make significant statements about the outcomes of a large number of similar experiments. We are thus led to a *statistical* description of the system, i.e., to a

---

† An act of observation can be regarded as an experiment in which the result of the observation constitutes the outcome of the experiment. Therefore we need make no distinction between experiments and observations.

‡ This is the situation, for example, in quantum mechanics where the outcome of a measurement on a microscopic system is ordinarily not predictable with certainty.

description in terms of *probabilities*. To achieve such a description, we proceed in the following way:

Instead of focusing attention on the single system A of interest, we contemplate an assembly (or an *ensemble*, in more customary terminology) consisting of some very large number $\mathcal{N}$ of "similar" systems. In principle, $\mathcal{N}$ is imagined to be arbitrarily large (i.e., $\mathcal{N} \to \infty$). The systems are supposed to be "similar" in the sense that each system satisfies the same conditions known to be satisfied by the system A. This means that each system is imagined to have been prepared in the same way as A and to be subjected to the same experiment as A. We can then ask in what fraction of cases a particular outcome of the experiment does occur. To be precise, we arrange matters so that we can enumerate in some convenient way all the possible mutually exclusive outcomes of the experiment. (The total number of such possible outcomes may be finite or infinite.) Suppose then that a particular outcome of the experiment is labeled by r and that there are, among the $\mathcal{N}$ systems of the ensemble, $\mathcal{N}_r$ systems which exhibit this outcome. Then the fraction

$$P_r \equiv \frac{\mathcal{N}_r}{\mathcal{N}} \qquad (\text{where } \mathcal{N} \to \infty) \tag{1}$$

is called the *probability of occurrence of the outcome r.* To the extent that $\mathcal{N}$ is made very large, a repetition of the same experiment on the ensemble is expected to lead with increasing reproducibility to the same ratio $\mathcal{N}_r/\mathcal{N}$. The definition (1) becomes thus unambiguous in the limit where $\mathcal{N}$ is made arbitrarily large.

The preceding discussion shows how the probability of occurrence of any possible outcome of an experiment on a system can be measured by repeating the experiment on a large number $\mathcal{N}$ of similar systems.†
Although the outcome of the experiment on a single system cannot be predicted, the task of a *statistical* theory is then that of predicting the *probability* of occurrence of each of the possible outcomes of the experiment. The predicted probabilities can then be compared against the probabilities actually measured by experiments on an ensemble of similar systems.

Several examples will serve to illustrate this statistical description in a number of concrete cases.

† *If* the situation being contemplated is independent of time, one could equally well repeat the same experiment $\mathcal{N}$ times in succession with the *one* particular system under consideration (being careful to start the experiment each time with the system in the same initial condition).

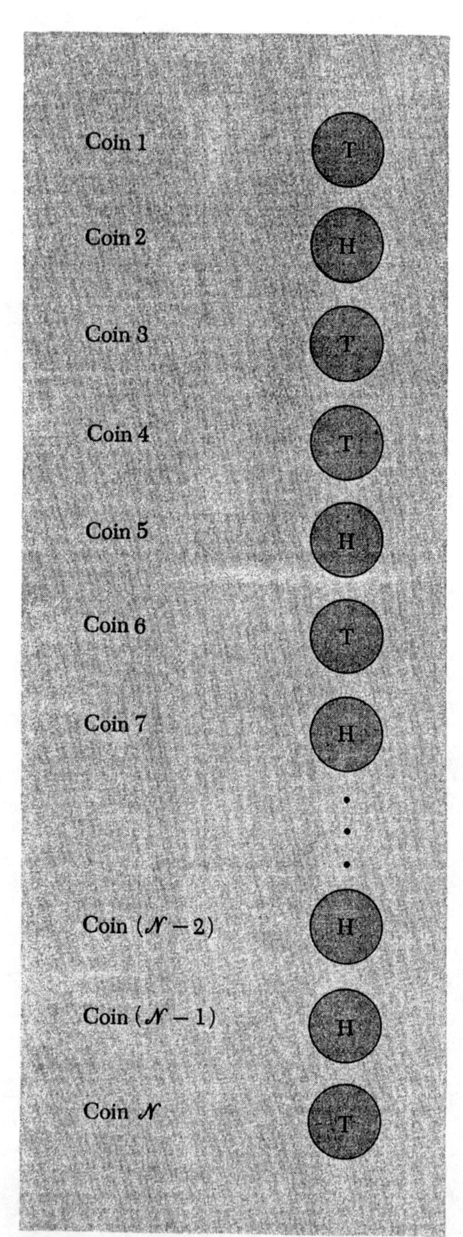

Fig. 2.1 In order to make probability statements about the tossing of a *single* coin, one considers an ensemble consisting of a large number $\mathcal{N}$ of similar coins. The diagram indicates schematically the appearance of this ensemble after each coin has been tossed. The symbol H denotes "head," the symbol T "tail."

Tossing of coins or dice

Consider the experiment of tossing a coin. There are only two possible outcomes of such an experiment, either "head" or "tail," depending on whether the coin comes to rest on a table with the engraved head uppermost or its other side uppermost.† In principle, the outcome of the experiment could be completely predicted if we knew exactly how the coin is tossed and the forces by which it interacts with the table. It would then only be necessary to make the requisite complicated calculations based on the laws of classical mechanics. But such detailed information about the tossing of the coin is not available in practice. It is, therefore, not possible to make a unique prediction about the outcome of a particular toss. (Indeed, even if all the requisite information could be obtained and all the requisite computations could be performed, we would ordinarily not be interested in making precise predictions at the expense of such enormous complexity.) The statistical formulation of the experiment is, however, very easy and familiar. We need only consider an ensemble consisting of a very large number $\mathcal{N}$ of similar coins. When these coins are tossed in similar ways, we can count the fraction of cases where the outcome is a head and that where it is a tail.‡ These fractions then give, respectively, the measured probability p of obtaining a head and probability q of obtaining a tail. A statistical theory would attempt to predict these probabilities. For example, if the center of mass of the coin coincides with its geometrical center, the theory might be based on the symmetry argument that there is nothing in the laws of mechanics which distinguishes between a head or a tail. In this case, half the experimental outcomes should be heads and half tails, so that $p = q = \frac{1}{2}$. Comparison with experiment might or might not verify the theory. For example, if heads were observed to occur more frequently than tails, we might conclude that the theory is not justified in assuming that the center of mass of the coin coincides with its geometrical center.

Consider now the somewhat more complicated experiment of tossing a set of N coins. Since the tossing of any one coin can have 2 possible outcomes, the tossing of the set of N coins can then have any of $2 \times 2 \times \cdots \times 2 = 2^N$ possible outcomes.§ The statistical formulation of the experiment requires again that, instead of dealing with a single set of N coins, we contemplate an ensemble consisting of $\mathcal{N}$ such sets of N coins, where each set is tossed in a similar way. A possible question of interest is then the probability that any particular one of the 2^N possible outcomes occurs in the ensemble. A less detailed question of interest might concern the probability of finding in the ensemble an outcome where any n of the coins show heads and the remaining $(N - n)$ coins show tails.

The problem of tossing a set of N dice is, of course, analogous. The only difference is that the toss of any one die can have 6 possible outcomes, depending on which of the six faces of the cubical die lands uppermost.

It is often convenient to use the simple word *event* to denote the outcome of an experiment or the result of an observation. Note that the probability of occurrence of an event depends crucially on the information available about the system under consideration. Indeed,

† We neglect the remote possibility that the coin comes to rest standing on its edge.

‡ Alternatively, we might take the same coin and toss it $\mathcal{N}$ times in succession, counting the fraction of cases showing heads or tails.

§ In the special case where $N = 4$, the $2^4 = 16$ possible outcomes are listed explicitly in Table 1.1 (p. 7) if one interprets the letter L as a head and the letter R as a tail.

this information determines the kind of statistical ensemble to be contemplated, since this ensemble must consist solely of systems which satisfy all the conditions satisfied by the particular system under consideration.

Example

Suppose that we are interested in the following question: What is the probability that a person living in the United States will be hospitalized at some time between the ages of 23 and 24 years? Then we must consider an ensemble consisting of a large number of persons in the United States and must ascertain the fraction of these persons hospitalized at some time between the ages of 23 and 24. But suppose that we are now also told that the person is of the female sex. Then the answer to our question changes because we must now contemplate an ensemble of *women* living in the United States and must ascertain how many of these women are hospitalized at some time between the ages of 23 and 24. (Indeed, women of this age tend to enter hospitals because of childbirth, a condition from which men are exempt.)

Application to systems of many particles

Consider a macroscopic system consisting of many particles. For example, the system might be an ideal gas of N molecules, a system of N spins, a liquid, or a piece of copper. In none of these cases is it possible to make a unique prediction about the behavior of each of the particles in the system,† nor is it of interest. We resort, therefore, to a statistical description of the system A under consideration. Instead of considering the single system A, we contemplate an ensemble consisting of a large number $\mathcal{N}$ of systems similar to A. To make a statistical statement about the system at the time t, we make observations on the $\mathcal{N}$ systems at this time t. Thus we can determine the probability $P_r(t)$ that the observation yields a particular outcome r at the time t. The procedure is most readily visualized by imagining that we take a motion picture of each system in the ensemble. We then end up with $\mathcal{N}$ film strips which contain the results of all observations on the systems of the ensemble. The behavior of any one system in the ensemble, say system number k, as a function of time can then be obtained by looking at the kth film strip (i.e., by looking along a particular horizontal line in Fig. 2.3). On the other hand,

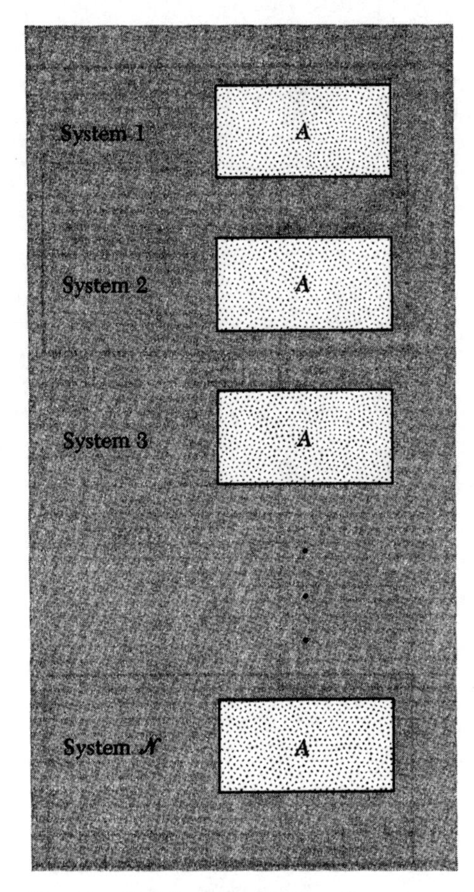

Fig. **2.2** Statistical description of a system A consisting of a gas in a box. The diagram illustrates schematically a statistical ensemble of $\mathcal{N}$ such systems similar to the system A under consideration.

† In the correct quantum-mechanical description of the system nonstatistical predictions are not possible, even in principle. In a classical description a unique prediction about a system would require knowledge of the position and velocity of each particle at one time, information which is not available to us.

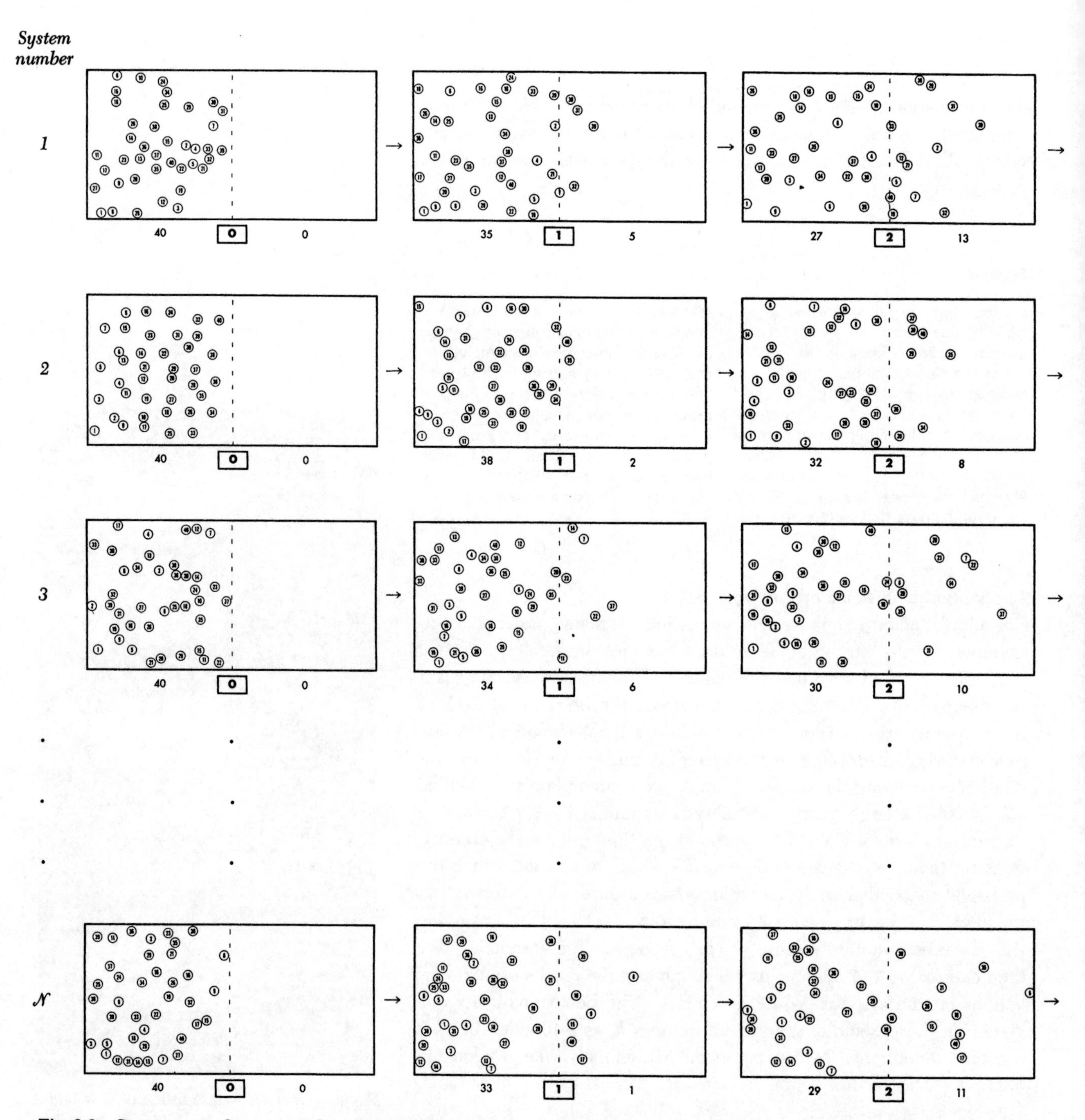

Fig. 2.3 Computer-made pictures showing a statistical ensemble of systems. This ensemble was constructed to represent a system consisting of 40 particles in a box when the information available about the system is the following: All particles are known to be in the left half of the box at some initial time corresponding to the frame $j = 0$, but nothing else is known about their positions or velocities.

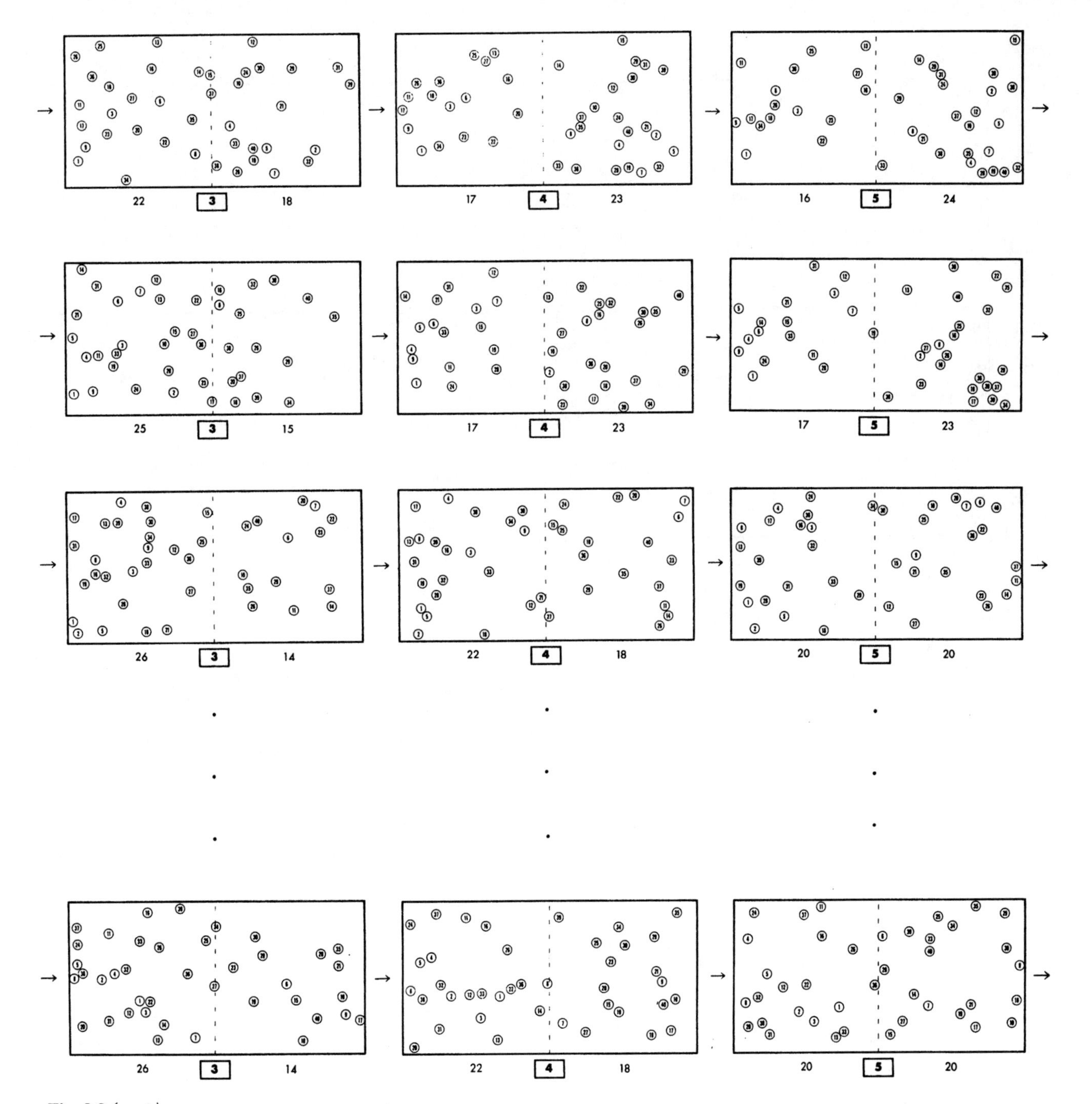

Fig. 2.3 (cont.)

The evolution in time of the kth system in the ensemble can be followed by looking horizontally at the successive frames $j = 0, 1, 2, \ldots$ for this system. Statistical statements about the system at any time corresponding to the jth frame can be made by looking vertically at all the systems in their jth frame and doing the counting necessary to determine probabilities.

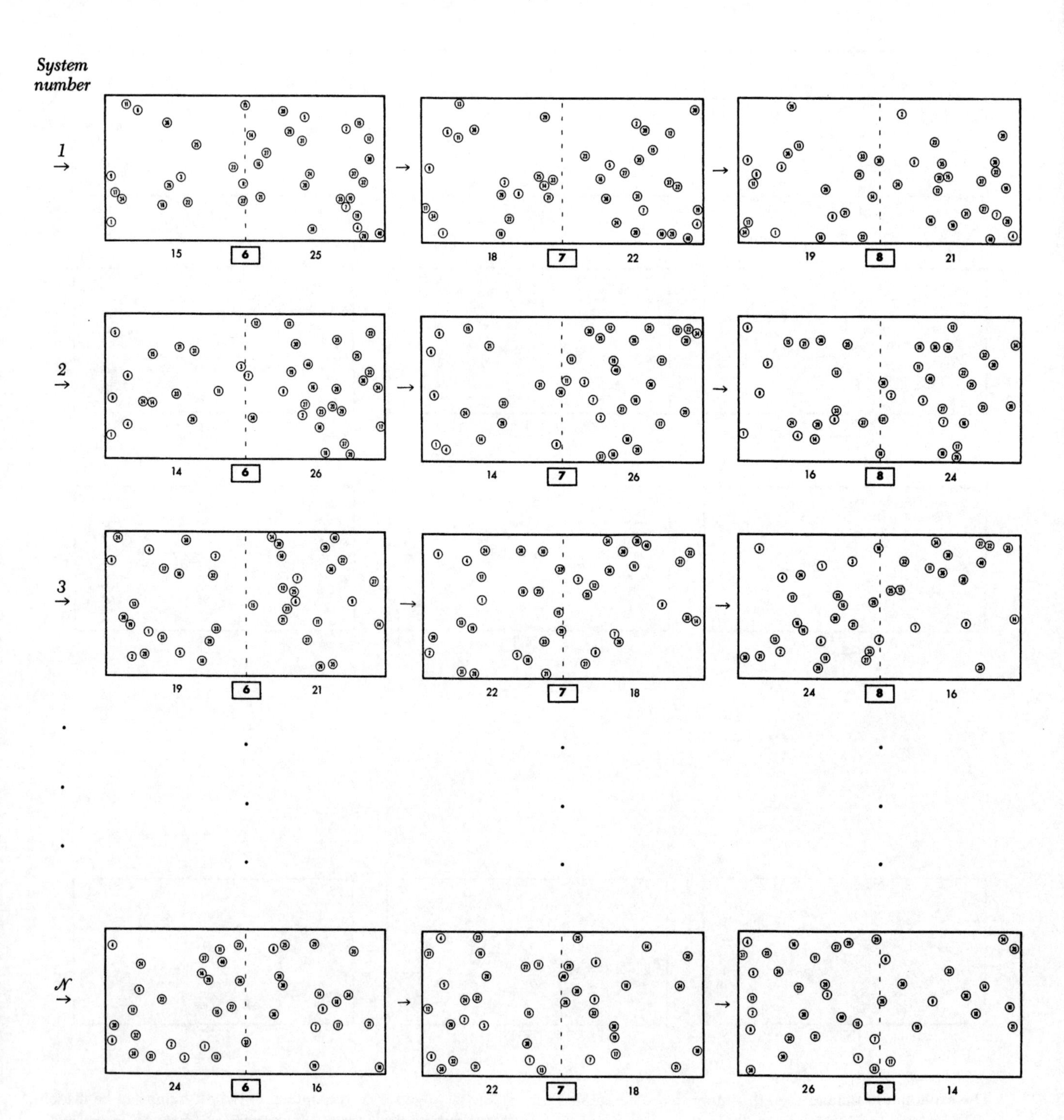

Fig. 2.4 Continuation of Fig. 2.3. The ensemble has by now become time-independent, i.e., the system has attained equilibrium.

probability statements about the system at any one time t are obtained by looking at all the film frames taken at the particular time t (i.e., by looking along a particular vertical line in Fig. 2.3 and by counting the fraction of systems exhibiting a given outcome at this time).

A statistical ensemble of systems is said to be time-independent if the number of its systems exhibiting any particular event is the same at every time (or equivalently, if the probability of occurrence of any particular event in this ensemble is independent of time). The statistical description then provides a very clear definition of equilibrium: *An isolated macroscopic system is said to be in equilibrium if a statistical ensemble of such systems is time-independent.*

Example

Consider an ideal gas of N molecules. At some initial time t_0, immediately after a partition has been removed, all the molecules of this gas are known to be in the left half of a box. How do we go about giving a statistical description of what happens at all subsequent times? We need only consider an ensemble consisting of some large number $\mathcal{N}$ of similar boxes of gas, each one of which has all its molecules concentrated in its left half at the time t_0. Such an ensemble is shown schematically in Fig. 2.3. We can then consider this ensemble at any time $t > t_0$ and ask various questions of interest. For example, focusing attention on any one molecule, what is the probability $p(t)$ that this molecule is in the left half of the container or the probability $q(t)$ that it is in the right half of the container? Or what is the probability $P(n,t)$ that, at any time t, n of the N molecules are located in the left half of the container? At the initial time t_0, we know that $p(t_0) = 1$ while $q(t_0) = 0$. [Similarly, $P(N,t_0) = 1$, while $P(n,t_0) = 0$ for $n \neq N$.] As time goes on, all these probabilities change until the molecules become uniformly distributed throughout the box so that $p = q = \frac{1}{2}$. Thereafter the probabilities remain unchanged in time, i.e., the ensemble has become time-independent and the system has attained equilibrium (see Fig. 2.4).† This time-independent situation is, of course, particularly simple. Indeed, the problem of the gas of N molecules then becomes analogous to the previously discussed problem of the set of N coins. In particular, the probability p of finding a molecule in the left half of the box is analogous to the probability p that a tossed coin shows a head; similarly, the probability q of finding a molecule in the right half of the box is analogous to the probability q that a coin shows a tail. Furthermore, as in the case of the coin, these probabilities are time-independent and such that $p = q = \frac{1}{2}$.

† Note that, despite the erratic fluctuations occurring in any *single* system as time goes on, the probability P in the *ensemble* at any one instant of time has always a unique well-defined value since the number $\mathcal{N}$ of systems in the ensemble is arbitrarily large. This remark illustrates the great conceptual simplicity of thinking in terms of ensembles rather than in terms of a single system.

Remark

In Eq. 1.4 of Chap. 1 we calculated probabilities by considering only a single system. This is ordinarily a valid procedure in the special case of a system in equilibrium. Since an ensemble of such systems is time-independent, a large number of successive observations on a single system is then equivalent to a large number of simultaneous observations on the many systems of the ensemble. In other words, suppose that one takes a film strip of a single system over a time τ and cuts it into $\mathcal{N}$ long pieces, each one of duration $\tau_1 \equiv \tau/\mathcal{N}$ (where τ_1 is so long that the behavior of the system in one piece is independent of its behavior in an adjacent piece). Then the collection of these $\mathcal{N}$ film strips of a single system is indistinguishable from a collection of $\mathcal{N}$ film strips taken of all the systems in the ensemble during any time interval of duration τ_1.

2.2 *Elementary Relations among Probabilities*

Probabilities satisfy some simple relations which are almost self-evident, but quite important. It will be worth deriving these relations by starting directly from the definition (1) of a probability. Throughout the following discussion the number $\mathcal{N}$ of systems in the ensemble will always be understood to be infinitely large.

Suppose that experiments on some system A can lead to any of α mutually exclusive outcomes. Let us label each such outcome or *event* by some index r; the index r can then refer to any of the α numbers $r = 1, 2, 3, \ldots,$ or α. In an ensemble of similar systems, $\mathcal{N}_1$ of them will be found to exhibit event 1, $\mathcal{N}_2$ of them event 2, . . . , and $\mathcal{N}_\alpha$ of them event α. Since these α events are mutually exclusive and exhaust all possibilities, it follows that

$$\mathcal{N}_1 + \mathcal{N}_2 + \cdots + \mathcal{N}_\alpha = \mathcal{N}.$$

Dividing by $\mathcal{N}$, this relation becomes

$$P_1 + P_2 + \cdots + P_\alpha = 1 \tag{2}$$

where $P_r \equiv \mathcal{N}_r/\mathcal{N}$ denotes the probability of occurrence of event r in accordance with the definition (1). The relation (2), which states merely that the sum of all possible probabilities adds up to unity, is called the *normalization condition* for probabilities. Using the summation symbol Σ defined in (M.1), this relation can also be written more compactly as

$$\sum_{r=1}^{\alpha} P_r = 1. \tag{3}$$

What is the probability of occurrence of *either* event r *or* event s? There are $\mathcal{N}_r$ systems in the ensemble exhibiting event r and $\mathcal{N}_s$ systems exhibiting event s. Hence there are $(\mathcal{N}_r + \mathcal{N}_s)$ systems exhibiting either event r or event s. Correspondingly the probability $P(r \text{ or } s)$ of occurrence of either alternative, event r or s, is simply given by

$$P(r \text{ or } s) = \frac{\mathcal{N}_r + \mathcal{N}_s}{\mathcal{N}}$$

so that

$$\boxed{P(r \text{ or } s) = P_r + P_s.} \tag{4}$$

Example

Suppose that we consider the tossing of a die which, by virtue of its symmetry, has equal probabilities $\frac{1}{6}$ of landing with any of its six faces uppermost. The probability that the die exhibits the number 1 is then $\frac{1}{6}$, as is the probability that it exhibits the number 2. The probability that the die exhibits either the number 1 or the number 2 is then, by (4), simply $\frac{1}{6} + \frac{1}{6} = \frac{1}{3}$. This is, of course, an obvious result since the events where 1 or 2 are uppermost represent one-third of all the six possible events 1, 2, 3, 4, 5, or 6.

The relation (4) is immediately generalizable to more than two alternative events. Thus the probability of occurrence of either one of several events is simply equal to the sum of their respective probabilities. In particular, we see that the normalization condition (2) merely states the obvious result that the sum of the probabilities on the left (i.e., the probabilities of occurrence of *either* event 1 *or* event 2 . . . *or* event α) is simply equal to unity (i.e., equivalent to certainty) since all possible events are exhausted in the enumeration of the α possible alternatives.

Joint probabilities

Suppose that the system under consideration can exhibit two different types of events, namely α possible events of a type labeled by r (where the index $r = 1, 2, 3, \ldots, \alpha$) and β possible events of a type labeled by s (where the index $s = 1, 2, 3, \ldots, \beta$). We shall denote by P_{rs} the probability of joint occurrence of *both* event r *and* event s. That is, in an ensemble consisting of a large number $\mathcal{N}$ of similar systems, $\mathcal{N}_{rs}$ of these are characterized by the joint occurrence of *both* an event r of the first type *and* an event s of the second type. Then $P_{rs} \equiv \mathcal{N}_{rs}/\mathcal{N}$. As usual, we denote by P_r the probability of occurrence of an event r (irrespective of the occurrence of events of type s). That is, if in the previous ensemble one pays no attention to events of type s and counts $\mathcal{N}_r$ systems in the ensemble exhibiting event r, then $P_r \equiv \mathcal{N}_r/\mathcal{N}$. Similarly, we denote by P_s the probability of occurrence of an event s (irrespective of the occurrence of events of type r).

A special, but important, case is that where the probability of occurrence of an event of type s is unaffected by the occurrence or nonoccurrence of an event of type r. The events of type r and s are then said to be *statistically independent* or *uncorrelated.* Now consider in the ensemble the $\mathcal{N}_r$ systems which exhibit any particular event r. Irrespective of the particular value of r, a fraction P_s of these systems will then also exhibit the event s. Thus the number $\mathcal{N}_{rs}$ of systems exhibiting jointly *both r and s* is simply

$$\mathcal{N}_{rs} = \mathcal{N}_r P_s.$$

Correspondingly the probability of joint occurrence of *both r and s* is given by

$$P_{rs} \equiv \frac{\mathcal{N}_{rs}}{\mathcal{N}} = \frac{\mathcal{N}_r P_s}{\mathcal{N}} = P_r P_s.$$

Hence we conclude that

if the events r and s are statistically independent,

$$P_{rs} = P_r P_s. \tag{5}$$

Note that the result (5) is not true if the events r and s are not statistically independent. The relation (5) can be immediately generalized; thus the joint probability of more than two statistically independent events is simply the product of their respective probabilities.

Example

Suppose that the system A under consideration consists of two dice A_1 and A_2. An event of type r might be the appearance uppermost of any one of the 6 faces of the die A_1; similarly an event of type s might be the appearance uppermost of any one of the 6 faces of die A_2. An event of system A is then specified by stating which face of die A_1 is uppermost and which face of die A_2 is uppermost. An experiment involving the tossing of the two dice would thus have $6 \times 6 = 36$ possible outcomes. Probability statements can be made about an ensemble consisting of a large number $\mathcal{N}$ of similar pairs of dice. We suppose that each die is symmetric so that it is equally likely to land with any one of its 6 faces uppermost. The probability P_r that each die lands with any particular face r uppermost is then simply $\frac{1}{6}$. If the dice do not interact with each other (e.g., if they are not both magnetized so as to exert forces tending to align each other) and are not tossed in precisely identical ways, they can be considered statistically independent. In this case the joint probability P_{rs} that die A_1 lands with a particular face r uppermost *and* that die A_2 lands with a particular face s uppermost is simply

$$P_{rs} = P_r P_s = \tfrac{1}{6} \times \tfrac{1}{6} = \tfrac{1}{36}.$$

This result is, of course, quite evident since the contemplated event represents one out of $6 \times 6 = 36$ possible outcomes.

2.3 The Binomial Distribution

We have now acquired sufficient familiarity with statistical methods to give a quantitative discussion of some physically important problems. Consider, for instance, an ideal system of N spins $\frac{1}{2}$, each having an associated magnetic moment μ_0. This system is of particular interest since it is a very simple system readily described in terms of quantum mechanics; it will therefore often be useful as a prototype of more complicated systems. For the sake of generality, we suppose that the spin system is located in an external magnetic field $\mathbf{B}$. Each magnetic moment can then point either "up" (i.e., parallel to field $\mathbf{B}$), or "down" (i.e., antiparallel to the field $\mathbf{B}$). We assume that the spin system is in equilibrium. A statistical ensemble consisting of $\mathcal{N}$ such spin systems is thus time-independent. Focusing attention on any one spin, we shall denote by p the probability that its magnetic moment points up and by q the probability that it points down. Since these two orientations exhaust all the possibilities, the normalization requirement (3) has the obvious consequence that

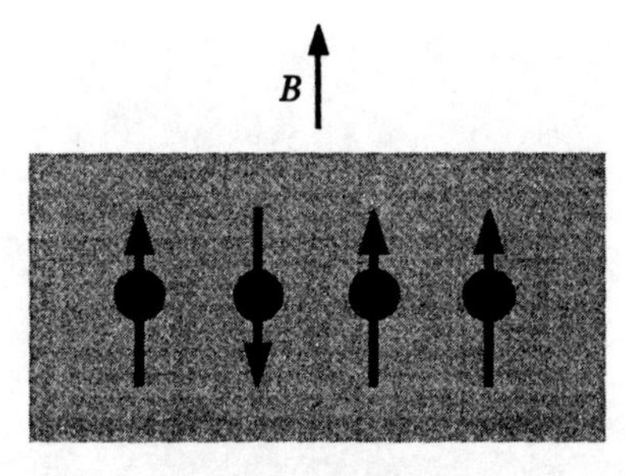

Fig. 2.5 A system consisting of N spins $\frac{1}{2}$ in the special case where $N = 4$. Each arrow indicates the direction of the magnetic moment of a spin. The external magnetic field is denoted by B.

$$p + q = 1 \tag{6}$$

or $q = 1 - p$. In the absence of the field, when $\mathbf{B} = 0$, there is no preferred direction in space so that $p = q = \frac{1}{2}$. But in the presence of the field, a magnetic moment will be more likely to point along the field than opposite to it so that $p > q$.† Since the spin system is ideal, the interaction between the spins is almost negligible so that their orientations can be considered statistically independent. The probability that any given moment points up is, therefore, unaffected by whether any other moment in the system points up or down.

Among the N magnetic moments of the spin system, let us denote by n the number of magnetic moments which point up and by n' the number of magnetic moments which point down. Of course,

$$n + n' = N \tag{7}$$

so that $n' = N - n$. Consider then the spin systems in the statistical ensemble. The number n of moments pointing up is not the same in each system, but can assume any of the possible values $n = 0, 1, 2, \ldots, N$. The question of interest is then the following: For each pos-

† We shall consider p and q as given quantities determined by experiment. In Chap. 4 we shall learn how to calculate p and q for any value of B if the spin system is known to be at a given temperature.

1	2	3	4	n	n'	$C_N(n)$
U	U	U	U	4	0	1
U	U	U	D	3	1	
U	U	D	U	3	1	4
U	D	U	U	3	1	
D	U	U	U	3	1	
U	U	D	D	2	2	
U	D	U	D	2	2	
U	D	D	U	2	2	6
D	U	U	D	2	2	
D	U	D	U	2	2	
D	D	U	U	2	2	
U	D	D	D	1	3	
D	U	D	D	1	3	4
D	D	U	D	1	3	
D	D	D	U	1	3	
D	D	D	D	0	4	1

Table 2.1. A table T which lists all the possible orientations of N magnetic moments in the special case where $N = 4$. The letter U indicates a moment which points up, the letter D a moment which points down. The number of moments pointing up is denoted by n, the number pointing down by n'. The number $C_N(n)$ of possible configurations in which n of the N moments point up is indicated in the last column. (Note that this table is analogous to Table 1.1.)

sible value of n, what is the probability $P(n)$ that n of the N magnetic moments point up?

The problem of finding the probability $P(n)$ is readily solved by the following argument. The probability that any one magnetic moment points up is p, the probability that it points down is $q = 1 - p$. Since all the magnetic moments are statistically independent, the general relation (5) allows us immediately to say that

$$\left[\begin{array}{l}\text{the probability of occurrence of}\\ \textit{one}\text{ particular configuration}\\ \text{where } n \text{ moments point up and}\\ \text{the remaining } n' \text{ point down}\end{array}\right] = \underbrace{pp\cdots p}_{n\text{ factors}}\;\underbrace{qq\cdots q}_{n'\text{ factors}} = p^n q^{n'}. \tag{8}$$

But a situation where any n moments point up can ordinarily be achieved in many alternative ways, as is illustrated in Table 2.1. Let us, therefore, introduce the notation

$$C_N(n) \equiv \text{the number of distinct configurations of } N \text{ moments where any } n \text{ of these moments point up (and the remaining } n' \text{ point down).}\dagger \tag{9}$$

The desired probability $P(n)$ that n of the N moments point up is equal to the probability that either the first, or the second, . . . , or the last of the $C_N(n)$ alternative possibilities is realized. In accordance with the general relation (4), the probability $P(n)$ is thus obtained by summing the probability (8) over the $C_N(n)$ possible configurations having n moments pointing up, i.e., by multiplying the probability (8) by $C_N(n)$. Thus we get

$$P(n) = C_N(n)p^n q^{N-n} \tag{10}$$

where we have put $n' = N - n$.

It only remains to calculate the number of configurations $C_N(n)$ in the general case of arbitrary N and n. Suppose then that we consider a table T, similar to Table 2.1, in which we list all the possible configurations of the N moments and designate by U each moment which points up and by D each moment which points down. The number $C_N(n)$ is then the number of entries in which the letter U appears n times. To examine how many such entries there are, let us consider n distinct moments which point up and label them by the letters $U_1, U_2, \ldots, U_n$. In how many ways can we list these in a table T' (such as Table 2.2 shown for the special case where $N = 4$ and $n = 2$)?

† The number $C_N(n)$ is sometimes called the number of combinations of N objects taken n at a time.

The letter U_1 can be listed in a line of the table in any one of N different places;
for each possible location of U_1, the letter U_2 can then be listed in any one of the remaining $(N - 1)$ places;
for each possible location of U_1 and U_2, the letter U_3 can then be listed in any one of the remaining $(N - 2)$ places;

. . . .

for each possible location of $U_1, U_2, \ldots, U_{n-1}$, the last letter U_n can then be listed in any one of the remaining $(N - n + 1)$ places.

The possible number $J_N(n)$ of distinct entries in the table T' is then obtained by multiplying together the number of possible locations of the letters $U_1, U_2, \ldots, U_n$; i.e.,

$$J_N(n) = N(N-1)(N-2)\cdots(N-n+1). \tag{11}$$

This can be written more compactly in terms of factorials. Thus†

$$J_N(n) = \frac{N(N-1)(N-2)\cdots(N-n+1)(N-n)\cdots(1)}{(N-n)\cdots(1)}$$

$$= \frac{N!}{(N-n)!}. \tag{12}$$

The symbols $U_1, U_2, \ldots, U_n$ were considered distinct in the preceding enumeration, while the subscripts are really irrelevant since all moments pointing up are equivalent; i.e., any U_i denotes a moment pointing up, irrespective of i. Those entries in the table T' which differ only by a permutation of the subscripts correspond, therefore, to physically indistinguishable situations (see, for example, Table 2.2). Since the possible number of permutations of the n subscripts is given by $n!$, the table T' contains $n!$ times too many entries if only distinct nonequivalent entries are to be considered.‡ The desired number $C_N(n)$ of distinct configurations of up and down moments is then given by dividing $J_N(n)$ by $n!$ so that

$$C_N(n) = \frac{J_N(n)}{n!} = \frac{N!}{n!(N-n)!}. \tag{13}$$

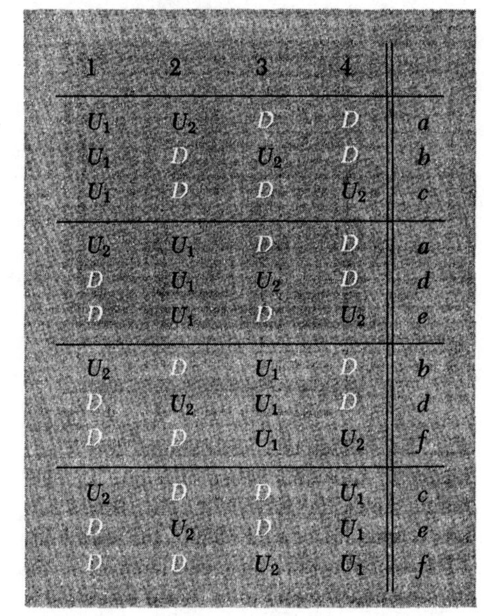

1	2	3	4	
U_1	U_2	D	D	a
U_1	D	U_2	D	b
U_1	D	D	U_2	c
U_2	U_1	D	D	a
D	U_1	U_2	D	d
D	U_1	D	U_2	e
U_2	D	U_1	D	b
D	U_2	U_1	D	d
D	D	U_1	U_2	f
U_2	D	D	U_1	c
D	U_2	D	U_1	e
D	D	U_2	U_1	f

Table 2.2. A table T' which lists all the possible arrangements of $N = 4$ identical moments, $n = 2$ of which point up. For ease of enumeration, these two moments have been labeled by U_1 and U_2, respectively, although they are physically indistinguishable. Hence entries which differ only in subscripts are equivalent. Such equivalent entries are indicated by identical letters in the last column. The table contains thus $n! = 2$ times too many entries if one is interested only in entries which are physically distinct.

† By definition, $N! \equiv N(N-1)(N-2)\cdots(1)$; in addition, $0! \equiv 1$.

‡ The first subscript can assume any of n possible values, the second subscript any of the remaining $(n-1)$ possible values, . . . , and the nth subscript the one remaining value. Hence the subscripts can be arranged in $n(n-1)\cdots(1) \equiv n!$ possible ways.

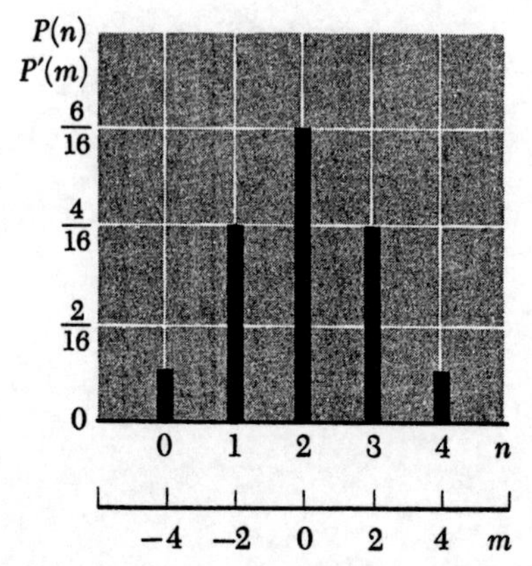

Fig. 2.6 Binominal distribution for $N = 4$ magnetic moments when $p = q = \frac{1}{2}$. The graph shows the probability $P(n)$ that n moments point up, or equivalently, the probability $P'(m)$ that the total magnetic moment in the up direction is equal to m (when measured in units of μ_0).

The desired probability (10) then becomes

$$\boxed{P(n) = \frac{N!}{n!(N-n)!} p^n q^{N-n}} \tag{14}$$

or, in more symmetrical form,

$$P(n) = \frac{N!}{n!n'!} p^n q^{n'}, \qquad \text{where } n' \equiv N - n. \tag{15}$$

In the special case

$$\text{when } p = q = \tfrac{1}{2}, \qquad P(n) = \frac{N!}{n!n'!}\left(\frac{1}{2}\right)^N. \tag{16}$$

For a given number N, the probability $P(n)$ is a function of n and is called the *binomial distribution.*

Remark

In expanding a binomial of the form $(p + q)^N$, the coefficient of the term $p^n q^{N-n}$ is simply equal to the number $C_N(n)$ of possible terms which involve the factor p exactly n times and the factor q exactly $(N - n)$ times. Hence one obtains the purely mathematical result known as the *binomial theorem*

$$(p+q)^N = \sum_{n=0}^{N} \frac{N!}{n!(N-n)!} p^n q^{N-n}. \tag{17}$$

Comparison with (14) shows that each term on the right is precisely the probability $P(n)$. This is the reason for the name "binomial distribution." Incidentally, since $p + q = 1$ when p and q are the probabilities of interest to us, Eq. (17) is equivalent to

$$1 = \sum_{n=0}^{N} P(n).$$

This verifies that the sum of the probabilities for all possible values of n is properly equal to unity, as required by the normalization condition (3).

Discussion

To examine the dependence of $P(n)$ on n, we investigate first the behavior of the coefficient $C_N(n)$ given by (13). Note first that $C_N(n)$ is symmetric under interchange of n and $N - n = n'$. Thus

$$C_N(n') = C_N(n). \tag{18}$$

In addition,

$$C_N(0) = C_N(N) = 1. \tag{19}$$

We also note that

$$\frac{C_N(n+1)}{C_N(n)} = \frac{n!(N-n)!}{(n+1)!(N-n-1)!} = \frac{N-n}{n+1}. \tag{20}$$

Starting with $n = 0$, the ratio of successive coefficients $C_N(n)$ is thus initially large, of the order of N; it then decreases monotonically with n, staying larger than (or at most becoming equal to) unity as long as $n < \frac{1}{2}N$, and becoming smaller than unity for $n \geq \frac{1}{2}N$. This behavior, combined with (19), shows that $C_N(n)$ has a maximum near $n = \frac{1}{2}N$ and that its value there is very large compared to unity, to the extent that N is large.

The behavior of the probability $P(n)$ is then apparent. By (16) it follows that,

$$\text{if } p = q = \tfrac{1}{2}, \qquad P(n') = P(n). \tag{21}$$

This result must, of course, be true by symmetry since there is no preferred spatial orientation if $p = q$ (i.e., in the absence of an applied magnetic field $\mathbf{B}$). In this case the probability $P(n)$ has a maximum† near $n = \frac{1}{2}N$. On the other hand, if $p > q$, the coefficient $C_N(n)$ still tends to produce a maximum of $P(n)$, but this maximum is now shifted to a value $n > \frac{1}{2}N$. Figures 2.6 and 2.7 illustrate the behavior of the probability $P(n)$ in some simple cases.

† The maximum is at $\frac{1}{2}N$ when N is even, and straddles this value otherwise.

Fig. 2.7 Binominal distribution for $N = 20$ magnetic moments when $p = q = \frac{1}{2}$. The graph shows the probability $P(n)$ that n moments point up, or equivalently, the probability $P'(m)$ that the total magnetic moment in the up direction is equal to m (when measured in units of μ_0).

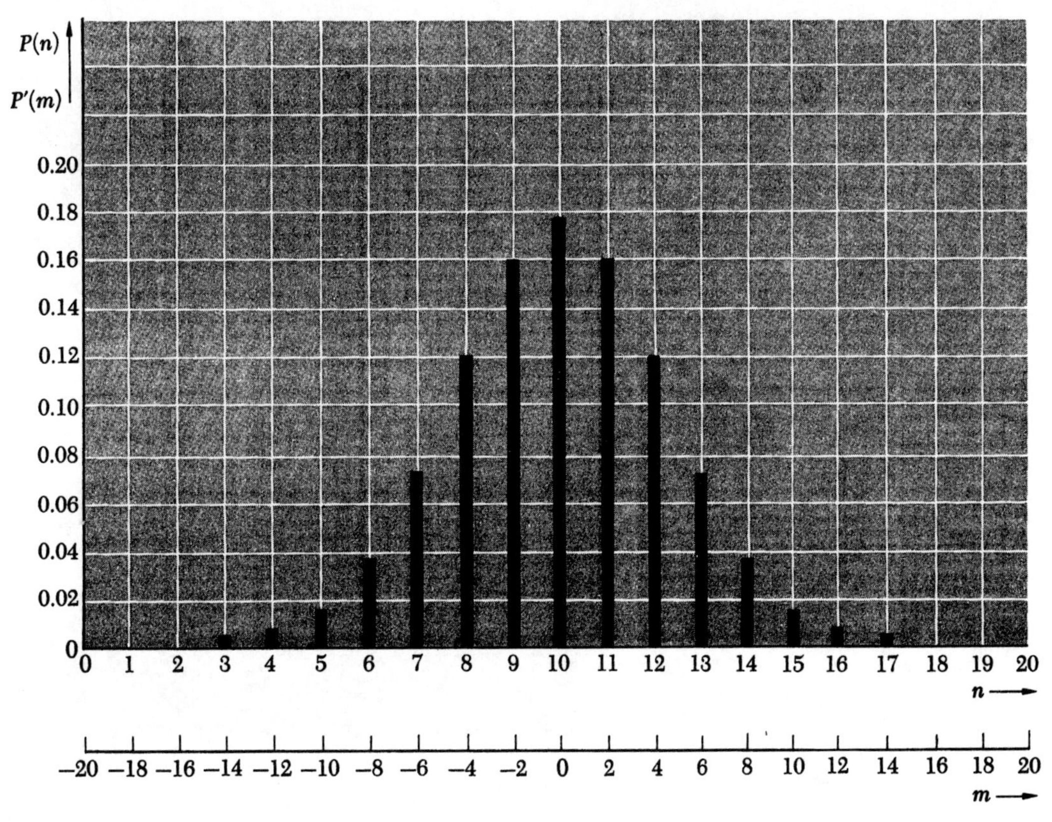

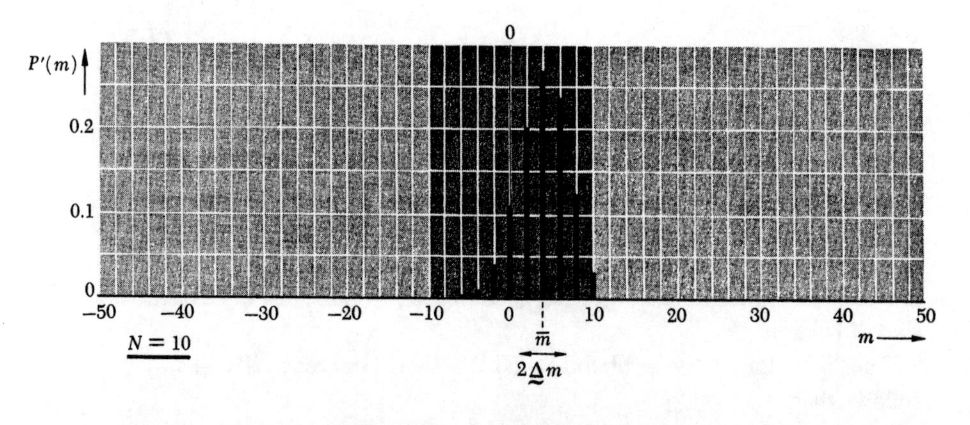

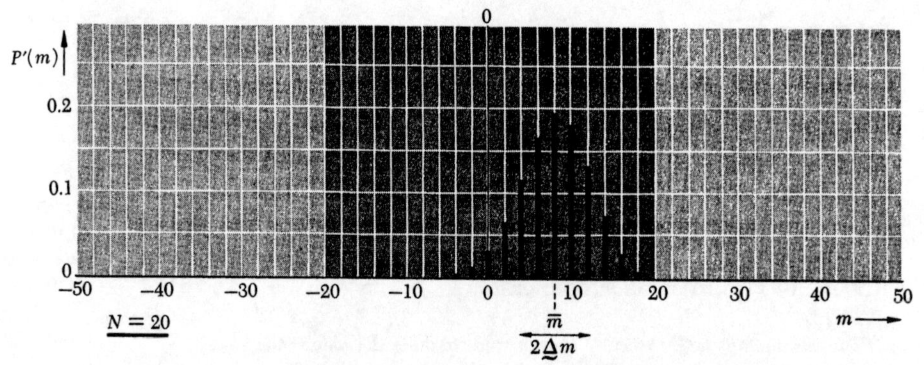

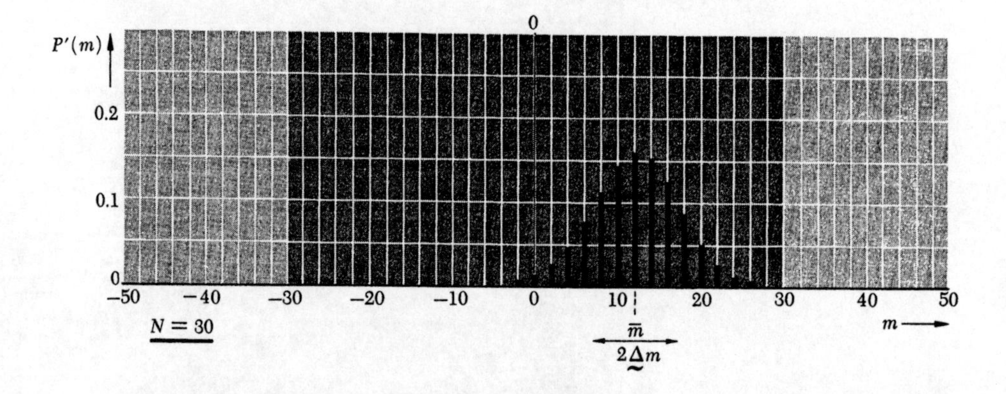

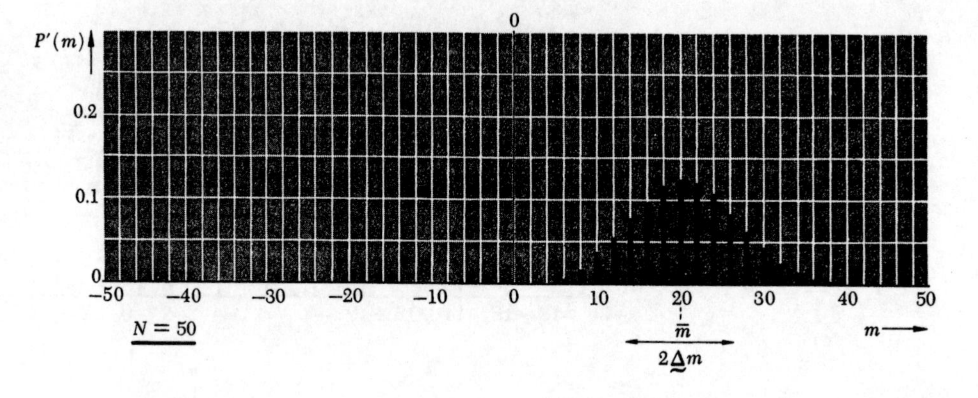

Fig. 2.8 The probability $P'(m)$ that the total magnetic moment of a system of N spins $\frac{1}{2}$ is equal to m (measured in units of μ_0). Because of the presence of an applied magnetic field, $p = 0.7$ and $q = 0.3$. The graphs show $P'(m)$ in four different cases corresponding to $N = 10$, $N = 20$, $N = 30$, and $N = 50$.

The total magnetic moment of a spin system is a quantity which is readily measured experimentally. Let us denote by M the total magnetic moment along the "up" direction. Since M is simply equal to the algebraic sum of the components along the "up" direction of the magnetic moments of all the N spins, it follows that

$$M = n\mu_0 - n'\mu_0 = (n - n')\mu_0$$

or

$$M = m\mu_0 \tag{22}$$

where

$$m \equiv n - n' \tag{23}$$

and where μ_0 denotes the magnitude of the magnetic moment of a spin. By (22), $m = M/\mu_0$ is simply the total magnetic moment measured in units of μ_0. The relation (23) can also be written as

$$m = n - n' = n - (N - n) = 2n - N. \tag{24}$$

This shows, incidentally, that the possible values of m must be odd if N is odd, and must be even when N is even. According to (24), a definite value of n corresponds to a unique value of m, and conversely

$$n = \tfrac{1}{2}(N + m). \tag{25}$$

The probability $P'(m)$ that m assumes a certain value must then be the same as the probability $P(n)$ that n assumes the corresponding value given by (25). Thus

$$P'(m) = P\left(\frac{N + m}{2}\right). \tag{26}$$

This expression gives the probability of occurrence of any possible value of the total magnetic moment of the spin system. In the special case where $p = q = \frac{1}{2}$, the expressions (16) and (26) yield thus explicitly

$$P'(m) = \frac{N!}{\left(\frac{N + m}{2}\right)!\left(\frac{N - m}{2}\right)!}\left(\frac{1}{2}\right)^N;$$

the most probable situation is then clearly at (or near) $m = 0$ where $M = 0$.

Generality of the binomial distribution

Although our discussion has dealt with the specific problem of a system of spins, it can be phrased more abstractly. Thus we have really solved the following general problem: Given N events which are statistically independent. Suppose that each such event occurs with a probability p; the probability that it does not occur is thus given by

$q = 1 - p$. What then is the probability $P(n)$ that any n out of these N events do occur (while the remaining $n = N - n$ events do not occur)? This question is immediately answered by the binomial distribution (14). Indeed, in our specific example of a system of N independent spins, the occurrence of an event is merely represented by a spin pointing up, while the nonoccurrence of an event is represented by a spin not pointing up, i.e., pointing down.

Some further examples will serve to illustrate some common problems which are immediately answered by the binomial distribution.

Ideal gas of N molecules

Consider an ideal gas of N molecules enclosed in a box of volume V_0. Since the molecules of an ideal gas interact with each other to an almost negligible extent, their motion is statistically independent. Suppose that the box is imagined to be subdivided into two parts of respective volumes V and V', where

$$V + V' = V_0. \tag{27}$$

Consider an ensemble of many such boxes of gas. Let p denote the probability that a given molecule is found in the volume V, and q the probability that it is found in the remaining volume V'. If the gas is in equilibrium, each molecule tends to be uniformly distributed throughout the box so that

$$p = \frac{V}{V_0} \quad \text{and} \quad q = \frac{V'}{V_0}. \tag{28}$$

Thus $p + q = 1$, as in (6). What then is the probability $P(n)$ in the ensemble that n of the N molecules are found in the volume V (while the remaining $n' = N - n$ molecules are found in the volume V')? The answer is given by the binomial distribution (14). In particular, if $V = V'$ so that $p = q = \frac{1}{2}$, we have thus solved explicitly the problem of Sec. 1.1 where we wanted to find the probability that n out of N molecules are found in the left half of a box.

Tossing of coins or dice

Consider the tossing of a set of N coins. The behavior of these coins can be considered statistically independent. Let p denote the probability that any given coin shows a head and q the probability that it shows a tail. By symmetry, we can assume that $p = q = \frac{1}{2}$. What then is the probability $P(n)$ that n of the N coins show heads?

The tossing of a set of N dice is similar. These again can be considered statistically independent. Let p denote the probability that any given die lands with a "6" uppermost and $q = 1 - p$ the probability that it does not. Since a die has 6 faces, we can assume by symmetry that $p = \frac{1}{6}$ while $q = 1 - p = \frac{5}{6}$. What then is the probability $P(n)$ that n of the N dice show a "6" uppermost? This question again is answered by the binomial distribution (14).

2.4 *Mean Values*

Suppose that a variable u of some system can assume any of α possible distinct values

$$u_1, u_2, \ldots, u_\alpha$$

with respective probabilities

$$P_1, P_2, \ldots, P_\alpha.$$

This means that, in an ensemble of $\mathcal{N}$ similar systems (where $\mathcal{N} \to \infty$), the variable u assumes the particular value u_r in a number $\mathcal{N}_r = \mathcal{N}P_r$ of these systems.

The specification of the probabilities P_r for all the α possible values u_r of the variable u constitutes the most complete statistical description of the system. It is, however, also convenient to define parameters that characterize in a less detailed way the distribution of the possible values of u in the ensemble. These parameters are certain *mean* (or *average*) values. The notion is quite familiar. For example, the result of an examination given to a group of students can be described most completely (if one does not wish to name individual students) by specifying the number of students receiving each of the possible grades given on this examination. But the result can also be characterized in a less detailed manner by computing the mean grade of the students. This is conventionally done by multiplying each possible grade by the number of students receiving this grade, adding all the resultant products, and then dividing this sum by the total number of students. In a similar manner, the mean value of u in the ensemble is defined by multiplying each possible value u_r by the number $\mathcal{N}_r$ of systems in the ensemble which exhibit this value, adding the resultant products for all α possible values of the variable u, and then dividing this sum by the total number $\mathcal{N}$ of systems in the ensemble. The *mean value* of u (or *ensemble average* of u), which we shall denote by $\bar{u}$, is thus *defined* by

$$\bar{u} \equiv \frac{\mathcal{N}_1 u_1 + \mathcal{N}_2 u_2 + \cdots + \mathcal{N}_\alpha u_\alpha}{\mathcal{N}} = \frac{\sum_{r=1}^{\alpha} \mathcal{N}_r u_r}{\mathcal{N}}. \tag{29}$$

But, since $\mathcal{N}_r/\mathcal{N} \equiv P_r$ is the probability of occurrence of the value u_r, the definition (29) becomes simply†

$$\bar{u} \equiv \sum_{r=1}^{\alpha} P_r u_r. \tag{30}$$

Similarly, if $f(u)$ is any function of u, the *mean value* (or *ensemble average*) of f is defined by the expression

$$\bar{f}(u) \equiv \sum_{r=1}^{\alpha} P_r f(u_r). \tag{31}$$

This definition implies that mean values have some very simple properties. For example, if $f(u)$ and $g(u)$ are any two functions of u,

$$\overline{f+g} \equiv \sum_{r=1}^{\alpha} P_r[f(u_r) + g(u_r)] = \sum_{r=1}^{\alpha} P_r f(u_r) + \sum_{r=1}^{\alpha} P_r g(u_r)$$

or

$$\overline{f+g} = \bar{f} + \bar{g}. \tag{32}$$

This result shows quite generally that the mean value of a sum of terms is equal to the sum of the mean values of these terms. Thus the successive operations of performing a sum and taking an average yield the same result irrespective of the order in which they are carried out.‡ Similarly, if c is any constant,

$$\overline{cf} = \sum_{r=1}^{\alpha} P_r[cf(u_r)] = c\sum_{r=1}^{\alpha} P_r f(u_r)$$

or

$$\overline{cf} = c\bar{f}. \tag{33}$$

Thus the operations of multiplying by a constant and taking an average can also be carried out in either order without affecting the result. If $f = 1$, the relation (33) makes the obvious assertion that the mean value of a constant is simply equal to this constant.

---

† The mean value $\bar{u}$ is time-dependent if the ensemble is time-dependent, i.e., if some of the probabilities P_r depend on time. Note also that the *mean value* or *ensemble average* $\bar{u}$ is an average over all systems in the ensemble at a particular time. It is ordinarily different from the time average defined in (1.6) for a single system, except in the special case of time-independent ensembles where the time average is taken over a very long interval of time.

‡ In mathematical language one would say that these operations "commute."

Example

Consider a system of 4 spins when $p = q = \frac{1}{2}$. The number of moments pointing up can then be $n = 0, 1, 2, 3, 4$. These numbers occur with probabilities $P(n)$ which follow immediately from (16) and which were already calculated very simply in (1.4*a*). As indicated in Fig. 2.6, these probabilities are respectively,

$$P(n) = \tfrac{1}{16}, \tfrac{4}{16}, \tfrac{6}{16}, \tfrac{4}{16}, \tfrac{1}{16}.$$

The mean number of moments pointing up is then

$$\begin{aligned}\bar{n} &= \sum_{n=0}^{4} P(n)n \\ &= (\tfrac{1}{16} \times 0) + (\tfrac{4}{16} \times 1) + (\tfrac{6}{16} \times 2) \\ &\quad + (\tfrac{4}{16} \times 3) + (\tfrac{1}{16} \times 4) \\ &= 2.\end{aligned}$$

Note that this result is simply equal to $Np = 4 \times \frac{1}{2}$.

Since $p = q$, there is no preferred direction in space. The mean number of moments pointing down must thus be equal to the mean number of moments pointing up, i.e.,

$$\bar{n'} = \bar{n} = 2.$$

This result also follows from the relation (32) which allows us to write

$$\overline{n'} = \overline{N - n} = \bar{N} - \bar{n} = 4 - 2 = 2.$$

Since there is no preferred direction in space, the mean magnetic moment must clearly vanish. Indeed, one has

$$\bar{m} = \overline{n - n'} = \bar{n} - \overline{n'} = 2 - 2 = 0.$$

The value of $\bar{m}$ could also be computed directly, of course, by using the probability $P'(m)$ that m assumes its possible values $m = -4, -2, 0, 2, 4$. Thus one has, by definition,

$$\begin{aligned}\bar{m} &= \sum_{m} P'(m)m \\ &= [\tfrac{1}{16} \times (-4)] + [\tfrac{4}{16} \times (-2)] \\ &\quad + [\tfrac{6}{16} \times 0] + [\tfrac{4}{16} \times 2] + [\tfrac{1}{16} \times 4] \\ &= 0.\end{aligned}$$

One last property of mean values is often of importance. Suppose that one is dealing with two variables, u and v, which can assume the values

$$u_1, u_2, \ldots, u_\alpha$$

and

$$v_1, v_2, \ldots, v_\beta,$$

respectively. Let us denote by P_r the probability that u assumes the value u_r, and by P_s the probability that v assumes the value v_s. *If* the probability that u assumes any of its values is independent of the value assumed by v (i.e., *if* the variables u and v are statistically independent), then the joint probability P_{rs} that u assumes the value u_r *and* that v assumes the value v_s is, by (5), simply equal to

$$P_{rs} = P_r P_s. \tag{34}$$

Suppose now that $f(u)$ is any function of u while $g(v)$ is any function of v. Then the mean value of the product fg is, by the definition (31), quite generally given by

$$\overline{f(u)g(v)} \equiv \sum_{r=1}^{\alpha} \sum_{s=1}^{\beta} P_{rs} f(u_r) g(v_s) \tag{35}$$

where the summation is over all possible values u_r and v_s of the variables. If the variables are statistically independent so that (34) is valid, Eq. (35) becomes

$$\begin{aligned}\overline{fg} &= \sum_r \sum_s P_r P_s f(u_r) g(v_s) \\ &= \sum_r \sum_s [P_r f(u_r)][P_s g(v_s)] \\ &= \left[\sum_r P_r f(u_r)\right]\left[\sum_s P_s g(v_s)\right].\end{aligned}$$

But the first of the factors on the right is simply the mean value of f, while the second is the mean value of g. Hence we arrive at the result that

if u and v are statistically independent,

$$\overline{fg} = \bar{f}\bar{g}; \tag{36}$$

i.e., the average of a product is then simply equal to the product of the averages.

Dispersion

Suppose that a variable u assumes its possible values u_r with respective probabilities P_r. Some general features of the probability distribution can then be characterized by a few useful parameters. One of these is simply the mean value of u itself, i.e., the quantity $\bar{u}$ defined in (30). This parameter indicates the central value of u about which the various values u_r are distributed. It is then often convenient to measure the possible values of u with respect to their mean value $\bar{u}$ by putting

$$\Delta u \equiv u - \bar{u} \tag{37}$$

where Δu is the deviation of u from the mean value $\bar{u}$. Note that the mean value of this deviation vanishes. Indeed, using the property (32),

$$\overline{\Delta u} = \overline{(u - \bar{u})} = \bar{u} - \bar{u} = 0. \tag{38}$$

It is also useful to define a parameter which measures the extent of the spread of the possible values of u about their mean value $\bar{u}$. The mean value of Δu itself does not provide such a measure since Δu is, on the average, as often positive as negative so that its mean value vanishes in accordance with (38). On the other hand, the quantity $(\Delta u)^2$ can never be negative. Its mean value, defined by

$$\overline{(\Delta u)^2} \equiv \sum_{r=1}^{\alpha} P_r (\Delta u_r)^2 \equiv \sum_{r=1}^{\alpha} P_r (u_r - \bar{u})^2 \tag{39}$$

is called the *dispersion* (or *variance*) of u and also can never be negative since each term in the sum of (39) is nonnegative.† Thus

$$\overline{(\Delta u)^2} \geq 0. \tag{40}$$

The dispersion can only vanish if *all* occurring values of u_r are equal to $\bar{u}$; it becomes increasingly large to the extent that these values have appreciable probability of occurring far from $\bar{u}$. The dispersion does, therefore, provide a convenient measure of the amount of scatter of the values assumed by u.

The dispersion $\overline{(\Delta u)^2}$ is a quantity which has the dimensions of the square of u. A linear measure of the spread of possible values of u is provided by the square root of the dispersion, i.e., by the quantity

$$\boxed{\underset{\sim}{\Delta} u \equiv [\overline{(\Delta u)^2}]^{1/2}} \tag{41}$$

which has the same dimensions as u itself and which is called the *standard deviation* of u. The definition (39) shows that even a few values of u occurring with appreciable probability far from $\bar{u}$ would make a large contribution to $\underset{\sim}{\Delta} u$. Most values of u must therefore occur within a range of the order of $\underset{\sim}{\Delta} u$ surrounding the mean value $\bar{u}$.

Example

Let us return to the previous example of four spins when $p = q = \frac{1}{2}$. Since $\bar{n} = 2$, the dispersion of n is, by definition,

$$\begin{aligned}\overline{(\Delta n)^2} &\equiv \sum_n P(n)(n-2)^2 \\ &= [\tfrac{1}{16} \times (-2)^2] + [\tfrac{4}{16} \times (-1)^2] \\ &\quad + [\tfrac{6}{16} \times (0)^2] + [\tfrac{4}{16} \times (1)^2] \\ &\quad + [\tfrac{1}{16} \times (2)^2] \\ &= 1.\end{aligned}$$

Hence the standard deviation of n is

$$\underset{\sim}{\Delta} n = \sqrt{1} = 1.$$

Similarly one can calculate the dispersion of the magnetic moment. Since $\bar{m} = 0$, one has, by definition,

$$\begin{aligned}\overline{(\Delta m)^2} &\equiv \sum_m P'(m)(m-0)^2 \\ &= [\tfrac{1}{16} \times (-4)^2] + [\tfrac{4}{16} \times (-2)^2] \\ &\quad + [\tfrac{6}{16} \times (0)^2] + [\tfrac{4}{16} \times (2)^2] \\ &\quad + [\tfrac{1}{16} \times (4)^2] \\ &= 4\end{aligned}$$

so that

$$\underset{\sim}{\Delta} m = \sqrt{4} = 2.$$

Let us verify that the preceding results are consistent. Since $\bar{m} = 0$ while $\bar{n} = \bar{n}' = 2$, one has for all values of m or n

$$\begin{aligned}\Delta m = m &= n - n' = n - (4 - n) \\ &= 2n - 4 = 2(n-2)\end{aligned}$$

or $$\Delta m = 2(n - \bar{n}) = 2\,\Delta n.$$

Hence $\overline{(\Delta m)^2} = 4\overline{(\Delta n)^2}$,

in agreement with what we found by explicit calculation.

† Note that $\overline{(\Delta u)^2}$ is different from the quantity $(\overline{\Delta u})^2$; i.e., it makes a great deal of difference whether one squares first and then takes the average, or whether one carries out these operations in the reverse order.

A knowledge of the probabilities P_r for all values u_r gives complete statistical information about the distribution of the values of u in the ensemble. On the other hand, a knowledge of a few mean values such as $\bar{u}$ and $\overline{(\Delta u)^2}$ provides only a partial knowledge of the characteristics of this distribution and is not sufficient to determine the probabilities P_r unambiguously. Such mean values, however, may often be calculated very simply without an explicit knowledge of the probabilities, even in cases where the actual computation of these probabilities would be a difficult task. We shall illustrate these remarks in the next section.

2.5 *Calculation of Mean Values for a Spin System*

Consider an ideal system of N spins $\frac{1}{2}$. The fact that these spins are statistically independent allows us to calculate various mean values quite simply under very general conditions. The calculation can be accomplished without the need of computing any probabilities such as the probability $P(n)$ obtained in (14).

Let us start from scratch then to investigate a physically interesting quantity of this spin system, its total magnetic moment M along the up direction. Let us denote by μ_i the component along the up direction of the ith spin. The total magnetic moment M is then simply equal to the sum of the magnetic moments of all the spins so that

$$M = \mu_1 + \mu_2 + \cdots + \mu_N$$

or, in more compact notation,

$$M = \sum_{i=1}^{N} \mu_i. \tag{42}$$

We should like to calculate the mean value and the dispersion of this total magnetic moment.

To compute the mean value of M, we need only take the mean values of both sides of (42). The general property (32), which allows us to interchange the order of averaging and summing, immediately yields the result

$$\bar{M} = \overline{\sum_{i=1}^{N} \mu_i} = \sum_{i=1}^{N} \bar{\mu}_i. \tag{43}$$

But the probability that any magnetic moment has a given orientation (either up or down) is the same for each moment; hence the mean magnetic moment is the same for each spin (i.e., $\bar{\mu}_1 = \bar{\mu}_2 = \cdots = \bar{\mu}_N$) and can be denoted simply by $\bar{\mu}$. The sum in (43) consists, therefore, of N equal terms so that (43) becomes simply

$$\boxed{\bar{M} = N\bar{\mu}.} \tag{44}$$

This result is almost self-evident; it asserts merely that the mean total magnetic moment of N spins is N times as large as the mean moment of one spin.

Now let us calculate the dispersion of M, i.e., the quantity $\overline{(\Delta M)^2}$ where

$$\Delta M \equiv M - \bar{M}. \tag{45}$$

Subtraction of (43) from (42) yields

$$M - \bar{M} = \sum_{i=1}^{N} (\mu_i - \bar{\mu})$$

or

$$\Delta M = \sum_{i=1}^{N} \Delta\mu_i, \tag{46}$$

where

$$\Delta\mu_i \equiv \mu_i - \bar{\mu}. \tag{47}$$

To find $(\Delta M)^2$, we need only multiply the sum in (46) by itself. Thus

$$\begin{aligned}(\Delta M)^2 &= (\Delta\mu_1 + \Delta\mu_2 + \cdots + \Delta\mu_N)\,(\Delta\mu_1 + \Delta\mu_2 + \cdots + \Delta\mu_N)\\ &= [(\Delta\mu_1)^2 + (\Delta\mu_2)^2 + (\Delta\mu_3)^2 + \cdots + (\Delta\mu_N)^2]\\ &\quad + [\Delta\mu_1\,\Delta\mu_2 + \Delta\mu_1\,\Delta\mu_3 + \cdots + \Delta\mu_1\,\Delta\mu_N\\ &\qquad + \Delta\mu_2\,\Delta\mu_1 + \Delta\mu_2\,\Delta\mu_3 + \cdots + \Delta\mu_N\,\Delta\mu_{N-1}]\end{aligned}$$

or

$$(\Delta M)^2 = \sum_{i=1}^{N} (\Delta\mu_i)^2 + \sum_{\substack{i=1\\ i\neq j}}^{N} \sum_{j=1}^{N} (\Delta\mu_i)(\Delta\mu_j). \tag{48}$$

The first term on the right represents all the squared terms arising from terms in the sum (46) which are multiplied by themselves; the second term represents all the cross terms arising from products of *different* terms in the sum (46). Taking the mean value of (48) and using again the property (32) which allows us to interchange the order of averaging and summing, we then obtain

$$\overline{(\Delta M)^2} = \sum_{i=1}^{N} \overline{(\Delta\mu_i)^2} + \sum_{\substack{i=1\\ i\neq j}}^{N} \sum_{j=1}^{N} \overline{(\Delta\mu_i)(\Delta\mu_j)}. \tag{49}$$

All products in the second sum, where $i \neq j$, refer to different spins. But since different spins are statistically independent, the property (36) implies that the mean value of each such product is simply equal to the product of the mean values of its factors. Thus,

for $i \neq j$,

$$\overline{(\Delta\mu_i)(\Delta\mu_j)} = (\overline{\Delta\mu_i})(\overline{\Delta\mu_j}) = 0 \tag{50}$$

since

$$\overline{\Delta\mu_i} = \bar{\mu}_i - \bar{\mu} = 0.$$

In short, each cross term in (49) vanishes on the average, being as often positive as negative. Hence (50) reduces simply to a sum of squared terms (none of which can be negative):

$$\overline{(\Delta M)^2} = \sum_{i=1}^{N} \overline{(\Delta\mu_i)^2} \tag{51}$$

The argument now becomes identical to that following Eq. (43). The probability that any moment has any given orientation is the same for each moment; hence the dispersion $\overline{(\Delta\mu_i)^2}$ is the same for each spin [i.e., $\overline{(\Delta\mu_1)^2} = \overline{(\Delta\mu_2)^2} = \cdots = \overline{(\Delta\mu_N)^2}$] and can be denoted simply by $\overline{(\Delta\mu)^2}$. The sum in (51) consists thus of N equal terms and reduces simply to

$$\boxed{\overline{(\Delta M)^2} = N\overline{(\Delta\mu)^2}.} \tag{52}$$

This relation asserts that the dispersion of the total magnetic moment is merely N times as large as the dispersion of the magnetic moment of an individual spin. Correspondingly (52) also implies that

$$\boxed{\underset{\sim}{\Delta} M = \sqrt{N}\, \underset{\sim}{\Delta}\mu} \tag{53}$$

where $\quad \underset{\sim}{\Delta} M \equiv [\overline{(\Delta M)^2}]^{1/2} \quad$ and $\quad \underset{\sim}{\Delta}\mu \equiv [\overline{(\Delta\mu)^2}]^{1/2}$

are, in accordance with the general definition (41), the standard deviations of the total magnetic moment and of the magnetic moment per spin, respectively.

The relations (44) and (53) show explicitly how $\bar{M}$ and $\underset{\sim}{\Delta} M$ depend on the total number N of spins in the system. When $\bar{\mu} \neq 0$, the mean total magnetic moment $\bar{M}$ increases proportionately to N. The standard deviation $\underset{\sim}{\Delta} M$ (which measures the width of the distribution of values of M about their mean value $\bar{M}$) also increases as N increases, but does so only proportionately to $N^{1/2}$. Hence the *relative* magnitude of $\underset{\sim}{\Delta} M$ compared to $\bar{M}$ *decreases* proportionately to $N^{-1/2}$; indeed (44) and (53) imply that

$$\text{for } \bar{\mu} \neq 0, \qquad \frac{\underset{\sim}{\Delta} M}{\bar{M}} = \frac{1}{\sqrt{N}}\left(\frac{\underset{\sim}{\Delta}\mu}{\bar{\mu}}\right). \tag{54}$$

Figure 2.8 illustrates these characteristic trends.

Note that the results (44) and (53) are very general. They depend only on the additive relation (43) and on the fact that the spins are statistically independent. All our considerations would thus remain equally valid even if the component μ_i of each magnetic moment could assume many possible values. (This would be the case if the spin of each particle were greater than $\frac{1}{2}$ so that it could exhibit more than two possible orientations in space.)

System of particles with spin $\frac{1}{2}$

The foregoing results are readily applied to the familiar special case where each particle has spin $\frac{1}{2}$. As usual we suppose that its magnetic moment then has probability p of pointing up so that $\mu_i = \mu_0$, and probability $q = 1 - p$ of pointing down so that $\mu_i = -\mu_0$. Its mean moment along the up direction is thus

$$\bar{\mu} \equiv p\mu_0 + q(-\mu_0) = (p - q)\mu_0 = (2p - 1)\mu_0. \tag{55}$$

As a check we note that, in the symmetrical case where $p = q$, $\bar{\mu} = 0$ as expected.

The dispersion of the magnetic moment of a spin is given by

$$\overline{(\Delta\mu)^2} \equiv \overline{(\mu - \bar{\mu})^2} \equiv p(\mu_0 - \bar{\mu})^2 + q(-\mu_0 - \bar{\mu})^2. \tag{56}$$

But $$\mu_0 - \bar{\mu} = \mu_0 - (2p - 1)\mu_0 = 2\mu_0(1 - p) = 2\mu_0 q,$$

and $$\mu_0 + \bar{\mu} = \mu_0 + (2p - 1)\mu_0 = 2\mu_0 p.$$

Thus (56) becomes

$$\overline{(\Delta\mu)^2} = p(2\mu_0 q)^2 + q(2\mu_0 p)^2 = 4\mu_0{}^2 pq(q + p)$$

or $$\overline{(\Delta\mu)^2} = 4pq\mu_0{}^2 \tag{57}$$

since $p + q = 1$.

The relations (44) and (52) yield therefore the results

$$\boxed{\bar{M} = N(p - q)\mu_0} \tag{58}$$

and $$\boxed{\overline{(\Delta M)^2} = 4Npq\,\mu_0{}^2.} \tag{59}$$

The standard deviation of M is accordingly

$$\boxed{\Delta M = 2\sqrt{Npq}\,\mu_0.} \tag{60}$$

If we write $M = m\mu_0$, so that the integer $m = M/\mu_0$ expresses the total magnetic moment in units of μ_0, the results (58) through (60) can also be expressed in the form

$$\bar{m} = N(p - q) = N(2p - 1), \tag{61}$$

$$\overline{(\Delta m)^2} = 4Npq, \tag{62}$$

$$\Delta m = 2\sqrt{Npq}. \tag{63}$$

These relations contain an appreciable amount of information about the distribution of the possible values of M or m in the ensemble of spin systems. Thus we know that only those values of m occur with appreciable probability which lie near $\bar{m}$ and do not differ from it by an amount much larger than $\underset{\sim}{\Delta} m$. Figure 2.8 provides a specific illustration.

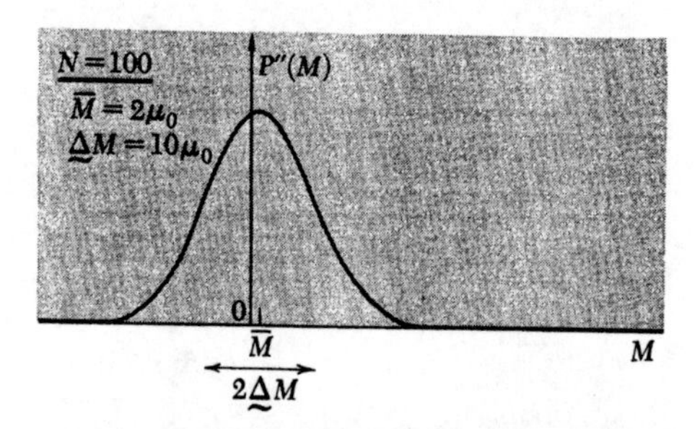

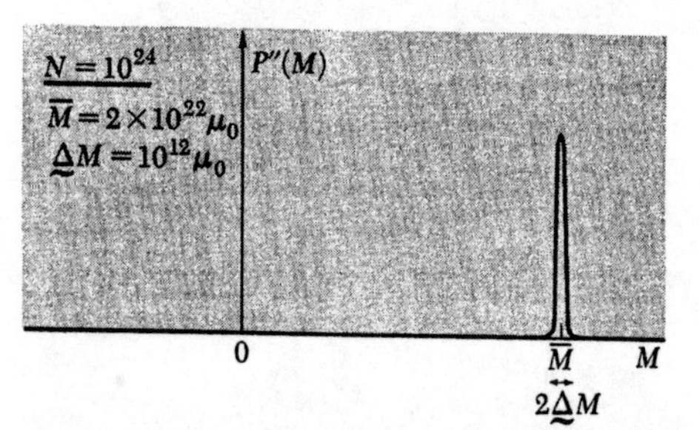

Fig. 2.9 The probability $P''(M)$ that the total magnetic moment of a spin system has a value M when $N = 100$ and when $N = 10^{24}$. The magnetic field is such that $p = 0.51$ and $q = 0.49$. The graphs indicate the envelope curves showing the possible values of $P''(M)$. The two graphs are not drawn to the same scale.

Example

Suppose that, in the presence of a certain applied magnetic field $\mathbf{B}$, the magnetic moment of each spin has a probability $p = 0.51$ of pointing parallel to $\mathbf{B}$ and a probability $q = 1 - p = 0.49$ of pointing antiparallel to $\mathbf{B}$. The mean total magnetic moment of a system of N spins is then

$$\bar{M} = 0.02N\mu_0.$$

The standard deviation of its total magnetic moment is given by (60) so that

$$\underset{\sim}{\Delta}M = 2\sqrt{Npq}\,\mu_0 \approx \sqrt{N}\,\mu_0.$$

Hence $\dfrac{\underset{\sim}{\Delta}M}{\bar{M}} \approx \dfrac{\sqrt{N}\,\mu_0}{0.02N\mu_0} = \dfrac{50}{\sqrt{N}}.$

Consider first a case where the total number of spins is fairly small. For example, suppose that $N = 100$. Then

$$\frac{\underset{\sim}{\Delta}M}{\bar{M}} \approx \frac{50}{\sqrt{100}} = 5$$

so that $\underset{\sim}{\Delta}M > \bar{M}$. The scatter in the possible values of M is then very pronounced. Indeed, it is quite likely that there occur values of M which differ widely from $\bar{M}$ and are even of opposite sign. (See Fig. 2.9.)

On the other hand, consider the case of a macroscopic system of spins where N is of the order of Avogadro's number, say $N = 10^{24}$. Then

$$\frac{\underset{\sim}{\Delta}M}{\bar{M}} \approx \frac{50}{\sqrt{10^{24}}} = 5 \times 10^{-11}$$

so that $\underset{\sim}{\Delta}M \lll \bar{M}$. The scatter in the possible values of M is then very small relative to the mean total magnetic moment. If we set out to measure the total magnetic moment of the system, we would thus almost always measure a value very close to $\bar{M}$. Indeed, unless our method of measurement were sufficiently precise to detect differences of magnetic moment smaller than about one part in 10^{10}, we would virtually always measure a magnetic moment equal to $\bar{M}$ without becoming aware of the existence of fluctuations about this value. This example illustrates concretely the general conclusion that the *relative* magnitude of fluctuations tends to become very small in a macroscopic system consisting of very many particles.

Distribution of molecules in an ideal gas

Consider an ideal gas of N molecules contained in a box of volume V_0. We are interested in investigating the number n of molecules found within any specified subvolume V of this box (see Fig. 2.10). If the gas is in equilibrium, then the probability p of finding a molecule in this volume V is simply equal to

$$p = \frac{V}{V_0} \tag{64}$$

as mentioned previously in (28).

It is very easy to calculate the mean value of n and its dispersion. We have already pointed out at the end of Sec. 2.3 that the problem of the ideal gas is analogous to that of the spin system. (Both problems are of the type leading to the binomial distribution.) Hence we can immediately apply the results (61) and (62) to find the desired information about n. Let n' denote the number of molecules in the remaining volume $V_0 - V$ of the box and let $m \equiv n - n'$. As shown in (25), it then follows that

$$n = \tfrac{1}{2}(N + m). \tag{65}$$

Using the result (61) for $\bar{m}$, we then obtain

$$\bar{n} = \tfrac{1}{2}(N + \bar{m}) = \tfrac{1}{2}N(1 + p - q)$$

or

$$\boxed{\bar{n} = Np} \tag{66}$$

since $q = 1 - p$. Furthermore, we obtain from (65) the relation

$$\Delta n \equiv n - \bar{n} = \tfrac{1}{2}(N + m) - \tfrac{1}{2}(N + \bar{m}) = \tfrac{1}{2}[m - \bar{m}]$$

or

$$\Delta n = \tfrac{1}{2}\,\Delta m.$$

Hence

$$\overline{(\Delta n)^2} = \tfrac{1}{4}\overline{(\Delta m)^2}$$

and (62) implies that†

$$\boxed{\overline{(\Delta n)^2} = Npq.} \tag{67}$$

The standard deviation of n is then

$$\boxed{\underset{\sim}{\Delta} n = \sqrt{Npq}} \tag{68}$$

so that

$$\frac{\underset{\sim}{\Delta} n}{\bar{n}} = \frac{\sqrt{Npq}}{Np} = \left(\frac{q}{p}\right)^{1/2} \frac{1}{\sqrt{N}}. \tag{69}$$

These relations show again that the standard deviation $\underset{\sim}{\Delta} n$ increases proportionately to $N^{1/2}$. Correspondingly, the relative value $\underset{\sim}{\Delta} n/\bar{n}$ of the standard deviation *de*creases proportionately to $N^{-1/2}$ and thus becomes very small when N is large. These statements are well illustrated by the special case of Chap. 1 where we considered the number n of molecules contained in one half of a box. In this case (64) implies that $p = q = \frac{1}{2}$ so that (66) reduces to the obvious result

$$\bar{n} = \tfrac{1}{2}N$$

while

$$\frac{\underset{\sim}{\Delta} n}{\bar{n}} = \frac{1}{\sqrt{N}}.$$

† The relations (66) and (67) could also be derived directly by the methods of this section without using the corresponding results for the quantity m (see Prob. 2.14).

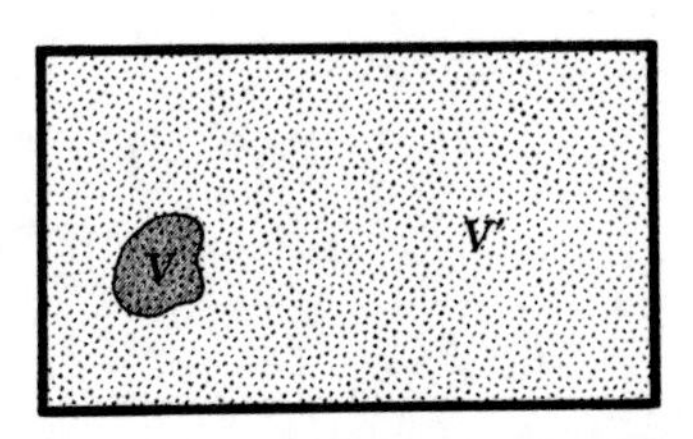

Fig. 2.10 A box of volume V_0 contains N molecules of an ideal gas. At any given time some number n of the molecules are located in the subvolume V, while the remaining number $n' = N - n$ are located in the rest of the volume $V' = V_0 - V$.

These relations put the discussion of fluctuations in Sec. 1.1 on a quantitative basis. The fact that the absolute magnitude of fluctuations, (measured by $\underset{\sim}{\Delta} n$) increases with increasing N while the relative magnitude of fluctuations (measured by $\underset{\sim}{\Delta} n/n$) decreases with increasing N is illustrated explicitly by the graphs of Figs. 1.5 and 1.6 for $N = 4$ and $N = 40$. When the box contains about a mole of gas, N is of the order of Avogadro's number so that $N \sim 10^{24}$. In this case the relative magnitude of fluctuations $\underset{\sim}{\Delta} n/\bar{n} \sim 10^{-12}$ becomes so small as to be almost always negligible.

2.6 *Continuous Probability Distributions*

Let us consider an ideal spin system consisting of a large number N of spins $\frac{1}{2}$. There are then many possible values of the total magnetic moment of this system. Indeed, by (22) and (24),

$$M = m\mu_0 = (2n - N)\mu_0 \tag{70}$$

so that M can assume any of the $(N + 1)$ possible values

$$M = -N\mu_0, -(N-2)\mu_0, -(N-4)\mu_0, \ldots, (N-2)\mu_0, N\mu_0. \tag{71}$$

The probability $P''(M)$ that the total magnetic moment assumes a particular value M is equal to the probability of occurrence of the corresponding value of m or n, i.e., to $P'(m)$ given by (26) or $P(n)$ given by (14).

Thus
$$P''(M) = P'(m) = P(n),$$
where
$$m = \frac{M}{\mu_0} \quad \text{and} \quad n = \tfrac{1}{2}(N + m). \tag{72}$$

Except when M is near its extreme possible values $\pm N\mu_0$ [where $P''(M)$ is negligibly small], the probability $P''(M)$ does not vary appreciably in going from one possible value of M to an adjacent one; i.e., $|P''(M + 2\mu_0) - P''(M)| \ll P''(M)$. The envelope of the possible values of $P''(M)$ then forms a smooth curve, as indicated in Fig. 2.11. Thus it is possible to regard $P''(M)$ as a smoothly varying function of the continuous variable M, although only the discrete values (71) of M are relevant.

Suppose that μ_0 is negligibly small compared to the smallest magnetic moment of interest in any macroscopic measurement. The fact that M can assume only discrete values separated by $2\mu_0$ is then unobservable within the precision of contemplated observations. Thus M can indeed be considered as a continuous variable. Furthermore, one can meaningfully talk of a range dM which is a "macroscopic infinitesimal," i.e., a quantity which is *macro*scopically very small although it is *micro*scopically large. (In other words, dM is supposed to be negligibly small compared to the smallest magnetic moment of interest in a macroscopic discussion, although it is much larger than μ_0.)† The following question is then of interest: What is the probability that the total magnetic moment of the system lies in a particular small range between M and $M + dM$? The magnitude of this probability depends obviously on the size of the range dM and must become vanishingly small as dM is made negligibly small. This probability is therefore expected to be simply proportional to dM so that it can be written in the form:

$$\begin{bmatrix}\text{Probability that the total magnetic} \\ \text{moment lies between } M \text{ and } M + dM\end{bmatrix} = \mathcal{P}(M)\, dM \tag{73}$$

where $\mathcal{P}(M)$ is independent of the magnitude of dM.‡ The quantity $\mathcal{P}(M)$ is called a *probability density;* it yields an actual probability when it is multiplied by the infinitesimal range dM.

The probability (73) is readily expressed explicitly in terms of the probability $P''(M)$ that the total magnetic moment assumes the particular discrete value M. Since (71) shows that the possible values of M are separated by amounts $2\mu_0$ and since $dM \gg 2\mu_0$, the range between M and $M + dM$ contains $dM/(2\mu_0)$ possible values of M. All of these occur with nearly the same probability $P''(M)$ since the probability is very slowly varying over the small range dM. Hence the probability that the total moment lies in the range between M and $M + dM$ is simply obtained by summing $P''(M)$ over all values of M lying in this range, i.e., by multiplying the nearly constant value $P''(M)$ by $dM/(2\mu_0)$. This probability is thus properly proportional to dM and is given explicitly by

$$\mathcal{P}(M)\, dM = P''(M)\frac{dM}{2\mu_0}. \tag{74}$$

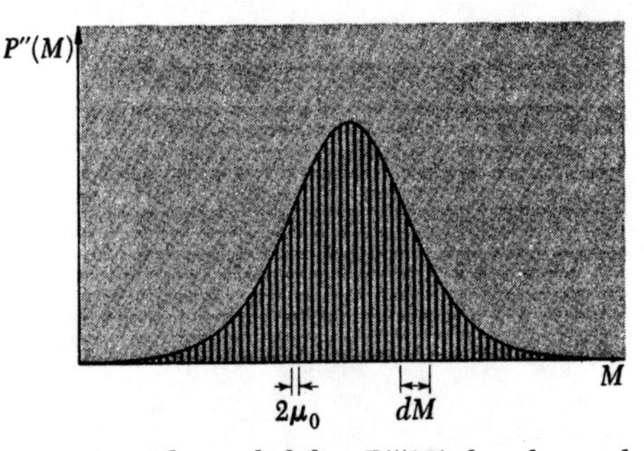

Fig. 2.11 The probability $P''(M)$ that the total magnetic moment of a spin system has a value M in a case where the number N of spins is large and the magnetic moment μ_0 of a spin is relatively small.

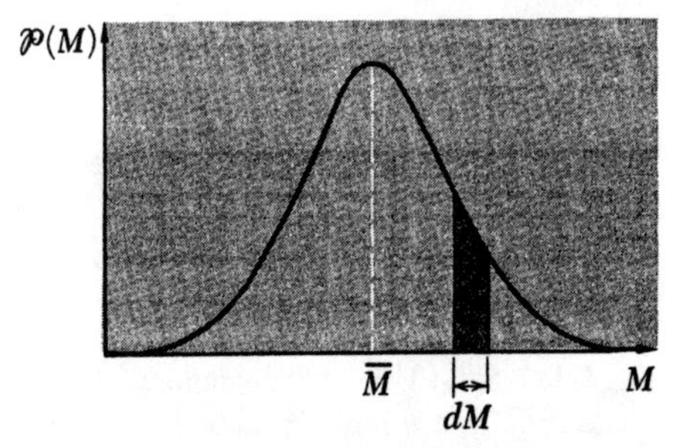

Fig. 2.12 The probability distribution of Fig. 2.11 is shown expressed in terms of the probability density $\mathcal{P}(M)$. Here $\mathcal{P}(M)\, dM$ [which is equal to the area under the curve in the small range between M and $M + dM$] is the probability that the total magnetic moment lies in the range between M and $M + dM$.

† It is worth noting that many differentials used in physics are macroscopic infinitesimals. For example, in the study of electricity, one often talks of the charge Q on a body and of a charge increment dQ. This differential description is valid if dQ is understood to be much larger than the discrete electronic charge e, although it is supposed to be negligibly small compared to Q itself.

‡ Since the probability is some smooth function of dM, it should be expressible near any value of M as a Taylor series in powers of dM when dM is small. Thus it should be of the form

$$\text{Probability} = a_0 + a_1\, dM + a_2(dM)^2 + \cdots$$

where the coefficients $a_0, a_1, \ldots$ depend on M. Here $a_0 = 0$ since the probability must approach zero as dM is made very small; furthermore, terms involving higher powers of dM are negligibly small compared to the leading term which is proportional to dM. Hence one is led to the result (73).

In practice, the actual calculation of $P''(M)$ may be laborious if M/μ_0 is large since the binomial distribution (14) then requires the evaluation of large factorials. These difficulties may, however, be circumvented by using the Gaussian approximation of Appendix A.1.

There are many problems where a variable of interest, call it u, is intrinsically continuous. For example, u might denote the angle between some vector in a plane and a fixed direction; this angle could then assume any value in the domain between 0 and 2π. In the general case u can assume any value in some domain $a_1 \leq u \leq a_2$. (This domain may be infinite in extent, i.e., $a_1 \to -\infty$, or $a_2 \to \infty$, or both.) Probability statements can be made about such a variable in a manner completely analogous to that discussed in the case of M. Thus one can focus attention on any infinitesimal range between u and $u + du$ and ask for the probability that the variable lies in this range. When du is sufficiently small, this probability must again be proportional to du so that it can be written in the form $\mathscr{P}(u)\, du$ where the quantity $\mathscr{P}(u)$ is a *probability density* independent of the size of du.

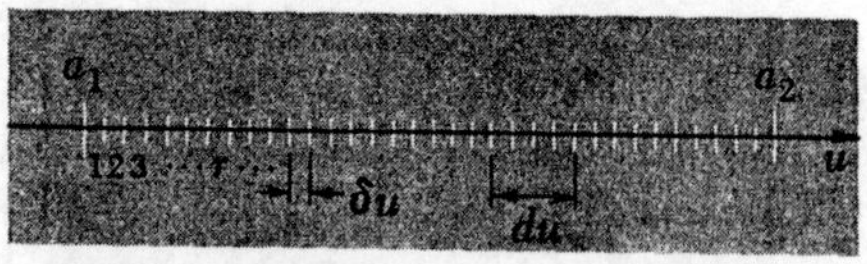

Fig. 2.13 Subdivision of the domain of a continuous variable u into a countable number of infinitesimal equal intervals of fixed size δu. Each such interval is labeled by an index r which can assume the values 1, 2, 3, 4, ... The size of a macroscopic infinitesimal range du is also shown.

Probability considerations involving a continuous variable u can readily be reduced to the simpler situation where the possible values of the variables are discrete and thus countable. It is only necessary to subdivide the domain of possible values of u into arbitrarily small equal intervals of fixed size δu. Each such interval can then be labeled by some index r. The value of u in this interval can be denoted by u_r and the probability that u lies in this interval by P_r or $P(u_r)$. This procedure allows us to deal with a countable set of values of the variable u, each such value corresponding to one of the infinitesimal intervals $r = 1, 2, 3, \ldots$. It also becomes apparent that relations involving probabilities of discrete variables remain equally valid for probabilities of continuous variables. For example, the simple properties (32) and (33) of mean values are also applicable if u is a continuous variable.

Note that the sums involved in calculating normalization conditions or mean values can be expressed as integrals if the variable is continuous. For example, the normalization condition asserts that the sum of the probabilities over all possible values of the variable must equal unity; in symbols

$$\sum_r P(u_r) = 1. \tag{75}$$

But, if the variable is continuous, one can first sum over all discrete intervals r for which u_r lies in the range between u and $u + du$; this gives the probability $\mathscr{P}(u)\, du$ that the variable lies in this range.† One

† Here the range du is understood to be large compared to the arbitrarily small interval δu (so that $du \gg \delta u$), but to be sufficiently small so that $P(u_r)$ does not vary appreciably within the range du.

can then complete the sum (75) by summing (i.e., integrating) over all such possible ranges du. Thus (75) is equivalent to

$$\int_{a_1}^{a_2} \mathcal{P}(u)\, du = 1 \tag{76}$$

which expresses the normalization condition in terms of the probability density $\mathcal{P}(u)$. Similarly, the general definition (31) of the mean value of a function $f(u)$ of discrete variables is given by

$$\overline{f(u)} \equiv \sum_r P(u_r) f(u_r). \tag{77}$$

In a continuous description one can again sum first over all intervals r for which u_r lies in the range between u and $u + du$; this contributes to the sum an amount $\mathcal{P}(u)\, du\, f(u)$. One can then complete the sum by integrating over all possible ranges du. Hence (77) is equivalent to the relation†

$$\overline{f(u)} = \int_{a_1}^{a_2} \mathcal{P}(u) f(u)\, du. \tag{78}$$

Generalization to the case of several variables

The generalization of the previous remarks to the case of more than one variable is immediate. Suppose, for example, that one is dealing with two continuous variables u and v. Then the joint probability that the variable u lies in the small range between u and $u + du$ *and* that the variable v lies in the small range between v and $v + dv$ is proportional to both du and dv and can be written in the form $\mathcal{P}(u,v)\, du\, dv$ where $\mathcal{P}(u,v)$ is a probability density independent of the size of du or dv. If desired, the situation can again be reduced to one of discrete probabilities by subdividing the variable u into very small fixed intervals δu, each of which can be labeled by some index r, and by subdividing the variable v into very small fixed intervals δv, each of which can be labeled by some other index s. In these terms the situation can then be described by specifying the probability P_{rs} that the variables have values lying in any given cell labeled by the pair of indices r and s.

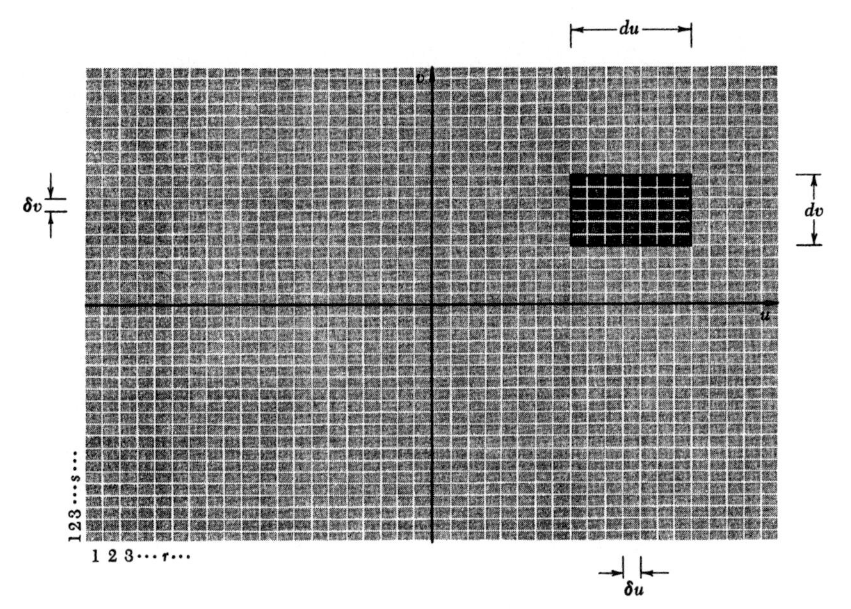

Fig. 2.14 Subdivision of the continuous variables u and v into small equal intervals of magnitude δu and δv labeled by r and s, respectively. The uv plane is thus subdivided into small cells, each labeled by the pair of indices r and s.

† Note that the probability *density* $\mathcal{P}(u)$ may become infinite for certain values of u. This does not lead to any difficulties as long as any integral $\int_{c_2}^{c_1} \mathcal{P}(u)\, du$ (which gives the probability that the value of u lies in an arbitrary range between c_1 and c_2) remains finite.

Summary of Definitions

statistical ensemble An assembly of a very large number of mutually noninteracting systems, each of which satisfies the same conditions as those known to be satisfied by a particular system under consideration.

time-independent ensemble An ensemble in which the number of systems exhibiting any particular property is the same at any time.

event The outcome of an experiment or result of an observation.

probability The probability P_r of occurrence of an event r in a system is defined with respect to a statistical ensemble of $\mathcal{N}$ such systems. If $\mathcal{N}_r$ systems in the ensemble exhibit the event r, then

$$P_r \equiv \frac{\mathcal{N}_r}{\mathcal{N}} \qquad (\text{where } \mathcal{N} \to \infty).$$

statistical independence Two events are statistically independent if the occurrence of one event does not depend on the occurrence or nonoccurrence of the other event.

mean value (or ensemble average) The mean value of u is denoted by $\bar{u}$ and defined as

$$\bar{u} \equiv \sum_r P_r u_r$$

where the sum is over all possible values u_r of the variable u and where P_r denotes the probability of occurrence of the particular value u_r.

dispersion (or variance) The dispersion of u is defined as

$$\overline{(\Delta u)^2} \equiv \sum_r P_r (u_r - \bar{u})^2.$$

standard deviation The standard deviation of u is defined as

$$\underset{\sim}{\Delta} u \equiv [\overline{(\Delta u)^2}]^{1/2}.$$

probability density The probability density $\mathcal{P}(u)$ is defined by the property that $\mathcal{P}(u)\,du$ yields the probability of finding the continuous variable u in the range between u and $u + du$.

Important Relations

Given N statistically independent events, each having probability p of occurring (and probability $q = 1 - p$ of not occurring).

Probability that n of these N events occur (binomial distribution):

$$P(n) = \frac{N!}{n!(N-n)!} p^n q^{N-n}. \tag{i}$$

Mean number of occurring events: $\bar{n} = Np.$ (ii)

Standard deviation of n: $\underset{\sim}{\Delta} n = \sqrt{Npq}.$ (iii)

Suggestions for Supplementary Reading

W. Weaver, *Lady Luck* (Anchor Books, Doubleday & Company, Inc., Garden City, N.Y., 1963). An elementary introduction to probability concepts.

F. Mosteller, R. E. K. Rourke, and G. B. Thomas, *Probability and Statistics* (Addison-Wesley Publishing Company, Reading, Mass., 1961).

F. Reif, *Fundamentals of Statistical and Thermal Physics,* chap. 1 (McGraw-Hill Book Company, New York, 1965). The random walk problem discussed in this book is analogous to the problem of the ideal spin system, but is treated at greater length.

H. D. Young, *Statistical Treatment of Experimental Data* (McGraw-Hill Book Company, New York, 1962). An elementary account of statistical methods, particularly as they are applied to problems of experimental measurement.

W. Feller, *An Introduction to Probability Theory and its Applications,* 2d ed. (John Wiley and Sons, New York, 1959). This book on probability theory is more advanced than the preceding ones, but discusses many concrete examples.

Problems

2.1 ***An elementary dice problem***

What is the probability of throwing a total of 6 points or less with 3 dice?

2.2 ***Random numbers***

A number is chosen at random between 0 and 1. What is the probability that exactly 5 of its first 10 decimal places consist of digits less than 5?

2.3 ***Tossing of dice***

Assume that each face of a die is equally likely to land uppermost. Consider a game which involves the tossing of 5 such dice. Find the probability that the number "6" appears uppermost

(*a*) in exactly one die,
(*b*) in at least one die,
(*c*) in exactly two dice.

2.4 ***Probability of survival***

In the macabre game of Russian roulette (*not* recommended by the author) one inserts a single cartridge into one of the 6 chambers of the cylinder of a revolver, leaving the other 5 chambers of the cylinder empty. One then spins the cylinder, aims at one's head, and pulls the trigger. What is the probability of still being alive after playing the game

(*a*) Once?
(*b*) Twice?
(*c*) N times?
(*d*) What is the probability of being shot when playing the game the Nth time?

2.5 *The random walk problem*

A man starts out from a lamppost in the middle of a street, taking steps of equal length l. The probability is p that any one of his steps is to the right, and $q = 1 - p$ that it is to the left. The man is so drunk that his behavior at any step shows no traces of memory of what he did at preceding steps. His steps are thus statistically independent. Suppose that the man has taken N steps.

(*a*) What is the probability $P(n)$ that n of these steps are to the right and the remaining $n' = (N - n)$ steps are to the left?

(*b*) What is the probability $P'(m)$ that the displacement of the man from the lamppost is equal to ml, where $m = n - n'$ is an integer?

2.6 *Probability of returning to starting point*

In the preceding problem, suppose that $p = q$ so that each step is equally likely to be to the right or to the left. What is the probability that the man will again be at the lamppost after taking N steps

(*a*) If N is even?

(*b*) If N is odd?

2.7 *One-dimensional diffusion of an atom*

Consider a thin copper wire stretched out along the x axis. A few of the copper atoms, located near $x = 0$, are made radioactive by bombardment with fast particles. When the temperature of the wire is raised, the atoms become more mobile. Each atom can then jump to an adjacent lattice site, either to the site on its right (i.e., in the $+x$ direction) or the site on its left (i.e., in the $-x$ direction). The possible lattice sites are separated by a distance l. Assume that one has to wait a time τ before a given atom jumps to an adjacent lattice site. This time τ is a rapidly increasing function of the absolute temperature of the wire. The process whereby the atoms move about by virtue of successive jumps to adjacent sites is known as *diffusion.*

Suppose that the wire is rapidly raised to some high temperature at some time $t = 0$ and is thereafter maintained at this temperature.

(*a*) Let $\mathcal{P}(x)\,dx$ denote the probability that a radioactive atom is found at a distance between x and $x + dx$ after a time t. [We assume that $t \gg \tau$ for all times t of physical interest, since τ is quite short when the temperature of the wire is high.] Make a rough sketch showing the behavior of $\mathcal{P}(x)$ as a function of x in the following three cases:

(1) shortly after the time $t = 0$;

(2) after a moderately long time t has elapsed;

(3) after an extremely long time t has elapsed.

(*b*) What is the mean displacement $\bar{x}$ from the origin of a radioactive atom after a time t?

(*c*) Find an explicit expression for the standard deviation Δx of the displacement of a radioactive atom after a time t.

2.8 *Computation of the dispersion*

Use the general properties of mean values to show that the dispersion of u can be calculated by the general relation

$$\overline{(\Delta u)^2} \equiv \overline{(u - \bar{u})^2} = \overline{u^2} - \bar{u}^2. \tag{i}$$

The last expression on the right provides a simple method for computing the dispersion.

Show also that (i) implies the general inequality

$$\overline{u^2} \geq \bar{u}^2. \tag{ii}$$

2.9 *Mean values for a single spin*

The magnetic moment of a spin $\frac{1}{2}$ is such that its component μ in the up direction has probability p of being equal to μ_0, and probability $q = 1 - p$ of being equal to $-\mu_0$.

(*a*) Calculate $\bar{\mu}$ and $\overline{\mu^2}$.

(*b*) Use the expression (i) of Prob. 2.8 to calculate $\overline{(\Delta\mu)^2}$. Show that your result agrees with that given in Eq. (57) of the text.

2.10 *The inequality* $\overline{u^2} \geq \bar{u}^2$

Suppose that the variable u can assume the possible values u_r with respective probabilities P_r.

(*a*) By using the definitions of $\bar{u}$ and $\overline{u^2}$, and remembering the normalization requirement $\sum_r P_r = 1$, show that

$$\overline{u^2} - \bar{u}^2 = \frac{1}{2} \sum_r \sum_s P_r P_s (u_r - u_s)^2 \tag{i}$$

where each sum extends over all possible values of the variable u.

(*b*) Since no term in the sum (i) can ever be negative, show that

$$\overline{u^2} \geq \bar{u}^2 \tag{ii}$$

where the equals sign applies only in the case where only one value of u occurs with nonzero probability. The result (ii) agrees with that derived in Prob. 2.8.

2.11 *The inequality* $(\overline{u^n})^2 \leq \overline{u^{n+1}}\,\overline{u^{n-1}}$

The result (i) of the preceding problem suggests an immediate generalization. Thus consider the expression

$$\sum_r \sum_s P_r P_s u_r^m u_s^m (u_r - u_s)^2 \tag{i}$$

where m is any integer. When m is even, this expression can never be negative; when m is odd, it can never be negative if the possible values of u are all nonnegative (or all nonpositive).

(*a*) By performing the indicated multiplications in (i), show the following results:

$$(\overline{u^n})^2 \leq \overline{u^{n+1}}\,\overline{u^{n-1}} \tag{ii}$$

where $n \equiv m + 1$. If n is odd, this inequality is always valid; if n is even, it is valid if the possible values of u are all nonnegative (or all nonpositive). The equals sign in (ii) applies only in the case where a single possible value of u occurs with nonzero probability.

(*b*) Show that (ii) implies, as a special case, the inequality

$$\overline{\left(\frac{1}{u}\right)} \geq \frac{1}{\bar{u}}, \tag{iii}$$

valid whenever the possible values of u are all positive (or all negative). The equals sign applies in the special case where only one value of u occurs with nonzero probability.

2.12 *Optimum investment method*

The following practical situation illustrates that different ways of averaging the same quantity can lead to significantly different results. Suppose that one desires to invest money by buying at the beginning of every month a certain number of shares of stock in a company. The cost c_r per share depends, of course, on the particular month r and varies from month to month in a fairly unpredictable fashion. Two alternative methods of regular investment suggest themselves: In method A, one buys the same number s of shares of stock every month; in method B, one buys stock for the same amount of money m every month. After N months, one will then have acquired a total number S of shares of stock and will have paid for them a total amount of money M. The best investment method is clearly that which yields the largest number of shares for the lowest amount of money, i.e., which yields the largest ratio S/M.

(*a*) Obtain an expression for the ratio S/M in the case of method A.

(*b*) Obtain an expression for the ratio S/M in the case of method B.

(*c*) Show that method B is the better investment method, no matter how the cost of shares fluctuates from month to month. [*Suggestion:* Exploit the inequality (iii) of the preceding problem.]

2.13 *System of nuclei with spin 1*

Consider a nucleus having spin 1 (i.e., spin angular momentum $\hbar$). Its component μ of magnetic moment along a given direction can then have *three* possible values, namely $+\mu_0$, 0, or $-\mu_0$. Suppose that the nucleus is not spherically symmetrical, but is ellipsoidal in shape. As a result it tends to be preferentially oriented so that its major axis is parallel to a given direction in the crystalline solid in which the nucleus is located. There is thus a probability p that $\mu = \mu_0$, and a probability p that $\mu = -\mu_0$; the probability that $\mu = 0$ is then equal to $1 - 2p$.

(*a*) Calculate $\bar{\mu}$ and $\overline{\mu^2}$.

(*b*) Calculate $\overline{(\Delta\mu)^2}$.

(*c*) Suppose that the solid under consideration contains N such nuclei which interact with each other to a negligible extent. Let M denote the total component of magnetic moment, along the specified direction, of all these nuclei. Calculate $\bar{M}$ and its standard deviation ΔM in terms of N, p, and μ_0.

2.14 ***Direct calculation of*** $\bar{n}$ ***and*** $\overline{(\Delta n)^2}$

Consider an ideal system of N identical spins $\frac{1}{2}$. The number n of magnetic moments which point in the up direction can then be written in the form

$$n = u_1 + u_2 + \cdots + u_N \tag{i}$$

when $u_i = 1$ if the ith magnetic moment points up and $u_i = 0$ if its points down. Use the expression (i) and the fact that the spins are statistically independent to establish the following results.

(*a*) Show that $\bar{n} = N\bar{u}$.

(*b*) Show that $\overline{(\Delta n)^2} = N\overline{(\Delta u)^2}$.

(*c*) Suppose that a magnetic moment has probability p of pointing up and probability $q = 1 - p$ of pointing down. What are $\bar{u}$ and $\overline{(\Delta u)^2}$?

(*d*) Calculate $\bar{n}$ and $\overline{(\Delta n)^2}$ and show that your results agree with the relations (66) and (67) found in the text by a less direct method.

2.15 ***Density fluctuations in a gas***

Consider an ideal gas of N molecules which is in equilibrium within a container of volume V_0. Denote by n the number of molecules located within any subvolume V of this container. The probability p that a given molecule is located within this subvolume V is then given by $p = V/V_0$.

(*a*) What is the mean number $\bar{n}$ of molecules located within V? Express your answer in terms of N, V_0, and V.

(*b*) Find the standard deviation Δn in the number of molecules located within the subvolume V. Hence calculate $\Delta n/\bar{n}$, expressing your answer in terms of N, V_0, and V.

(*c*) What does the answer to part (*b*) become when $V \ll V_0$?

(*d*) What value should the standard deviation Δn assume when $V \to V_0$? Does the answer to part (*b*) agree with this expectation?

2.16 ***Shot effect***

Electrons of charge e are emitted at random from the hot filament of a vacuum tube. To good approximation the emission of any one electron does not affect the likelihood of emission of any other electron. Consider any very small time interval Δt. Then there is some probability p that an electron will be emitted from the filament during this time interval (and probability $q = 1 - p$ that an electron will not be emitted). Since Δt is very small, the probability p of emission during this time interval is very small (i.e., $p \ll 1$) and the probability that more than one electron will be emitted during the time Δt is negligible.

Consider any time t which is much larger than Δt. During this time there are then $N = t/\Delta t$ possible time intervals Δt during which an electron can be emitted. The total charge emitted in time t can then be written as

$$Q = q_1 + q_2 + q_3 + \cdots + q_N$$

where q_i denotes the charge which is emitted during the ith interval Δt; thus $q_i = e$ if an electron is emitted and $q_i = 0$ if it is not.

(*a*) What is the mean charge $\overline{Q}$ emitted from the filament during the time t?

(*b*) What is the dispersion $\overline{(\Delta Q)^2}$ of the charge Q emitted from the filament during the time t? Make use of the fact that $p \ll 1$ to simplify the answer to this question.

(*c*) The current I emitted during the time t is given by Q/t. Relate thus the dispersion $\overline{(\Delta I)^2}$ of the current to the mean current $\bar{I}$, showing that

$$\overline{(\Delta I)^2} = \frac{e}{t}\bar{I}.$$

(*d*) The fact that the current measured during any interval t exhibits fluctuations (which become more pronounced the shorter the time interval, i.e., the fewer the total number of individual electrons involved in the emission process) is known as the *shot effect.* Calculate the standard deviation ΔI of the current if the mean current $\bar{I} = 1$ microampere and the measuring time is 1 second.

2.17 ***Calculation of a mean square value***

A battery of total emf V is connected to a resistor R. As a result, an amount of power $P = V^2/R$ is dissipated in this resistor. The battery itself consists of N individual cells connected in series so that V is just equal to the sum of the emf's of all these cells. The battery is old, however, so that not all the cells are in perfect condition. Thus there is only a probability p that the emf of any individual cell has its normal value v; and a probability $(1 - p)$ that the emf of any individual cell is zero because the cell has become internally shorted. The individual cells are statistically independent of each other. Under these conditions, calculate the *mean* power P dissipated in the resistor. Express the result in terms of N, v, p, and R.

2.18 ***Estimate of error of measurement***

A man attempts to lay off a distance of 50 meters by placing a meter stick end to end 50 times in succession. This procedure is necessarily accompanied by some error. Thus the man cannot guarantee a distance of precisely one meter between the two chalk marks which he makes each time he places the meter stick on the ground. He knows, however, that the distance between the two marks is equally likely to lie anywhere between 99.8 and 100.2 cm, and that it certainly does not lie outside these limits. After repeating the operation 50 times, the man will indeed have laid off a mean distance of 50 meters. To estimate his total error, calculate the standard deviation of his measured distance.

2.19 *Diffusion of a molecule in a gas*

A molecule in a gas is free to move in three dimensions. Let $\mathbf{s}$ denote its displacement between successive collisions with other molecules. Displacements of the molecule between successive collisions are, to fair approximation, statistically independent. Furthermore, since there is no preferred direction in space, a molecule is as likely to move in a given direction as in the opposite direction. Thus its mean displacement $\bar{\mathbf{s}} = 0$ (i.e., each component of this displacement vanishes on the average so that $\bar{s}_x = \bar{s}_y = \bar{s}_z = 0$).

The total displacement $\mathbf{R}$ of the molecule after N successive displacements can then be written as

$$\mathbf{R} = \mathbf{s}_1 + \mathbf{s}_2 + \mathbf{s}_3 + \cdots + \mathbf{s}_N$$

where $\mathbf{s}_i$ denotes the ith displacement of the molecule. Use reasoning similar to that of Sec. 2.5 to answer the following questions:

(*a*) What is the mean displacement $\overline{\mathbf{R}}$ of the molecule after N displacements?

(*b*) What is the standard deviation $\Delta R \equiv \overline{(\mathbf{R} - \overline{\mathbf{R}})^2}$ of this displacement after N collisions? In particular, what is ΔR if the *magnitude* of each displacement $\mathbf{s}$ has the same length l?

2.20 *Displacement distribution of random oscillators*

The displacement x of a classical simple harmonic oscillator as a function of the time t is given by

$$x = A \cos(\omega t + \varphi)$$

where ω is the angular frequency of the oscillator, A is its amplitude of oscillation, and φ is an arbitrary constant which can have any value in the range $0 \leq \varphi < 2\pi$. Suppose that one contemplates an ensemble of such oscillators all of which have the same frequency ω and amplitude A, but which have random phase relationships so that the probability that φ lies in the range between φ and $\varphi + d\varphi$ is given simply by $d\varphi/(2\pi)$. Find the probability $\mathcal{P}(x)\,dx$ that the displacement of an oscillator, at any given time t, is found to lie in the range between x and $x + dx$.

Chapter 3

Statistical Description of Systems of Particles

Chapter 3 *Statistical Description of Systems of Particles*

The preceding review of basic probability concepts has prepared us to turn the qualitative considerations of the first chapter into a systematic quantitative theory of systems consisting of very many particles. Our aim will be to combine statistical considerations with our knowledge of the laws of mechanics applicable to the particles constituting a macroscopic system. The resulting theory, therefore, is called *statistical mechanics*. The reasoning that leads to this theory is very simple and uses only the most primitive ideas of mechanics and probability. The beauty of the subject lies precisely in the fact that arguments of great simplicity and apparent innocence are capable of yielding results of impressive generality and predictive power.

Indeed, the arguments used to discuss a macroscopic system are completely analogous to those used to discuss the familiar experiment of tossing a set of coins. The essential ingredients for an analysis of this experiment are the following:

(i) Specification of the state of the system

We must have available a method for specifying any of the possible outcomes of an experiment involving the system. For example, the state of the set of coins after any toss can be described by specifying which particular face of each coin is uppermost.

(ii) Statistical ensemble

We have far too little information about the precise manner in which the coins are tossed to be able to use the laws of mechanics to make a unique prediction about the outcome of any particular experiment. Hence we resort to a statistical description of the situation. Instead of considering the particular set of coins of interest, we thus focus attention on an ensemble consisting of a very large number of similar sets of coins subjected to the same experiment. We can then ask for the probability of occurrence of any experimental outcome. This probability can be measured by observing the ensemble and determining the fraction of systems exhibiting the particular outcome. Our theoretical aim is the prediction of any such probability.

(iii) Statistical postulates
To make theoretical progress, it is necessary to introduce some postulates. In the case of ordinary coins of uniform density, there is nothing intrinsic in the laws of mechanics which implies that one face of a coin should land uppermost more frequently than the other. Hence we are led to introduce the *postulate* that "a priori" (i.e., based on our prior notions as yet unverified by actual observations) a coin has equal probability of landing on each of its two faces. This postulate is eminently reasonable and certainly does not contradict any of the laws of mechanics. The actual validity of the postulate can, however, only be decided by using it to make theoretical predictions and by checking that these predictions are confirmed by experimental observations. To the extent that such predictions are consistently verified, the validity of this postulate can be accepted with confidence.

(iv) Probability calculations
Once the basic postulate has been adopted, we can calculate the probability of occurrence of any particular outcome involving the set of coins under consideration. We can also compute various mean values of interest. Thus we can answer all questions that can meaningfully be asked in a statistical theory.

In studying systems consisting of a large number of particles, our considerations will be very similar to those used in formulating the preceding problem of a set of coins. The next four sections will make this analogy explicit.

3.1 *Specification of the State of a System*

The study of atomic particles has shown that any system of such particles is described by the laws of quantum mechanics. These laws, whose validity is supported by an overwhelming amount of experimental evidence, will thus form the conceptual basis for our entire discussion.

In a quantum-mechanical description the most precise possible measurement on a system always shows this system to be in some one of a set of discrete *quantum states* characteristic of the system. The microscopic state of a system can thus be described completely by specifying the particular quantum state in which the system is found.

Fig. 3.1 Highly schematic diagram illustrating the first few energy levels of an arbitrary system. Each line denotes a possible quantum state of the system, while the vertical position of the line indicates the energy E of the system in this state. Note that there are many states which have the same energy.

Each quantum state of an isolated system is associated with a definite value of its energy and is called an *energy level.*† There may be several quantum states corresponding to the same energy of the system. (These quantum states are then said to be *degenerate.*) Every system has a lowest possible energy. There is usually only one possible quantum state of the system corresponding to this lowest energy; this state is said to be the *ground state* of the system.‡ In addition there are, of course, many (indeed, ordinarily infinitely many) possible states with higher energies; these are called the *excited states* of the system.

The preceding comments are completely general and applicable to any system, no matter how complex. They are best illustrated by some simple examples of great practical interest.

r	σ	M	E
1	$+1$	μ_0	$-\mu_0 B$
2	-1	$-\mu_0$	$+\mu_0 B$

Table 3.1 Quantum states of a single spin $\frac{1}{2}$ having a magnetic moment μ_0 and located in a magnetic field **B**. Each state of the system can be labeled by an index r, or alternatively, by the quantum number σ. The magnetic moment (along the "up" direction specified by the field **B**) is denoted by M; the total energy of the system is denoted by E.

(i) Single spin

Consider a single particle, assumed to be fixed in position, which has spin $\frac{1}{2}$ and a magnetic moment of magnitude μ_0. As already discussed in Sec. 1.3, this moment will be found to point either "up" or "down" (i.e., parallel or antiparallel) with respect to any specified direction. The system consisting of this single spin thus has only two quantum states which we shall label by a *quantum number* σ. We can then denote the state where the magnetic moment of the particle points up by $\sigma = +1$, and the state where it points down by $\sigma = -1$.

If the particle is in the presence of a magnetic field **B**, this field specifies the direction of physical interest in the problem. The energy E of the system then is lower when the magnetic moment is aligned parallel to the field rather than antiparallel to it. The situation is analogous to that of a bar magnet located in an external magnetic field. Thus, when the magnetic moment points up (i.e., parallel to the field **B**), its magnetic energy is simply $-\mu_0 B$. Conversely, when the moment points down (i.e., antiparallel to the field **B**), its magnetic energy is simply $\mu_0 B$. The two quantum states (or energy levels) of the system then correspond to different energies.

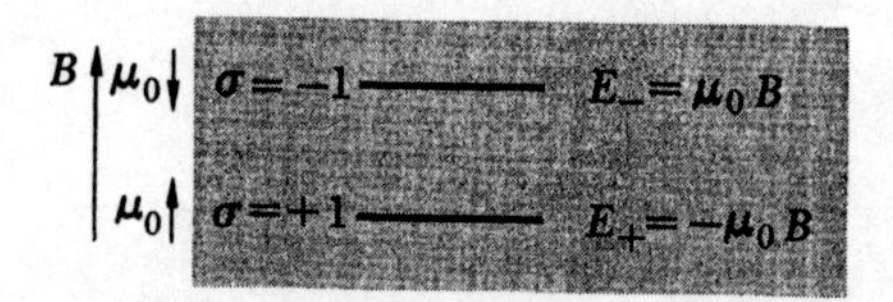

Fig. 3.2 Diagram showing the two energy levels of a spin $\frac{1}{2}$ having a magnetic moment μ_0 and located in a magnetic field **B**. The state where the magnetic moment points "up," so that its direction is parallel to **B**, is denoted by $\sigma = +1$ (or simply by $+$); that where it points "down" is denoted by $\sigma = -1$ (or simply by $-$).

† The hydrogen atom is likely to be a familiar example of a system described in terms of discrete energy levels. Transitions of the atom between states of different energy give rise to the sharp spectral lines emitted by the atom. A description in terms of energy levels is, of course, equally applicable to any atom, molecule, or system consisting of many atoms.

‡ In some cases there may be a relatively small number of quantum states of the same energy equal to the lowest possible energy of the system. The ground state of the system is then said to be *degenerate*.

(ii) Ideal system of N spins

Consider a system consisting of N particles, assumed to be fixed in position, where each particle has spin $\frac{1}{2}$ and a magnetic moment μ_0. The system is located in an applied magnetic field $\mathbf{B}$. The interaction between the particles is assumed to be almost negligible.†

The magnetic moment of each particle can point either up or down with respect to the field $\mathbf{B}$. The orientation of the ith moment can thus be specified by the value of its quantum number σ_i so that $\sigma_i = +1$ when this moment points up and $\sigma_i = -1$ when it points down. A particular state of the whole system can then be specified by stating the orientation of each of the N moments, i.e., by specifying the values assumed by the set of quantum numbers $\{\sigma_1, \sigma_2, \ldots, \sigma_N\}$. Thus one can enumerate, and label by some index r, all the possible states of the whole system. This is done in Table 3.2 for the special case where $N = 4$. The total magnetic moment of the system is simply equal to the sum of the magnetic moments of the individual spins. Since the interaction between these spins is almost negligible, the total energy E of the system is also simply equal to the sum of the energies of the individual spins.

r	σ_1	σ_2	σ_3	σ_4	M	E
1	+	+	+	+	$4\mu_0$	$-4\mu_0 B$
2	+	+	+	−	$2\mu_0$	$-2\mu_0 B$
3	+	+	−	+	$2\mu_0$	$-2\mu_0 B$
4	+	−	+	+	$2\mu_0$	$-2\mu_0 B$
5	−	+	+	+	$2\mu_0$	$-2\mu_0 B$
6	+	+	−	−	0	0
7	+	−	+	−	0	0
8	+	−	−	+	0	0
9	−	+	+	−	0	0
10	−	+	−	+	0	0
11	−	−	+	+	0	0
12	+	−	−	−	$-2\mu_0$	$2\mu_0 B$
13	−	+	−	−	$-2\mu_0$	$2\mu_0 B$
14	−	−	+	−	$-2\mu_0$	$2\mu_0 B$
15	−	−	−	+	$-2\mu_0$	$2\mu_0 B$
16	−	−	−	−	$-4\mu_0$	$4\mu_0 B$

Table 3.2 Quantum states of an ideal system of 4 spins $\frac{1}{2}$, each having a magnetic moment μ_0 and located in a magnetic field $\mathbf{B}$. Each quantum state of the whole system is labeled by the index r, or equivalently, by the set of 4 numbers $\{\sigma_1, \sigma_2, \sigma_3, \sigma_4\}$. For the sake of brevity, the symbol + indicates $\sigma = +1$ and the symbol − indicates $\sigma = -1$. The total magnetic moment (along the "up" direction specified by $\mathbf{B}$) is denoted by M; the total energy of the system is denoted by E.

(iii) Particle in a one-dimensional box

Consider a single particle, of mass m, free to move in one dimension. The particle is supposed to be confined within a box of length L, so that the particle's position coordinate x must lie in the range $0 \leq x \leq L$. Within this box the particle is subject to no forces.

In a quantum-mechanical description, the particle has wave properties associated with it. The particle confined within the box and bouncing back and forth between its walls is thus represented by a wave function ψ in the form of a standing wave whose amplitude must vanish at the boundaries of the box (since ψ itself must vanish outside the box).‡ The wave function thus must be of the form

$$\psi(x) = A \sin Kx \tag{1}$$

(where A and K are constants) and must satisfy the boundary conditions

$$\psi(0) = 0 \quad \text{and} \quad \psi(L) = 0. \tag{2}$$

The expression (1) obviously satisfies the condition $\psi(0) = 0$. In order that it also satisfy the condition $\psi(L) = 0$, the constant K must be such that

$$KL = \pi n$$

or

$$K = \frac{\pi}{L} n, \tag{3}$$

where n can assume any of the integral values§

$$n = 1, 2, 3, 4, \ldots. \tag{4}$$

† This assumption implies that one can virtually neglect the magnetic field at the position of any one particle caused by the magnetic moments of the other particles.

‡ The physical interpretation of the wave function is that $|\psi(x)|^2\,dx$ represents the probability that the particle is found in the range between x and $x + dx$.

§ The value $n = 0$ is irrelevant since it leads to $\psi = 0$, i.e., to no wave function (or no particle) existing within the box. Negative integral values of n lead to no distinct new wave functions since a change of sign of n, and thus of K, leads merely to a change of sign of ψ in (1) and leaves the probability $|\psi|^2\,dx$ unchanged. Hence the positive integral values of n yield all the distinct possible wave functions of the form (1). Physically this means that only the *magnitude* $\hbar K$ of the particle's momentum is relevant, since this momentum is equally likely to be positive or negative as a result of successive reflections of the particle by the walls.

The constant K in (1) is the *wave number* associated with the particle; it is related to the wavelength λ (the so-called *de Broglie wavelength* associated with the particle) by the relation

$$K = \frac{2\pi}{\lambda}. \tag{5}$$

Hence (3) is equivalent to

$$L = n\frac{\lambda}{2}$$

and represents merely the familiar condition that standing waves are obtained when the length of the box is equal to some integral multiple of half-wavelengths.

The momentum p of the particle is related to K (or λ) by the famous de Broglie relation

$$p = \hbar K = \frac{h}{\lambda} \tag{6}$$

where $\hbar \equiv h/2\pi$ and h is Planck's constant. The energy E of the particle is simply its kinetic energy, since there is no potential energy due to external forces. Hence E can be expressed in terms of the velocity v or momentum $p = mv$ of the particle as

$$E = \frac{1}{2}mv^2 = \frac{1}{2}\frac{p^2}{m} = \frac{\hbar^2 K^2}{2m}. \tag{7}$$

The possible values (3) of K then yield the corresponding energies

$$E = \frac{\hbar^2}{2m}\left(\frac{\pi}{L}n\right)^2 = \frac{\pi^2\hbar^2}{2m}\frac{n^2}{L^2}. \tag{8}$$

Equivalently, we could have discussed the whole problem from a more mathematical point of view by starting from the fundamental Schrödinger equation for the wave function ψ. For a free particle in one dimension this equation is

$$-\frac{\hbar^2}{2m}\frac{\partial^2\psi}{\partial x^2} = E\psi.$$

The functional form (1) satisfies this equation provided that the energy E is related to K by (7). The condition (2) that the wave function must vanish at the boundaries of the box leads again to (3) and hence to the expression (8) for the energy.

The possible quantum states of the particle in the box thus can be specified by the possible values (4) of the quantum number n. The corresponding discrete energies of these states (i.e., the corresponding energy levels of the particle) are then given by (8).

The relation (8) shows that the separation in energy between successive quantum states of the particle is very small if the length L of the box is of macroscopic size. The lowest possible energy of the particle, i.e., its ground-state energy, corresponds to the state $n = 1$. Note that this ground-state energy does not vanish.†

(iv) Particle in a three-dimensional box

The generalization of the preceding problem to the case of a particle free to move in three dimensions is straightforward. Suppose that the particle is confined within a box in the shape of a rectangular parallelepiped of edge lengths L_x, L_y, and L_z. The position coordinates x, y, z of the particle then can be assumed to lie in the respective ranges

$$0 \le x \le L_x, \quad 0 \le y \le L_y, \quad 0 \le z \le L_z.$$

The particle has mass m and is subject to no forces within the box.

Fig. 3.3 A box in the shape of a rectangular parallelepiped with edge lengths L_x, L_y, and L_z.

† This conclusion is consistent with the Heisenberg uncertainty principle ($\Delta x\,\Delta p > \hbar$) according to which a particle confined within a linear dimension of length L (so that $\Delta x \sim L$) must have associated with it a minimum momentum p of the order of $p \sim \hbar/L$. Hence the lowest possible energy of the particle in the box is a kinetic energy of the order of $p^2/2m = \hbar^2/2mL^2$.

The wave function of the particle now represents a standing wave in three dimensions. Thus it is of the form

$$\psi = A(\sin K_x x)(\sin K_y y)(\sin K_z z) \quad (9)$$

where the constants K_x, K_y, K_z can be regarded as the three components of a vector $\mathbf{K}$, the *wave vector* of the particle. According to the de Broglie relation the momentum of the particle is then given by

$$\mathbf{p} = \hbar\mathbf{K} \quad (10)$$

so that the relationship between the magnitude of p and the magnitude of K (or the wavelength λ) is the same as in (6). The energy of the particle is then given by

$$E = \frac{\mathbf{p}^2}{2m} = \frac{\hbar^2\mathbf{K}^2}{2m} = \frac{\hbar^2}{2m}(K_x{}^2 + K_y{}^2 + K_z{}^2). \quad (11)$$

Equivalently, it can be immediately verified that ψ in (9) is indeed a solution of the time-independent Schrödinger equation for a free particle in three dimensions,

$$-\frac{\hbar^2}{2m}\left(\frac{\partial^2\psi}{\partial x^2} + \frac{\partial^2\psi}{\partial y^2} + \frac{\partial^2\psi}{\partial z^2}\right) = E\psi,$$

provided that E is related to $\mathbf{K}$ by (11).

The fact that ψ must vanish at the boundaries of the box imposes the conditions that

$\psi = 0$ at the planes

$$\left.\begin{array}{lll} x = 0, & y = 0, & z = 0, \\ x = L_x, & y = L_y, & z = L_z. \end{array}\right\} \quad (12)$$

The expression (9) vanishes properly when $x = 0$, $y = 0$, or $z = 0$. To make it vanish for $x = L_x$, $y = L_y$, or $z = L_z$, the constants K_x, K_y, K_z must satisfy the respective conditions

$$K_x = \frac{\pi}{L_x}n_x, \quad K_y = \frac{\pi}{L_y}n_y, \quad K_z = \frac{\pi}{L_z}n_z, \quad (13)$$

where each of the numbers n_x, n_y, and n_z can assume any of the positive integral values

$$n_x, n_y, n_z = 1, 2, 3, 4, \ldots . \quad (14)$$

Any particular quantum state of the particle can then be designated by the values assumed by the set of quantum numbers $\{n_x, n_y, n_z\}$. Its corresponding energy is, by (11) and (13), equal to

$$\boxed{E = \frac{\pi^2\hbar^2}{2m}\left(\frac{n_x{}^2}{L_x{}^2} + \frac{n_y{}^2}{L_y{}^2} + \frac{n_z{}^2}{L_z{}^2}\right).} \quad (15)$$

(v) Ideal gas of N particles in a box

Consider a system consisting of N particles, each of mass m, confined within the box of the preceding example. The interaction between the particles is assumed to be almost negligible so that the particles constitute an ideal gas. The total energy E of the gas is then just equal to the sum of the energies of the individual particles, i.e.,

$$E = \epsilon_1 + \epsilon_2 + \cdots + \epsilon_N \quad (16)$$

where ϵ_i denotes the energy of the ith particle. The state of each such particle can be specified, as in the preceding example, by the values of its 3 quantum numbers n_{ix}, n_{iy}, n_{iz}; its energy ϵ_i is then given by an expression analogous to (15). Each possible quantum state of the *entire* gas can thus be specified by the values assumed by the $3N$ quantum numbers

$$\{n_{1x}, n_{1y}, n_{1z}; n_{2x}, n_{2y}, n_{2z}; \ldots \ldots; n_{Nx}, n_{Ny}, n_{Nz}\}.$$

Its corresponding energy is given by (16), where each term in the sum has the form (15).

The preceding examples are typical of the quantum-mechanical description and serve to illustrate the general comments made at the beginning of this section. We can summarize our remarks as follows: Each possible quantum state of a system can be specified by some set of f quantum numbers. This number f, called the number of *degrees of freedom* of the system, is equal to the number of independent coordinates (including spin coordinates) needed to describe the system.† Any one quantum state of the system can be specified by the particular values assumed by all of its quantum numbers. For simplicity, each such state can be labeled by some index r so that the possible quantum states can be listed and enumerated in some convenient order $r = 1, 2, 3, 4, \ldots$. Our question about the most detailed quantum-mechanical description of a system is then answered by the following statement:

> The microscopic state of a system can be described by specifying the particular quantum state r in which the system is found.

A completely precise description of an isolated system of particles would have to take into account *all* the interactions between its particles and would determine the rigorously exact quantum states of the system. If the system is in any one of these exact states, it would then remain in this state forever. But no system is ever so completely isolated in practice that it does not interact at all with its environment; furthermore, it would be neither possible, nor indeed useful, to strive for precision so great that all interactions between particles are rigorously taken into account. Thus, the quantum states actually used to describe a system are in practice always its approximate quantum states, determined by taking into account all the important dynamical properties of its particles while neglecting small residual interactions. A system initially known to be in one of its *approximate* quantum states does not remain in this state forever. In the course of time it will instead, under the influence of the small residual interactions, make transitions to its other quantum states (except to those to which it cannot pass without violating known restrictions imposed by the laws of mechanics).

† For instance, in the example of N particles without spin, the number of degrees of freedom $f = 3N$.

The hydrogen atom provides a familiar illustration of the preceding comments. The quantum states ordinarily used to describe this atom are those determined by taking into account only the Coulomb attraction between the nucleus and electron. The residual interaction of the atom with the surrounding electromagnetic field then causes transitions between these states. The result is the emission or absorption of electromagnetic radiation, giving rise to the observed spectral lines.

Illustrations of greater pertinence to us are provided by an isolated ideal spin system or an isolated ideal gas. If the particles in such a system would not interact with each other at all, the quantum states calculated in examples (ii) or (v) of this section would be exact, and hence no transitions would occur. But this situation does not correspond to reality. Indeed, we have been careful to stress that, even when a spin system or a gas is ideal, the interaction between its constituent particles is to be considered as *almost* negligible rather than as completely negligible. Thus small interactions exist in a spin system because each magnetic moment produces a small magnetic field at the positions of neighboring moments. Similarly, small interactions exist in the gas because forces exerted by one particle on another particle come into play whenever these particles approach each other sufficiently closely (and thus "collide" with each other). If these interactions are taken into account, the quantum states calculated in the examples (ii) and (v) become approximate quantum states. The effect of the interactions then is to cause occasional transitions between these states (the occurrence of such transitions being less frequent the smaller the magnitude of the interactions). Consider, for instance, the system consisting of 4 spins whose quantum states are listed in Table 3.2. Suppose that this system is originally known to be in the state $\{+-++\}$. As a result of the small interactions between the spins, there is a nonvanishing probability that the system will be found at some later time in any other state, such as $\{++-+\}$, to which it can pass without violating the condition of conservation of energy.

Our discussion of the specification of the state of a system has been given in terms of quantum ideas since the atoms and molecules constituting any system are known to be described by the laws of quantum mechanics. Under appropriate conditions it may sometimes be a convenient approximation to describe a system in terms of classical mechanics. The utility and validity of such approximations will be considered in Chap. 6.

3.2 Statistical Ensemble

A precise knowledge of the particular microscopic state in which a system of particles is found at any one time would, in principle, allow us to use the laws of mechanics to calculate in the greatest possible detail all of the system's properties at any arbitrary time. Ordinarily, however, we do not have available such precise microscopic knowledge about a macroscopic system, nor are we interested in such an excessively detailed description. Hence we are led to discuss the system in terms of probability concepts. Instead of considering the single macroscopic system of interest, we then focus attention on an ensemble consisting of a very large number of systems of this kind, all satisfying the same conditions as those known to be satisfied by the particular system under consideration. With respect to this ensemble, we can then make various probability statements about the system.

A complete macroscopic description of a system of many particles defines the so-called *macro*scopic state or *macrostate* of the system. Since such a description is based entirely on the specification of quantities which can be readily ascertained by macroscopic measurements alone, it provides only very limited information about the particles in the system. This information is typically the following:

(*i*) *Information about the external parameters of the system*
There are certain parameters of the system which are macroscopically measurable and which affect the motion of the particles in the system. These parameters are called *external parameters* of the system. For example, the system may be known to be located in a given external magnetic field **B** or a given external electric field $\boldsymbol{\mathcal{E}}$. Since the presence of such fields affects the motion of the particles in the system, **B** or $\boldsymbol{\mathcal{E}}$ are external parameters of the system. Similarly, suppose that a gas is known to be contained within a box of dimensions L_x, L_y, and L_z. Then each molecule of the gas must move so as to remain confined within this box. The dimensions L_x, L_y, L_z are thus external parameters of the gas.

Since the external parameters affect the equations of motion of the particles in the system, they must also affect the energy levels of these particles. Thus the energy of each quantum state of a system is ordinarily a function of the external parameters of the system. For example, in the case of a spin system, Table 3.1 shows explicitly that the energies of the quantum states depend on the value of the external magnetic field B. Similarly, in the case of a particle in a box, the expression (15) shows explicitly that any quantum state specified by the quantum numbers $\{n_x,n_y,n_z\}$ has an energy which depends on the dimensions L_x, L_y, L_z of the box.

A knowledge of all the external parameters of a system thus serves to determine the actual energies of its quantum states.

(ii) Information about the initial preparation of the system
In view of the conservation laws of mechanics, the initial preparation of a system implies certain general restrictions on the subsequent motion of the particles in the system. For example, suppose that we are dealing with a system which is isolated so that it does not interact with any other system. Then the laws of mechanics require that the total energy of this system (i.e., the total kinetic and potential energy of all the particles in the system) remain constant. The system, when originally prepared for observation, must have some total energy which may be determined to within some finite precision, i.e., this energy may be known to lie within some small range between E and $E + \delta E$. Then the conservation of energy implies that the total energy of the system must *always* be between E and $E + \delta E$. As a consequence of this restriction, the system can be found only in those of its quantum states which have an energy lying in this range.†

We shall call the *accessible states* of a system those of its quantum states in which the system can be found without violating any conditions imposed by the information available about the system. A statistical ensemble selected in accordance with the information available about the system must, therefore, comprise systems all of which are in their accessible states. As pointed out previously, the specification of the macrostate of a system consisting of very many particles provides only very limited information about the system. If the system is known to be in a given *macro*state, the number of quantum states accessible to the system is thus ordinarily very large (since the number

---

† In some cases one might also want to take into account other restrictions, such as those imposed by the conservation of total momentum. Although this could be done, it is ordinarily not of interest for the following reason: In the case of most laboratory experiments, it can be imagined that the system under investigation is enclosed in some container fastened to the floor of the laboratory and thus to the large mass of the earth. Any collision of the particles in the system with the container results then in a nearly negligible velocity change of the earth, although the latter can absorb any amount of momentum from the system while acquiring negligible energy. (The situation is similar to that of a ball being bounced off the earth.) Under these circumstances, there is no restriction on the possible momentum of the system, although its energy remains constant. Thus the system can be considered isolated with respect to energy transfer, while it is not isolated with respect to momentum transfer.

of particles in the system is very large). For instance, in the case of an isolated system merely known to have an energy between E and $E + \delta E$, all quantum states having energies in this range are accessible to the system.

It is conceptually simplest to discuss the case of a system which is *isolated* in the sense that it does not interact with any other systems so as to exchange energy with them.† Suppose that the macrostate of such an isolated system is specified by stating the values of its external parameters and the particular small range in which its energy is known to lie. This information determines then, respectively, the energies of the various quantum states of the system and the subset of those quantum states which are actually accessible to the system.

The essential content of the preceding remarks can be illustrated most simply by some systems with very few particles.

Example (i)

Consider a system of four spins $\frac{1}{2}$ (each having a magnetic moment μ_0) located in an applied magnetic field **B**. The possible quantum states and associated energies of this system are listed in Table 3.2. Suppose that this system is isolated and known to have a total energy which is equal to $-2\mu_0 B$. The system can then be found in any one of the following four states accessible to it:

$$\{+++-\}, \quad \{++-+\},$$
$$\{+-++\}, \quad \{-+++\}.$$

Example (ii)

Consider a system A^* which consists of two subsystems A and A' which can interact to a small extent and thus exchange energy with each other. System A consists of three spins $\frac{1}{2}$, each having a magnetic moment μ_0. System A' consists of two spins $\frac{1}{2}$, each having a magnetic moment $2\mu_0$. The system A^* is located in an applied magnetic field **B**. We shall denote by M the total magnetic moment of A along the direction of **B**, and by M' the total magnetic moment of A' in this direction. The interaction between the spins is assumed to be almost negligible. The total energy E^* of the entire system A^* is then given by

$$E^* = -(M + M')B.$$

The system A^* consists of 5 spins and thus has a total of $2^5 = 32$ possible quantum states. Each of these can be labeled by five quantum numbers, the three numbers σ_1, σ_2, σ_3 specifying the orientations of the three magnetic moments of A, and the two quantum numbers σ_1', σ_2' specifying the orientations of the two magnetic moments of A'. Suppose that the isolated system A^* is known to have a total energy E^* equal to $-3\mu_0 B$. Then A^* must be found in any one of the five accessible states, listed in Table 3.3, which are compatible with this total energy.

r	σ_1	σ_2	σ_3	σ_1'	σ_2'	M	M'
1	+	+	+	+	−	$3\mu_0$	0
2	+	+	+	−	+	$3\mu_0$	0
3	+	−	−	+	+	$-\mu_0$	$4\mu_0$
4	−	+	−	+	+	$-\mu_0$	$4\mu_0$
5	−	−	+	+	+	$-\mu_0$	$4\mu_0$

Table 3.3 Systematic enumeration of all the states, labeled by some index r, which are accessible to the system A^* when its total energy in a magnetic field **B** is equal to $-3\mu_0 B$. The system A^* consists of a subsystem A with three spins $\frac{1}{2}$, each having magnetic moment μ_0, and a subsystem A' with two spins $\frac{1}{2}$, each having magnetic moment $2\mu_0$.

† Any system which is *not* isolated can then be treated as a part of a larger system which *is* isolated.

The task of giving a statistical description of a macroscopic system can now be formulated very precisely. In a statistical ensemble of such systems, every system is known to be in one of its accessible quantum states. We should then like to predict the probability that the system is found in any given one of these accessible states. In particular, various macroscopic parameters of the system (e.g., its total magnetic moment or the pressure which it exerts) have values which depend on the particular quantum state of the system. Knowing the probability of finding the system in any one of its accessible states, we should then be able to answer the following questions of physical interest: What is the probability that any parameter of the system assumes any specified value? What is the mean value of such a parameter? What is its standard deviation?

3.3 *Statistical Postulates*

In order to make theoretical predictions concerning various probabilities and mean values, we must introduce some statistical postulates. Let us, therefore, consider the simple case of an *isolated* system (with given external parameters) whose energy is known to lie in a specified small range between E and $E + \delta E$. As already mentioned, this system can be found in any one of a large number of accessible states. Focusing attention on a statistical ensemble of such systems, what can we say about the *probability* of finding the system in any one of these accessible states?

To elucidate this question, we shall exploit some simple physical arguments similar to those used in Secs. 1.1 and 1.2. There we considered the example of an ideal gas and talked about the distribution of molecules in a box over their possible positions in space. Analogously, our more abstract general arguments will now deal with the distribution of systems in an ensemble over their accessible states. Our discussion will then readily lead to the formulation of general postulates suitable as a basis for a statistical theory.

Let us examine first the simple situation where the system under consideration is known at some time to have equal probability of being found in each one of its accessible states. In other words, we consider the case where the systems in the statistical ensemble are known at some time to be uniformly distributed over all their accessible states. What happens then as time goes on? A system in a given state will, of course, not remain there forever; as we pointed out at the end of Sec. 3.1, it will continually make transitions between the various states accessible to it. The situation is thus a dynamic one. But there is nothing intrinsic in the laws of mechanics which gives preference to any one of the accessible states of a system compared to any other one. Thus, contemplating the ensemble of systems as time goes on, we do not expect that the number of systems in some particular subset of the accessible states will become less while that in some other subset will become greater.† Indeed, the laws of mechanics can be applied to an ensemble of isolated systems to show explicitly that, if the systems are initially uniformly distributed over all their accessible states, then they will remain uniformly distributed over these states forever.‡ Such a uniform distribution remains, therefore, unchanged in time.

Example

To give a very simple illustration, we may return to Example (i) of Sec. 3.2. There we dealt with an isolated system of four spins having a total energy $-2\mu_0 B$. Suppose that this system is known at some time to have equal probability of being found in each one of its four accessible states

$$\{++\,+-\},\quad \{++-+\},$$
$$\{+-++\},\quad \{-+++\}.$$

In accord with our preceding arguments, there is nothing intrinsic in the laws of mechanics which would give preferred status to one of these four states compared to any other one. Hence, we do not expect that the system would at some later time be more likely to be found in one particular state, say the state $\{+++-\}$, rather than in any other one of its four accessible states. The contemplated situation, therefore, does not change in time. Hence the system will continue to have equal probability of being found in each one of its four accessible states.

The preceding argument thus leads to the following conclusion about an ensemble of isolated systems: If the systems in such an ensemble are uniformly distributed over their accessible states, the ensemble is time-independent. In terms of probabilities, this statement can be phrased as follows: If an isolated system has equal probability of being found in each one of its accessible states, then the probability of finding the system in each one of its states is time-independent.

---

† This argument is merely a more general version of that used in Sec. 1.1 where we considered an ideal gas. When the molecules of the gas are initially distributed uniformly throughout a box, they are not expected to concentrate themselves preferentially in one part of the box as time goes on since there is nothing in the laws of mechanics which gives preference to one part of the box compared to any other.

‡ This result is a consequence of the so-called "Liouville theorem." Proofs of this theorem, which require a knowledge of advanced mechanics far beyond the level of this book, can be found in R. C. Tolman, *The Principles of Statistical Mechanics,* chaps. 3 and 9 (Oxford University Press, Oxford, 1938).

An isolated system, by definition, is said to be *in equilibrium* if the probability of finding the system in each one of its accessible states is independent of time. In this case the mean value of every measurable macroscopic parameter of the system is, of course, also independent of time.† In terms of this definition of equilibrium, the conclusion of the preceding paragraph can be summarized by the following statement:

If an isolated system is found with equal probability in each one of its accessible states, it is in equilibrium.	(17)

Let us next examine the general case where the isolated system under consideration is known at some initial time to be in some *sub*set of the states actually accessible to it. A statistical ensemble of such systems at this time would then contain many systems in some subset of their accessible states, and no systems at all in the remaining accessible states. What happens now as time goes on? As we mentioned before, there is nothing intrinsic in the laws of mechanics which would tend to favor some of the accessible states over others. By their definition, the accessible states are also such that the laws of mechanics impose no restrictions preventing the system from being found in any one of these states. It is, therefore, exceedingly unlikely that a system in the ensemble remains indefinitely in the subset of states in which it finds itself initially and avoids the other states which are equally accessible to it.‡ Indeed, by virtue of small interactions between its constituent particles, the system will in the course of time continually make transitions between all of its accessible states. As a result, each system in the ensemble will ultimately pass through essentially all the states in which it can possibly be found. The net effect of these continual transitions is analogous to that resulting from the repeated shuffling of a deck of cards. If the shuffling is kept up long enough, the cards get so mixed up that each one is equally likely to occupy any position in the deck irrespective of how the deck was arranged initially. Similarly, in the case of the ensemble of systems, one expects that the systems will

---

† Indeed, to ascertain experimentally that a system is in equilibrium, one verifies that the mean values of *all* its observable macroscopic parameters are independent of time. One then presumes that the probability of finding the system in any one state is independent of time so that the system is in equilibrium.

‡ The present discussion in terms of quantum states and transitions between them is again merely a generalization of the arguments applied in Sec 1.2 to the case of an ideal gas. When all the molecules of the gas are in the left half of a box, the resulting situation is very unlikely and the molecules become quickly distributed throughout the entire box.

ultimately become uniformly (i.e., randomly) distributed among all their accessible states.† Once this situation has been attained, the distribution remains uniform in accordance with (17). This final situation thus corresponds to a time-independent equilibrium situation.

The conclusion of the preceding argument can then be summarized as follows:

> If an isolated system is not found with equal probability in each one of its accessible states, it is *not* in equilibrium. It then tends to change in time until it attains ultimately the equilibrium situation where it is found with equal probability in each one of its accessible states. (18)

Note that this conclusion is analogous to the statement (1.7) of Chap. 1. It merely formulates in more precise and general terms the tendency of an isolated system to approach its most random situation.

Example

The preceding comments can again be illustrated with our preceding example of the isolated system of four spins. Suppose that this system has been prepared in such a way that it is initially known to be in the state $\{+++-\}$. The total energy of the system is then $-2\mu_0 B$ and remains constant at this value. There are, however, three other states

$$\{++-+\}, \quad \{+-++\}, \quad \{-+++\}$$

which also have this energy and are thus equally accessible to the system. Indeed, as a result of small interactions between the magnetic moments, processes occur in which one moment flips from the "up" direction to the "down" direction while another moment does the reverse (the total energy remaining, of course, unchanged). As a result of any such process, the system goes from some initial accessible state to some other accessible state. After many repeated transitions of this kind, the system will ultimately be found with equal probability in any one of its four accessible states

$$\{+++-\}, \quad \{++-+\},$$
$$\{+-++\}, \quad \{-+++\}.$$

We shall adopt the statements (17) and (18) as the fundamental postulates of our statistical theory. Both (17) and (18) can actually be deduced from the laws of mechanics, (17) rigorously and (18) with the aid of some assumptions. The statement (18) is of special importance since it leads, in particular, to the following assertion:

† With some assumptions inherent in a statistical description, this expectation can be deduced from the laws of mechanics as a consequence of the so-called "H-theorem." A simple proof of this theorem and further references can be found in F. Reif, *Fundamentals of Statistical and Thermal Physics*, Appendix A.12 (McGraw-Hill Book Company, New York, 1965).

If an isolated system is in equilibrium, it is found with equal probability in each one of its accessible states.	(19)

This statement is the converse of (17). The validity of (19) follows immediately from (18). Indeed, if the conclusion of (19) were false, (18) implies that the premise of (19) would be violated.

The statistical situation of greatest simplicity is clearly that which is time-independent, i.e., that involving an isolated system in equilibrium. In this case, the statement (19) makes an unambiguous assertion about the probability of finding the system in each one of its accessible states. The statement (19) is, therefore, the fundamental postulate upon which we can erect the entire theory of macroscopic systems in equilibrium. This fundamental postulate of equilibrium statistical mechanics is sometimes called the *postulate of equal a priori probabilities*. Note that this postulate is exceedingly simple. Indeed, it is completely analogous to the simple postulate (assigning equal probabilities to head or tail) used in discussing the experiment of tossing a coin. The ultimate validity of the postulate (19) can only be established, of course, by checking its predictions against experimental observations. Since a large body of calculations based on this postulate have yielded results in very good agreement with experiment, the validity of this postulate can be accepted with great confidence.

The theoretical problems become much more complex when we are dealing with a statistical situation which *does* change with time, i.e., with a system which is *not* in equilibrium. In this case the only general statement which we have made is embodied in (18). This postulate makes an assertion about the *direction* in which the system tends to change (i.e., this direction is such that the system tends to approach the equilibrium situation of uniform statistical distribution over all accessible states). The postulate provides *no* information, however, about the actual time required for the system to attain its ultimate equilibrium condition (the so-called *relaxation time*). This time might be shorter than a microsecond or longer than a century, depending on the detailed nature of the interactions between the particles of the system and on the resultant frequency with which transitions actually occur between the accessible states of this system. A quantitative description of a nonequilibrium situation may thus become quite difficult since it requires a detailed analysis of how the probability of finding a system in each of its states changes with time. On the other hand, a problem involving a system in equilibrium merely requires use of the simple postulate (19) of equal a priori probabilities.

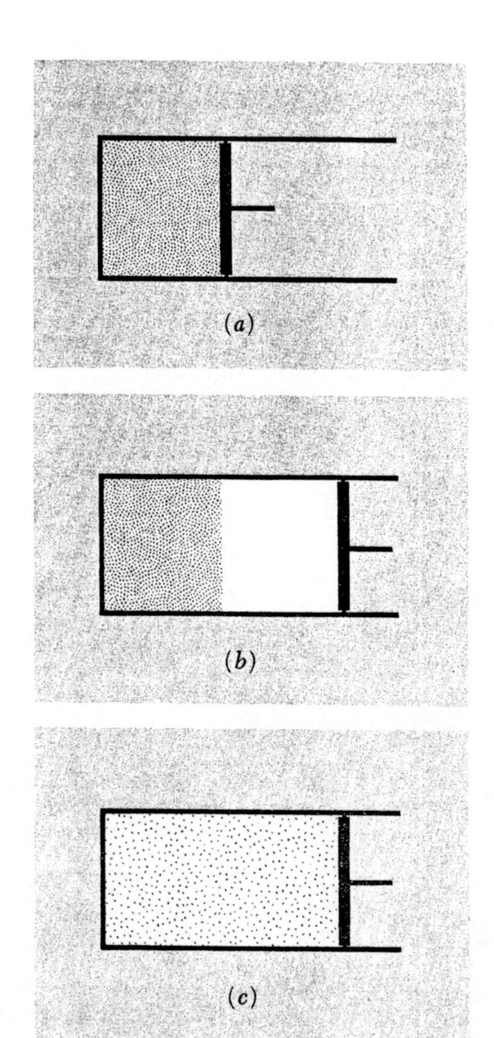

Fig. 3.4 Sudden expansion of a gas. (*a*) Initial situation. (*b*) Situation immediately after the piston has been moved. (*c*) Final situation.

Remarks concerning the applicability of equilibrium arguments

It is worth pointing out that the idealized concept of equilibrium is, in practice, a relative one. The important comparison is always between the relaxation time τ_r (the characteristic time necessary for the system to attain equilibrium when it is not in equilibrium initially) and the time τ_e of experimental interest in a given discussion.

Imagine, for example, that when the piston in Fig. 3.4 is suddenly pulled to the right, it takes about 10^{-3} seconds before equilibrium is attained and the whole gas is uniformly distributed throughout the entire volume of the box. Thus $\tau_r \sim 10^{-3}$ sec. Suppose now that, instead of doing the preceding experiment, we proceed as in Fig. 3.5 by moving the piston to the right quite slowly so that it takes a time $\tau_e = 100$ sec to complete the task. Strictly speaking, the gas is not in equilibrium during this time since its volume is changing. But since $\tau_e \gg \tau_r$, the molecules at any instant have had sufficient time to become essentially uniformly distributed throughout the entire volume available to them at that instant. The situation in the gas would then remain effectively unchanged if one imagined the piston to be really stopped at any instant. In practice, the gas can thus be considered to be in equilibrium at all times.

As an example of the opposite limit where $\tau_e \ll \tau_r$, consider a piece of iron which is rusting at a very slow rate; in a time $\tau_r = 100$ years it would be completely transformed into iron oxide. Strictly speaking, this again is not an equilibrium situation. But during a time τ_e of experimental interest, say $\tau_e = 2$ days, the situation would be effectively unchanged if one imagined rusting was prevented (e.g., by removing all the surrounding oxygen gas). The iron would then be in strict equilibrium. Thus, in practice, the piece of iron can again be discussed by equilibrium arguments.

It is, therefore, only in the case when $\tau_e \sim \tau_r$ (i.e., when the time of experimental interest is comparable to the time required to reach equilibrium) that the time-dependence of the system is of essential significance. The problem is then more difficult and cannot be reduced to a discussion of equilibrium, or near-equilibrium, situations.

3.4 *Probability Calculations*

The fundamental postulate (19) of equal a priori probabilities permits the statistical calculation of all time-independent properties of any system in equilibrium. In principle, such a calculation is very simple. Consider thus an isolated system in equilibrium and denote by Ω the total number of states accessible to it. According to our postulate, the probability of finding the system in each one of its accessible states is then the same and hence simply equal to $1/\Omega$. (The probability of finding the system in a state which is not accessible to it is, of course, zero.) Suppose now that we are interested in some parameter y of the system; for example, y might be the magnetic moment of the system or the pressure exerted by it. When the system is in any particular state, the parameter y assumes correspondingly some definite value.

Let us enumerate the possible values which y can assume† and designate them by $y_1, y_2, \ldots, y_n$. Among the Ω states accessible to the system, there will then be some Ω_i such states in which the parameter assumes the particular value y_i. The probability P_i that the parameter assumes the value y_i is then simply the probability that the system is found among the Ω_i states characterized by this value y_i. Thus P_i is obtained by summing $1/\Omega$ (the probability of finding the system in any single one of its accessible states) over the Ω_i states where y assumes the value y_i; i.e., P_i is merely Ω_i times as large as the probability $1/\Omega$ of finding the system in each one of its accessible states. Hence‡

$$\boxed{P_i = \frac{\Omega_i}{\Omega}.} \tag{20}$$

The mean value of the parameter y is then, by its definition, given by

$$\bar{y} \equiv \sum_{i=1}^{n} P_i y_i = \frac{1}{\Omega} \sum_{i=1}^{n} \Omega_i y_i \tag{21}$$

where the summation is over all possible values of y. The dispersion of y can be calculated in a similar way. All statistical calculations are thus basically just as simple as those involved in discussing the tossing of a set of coins.

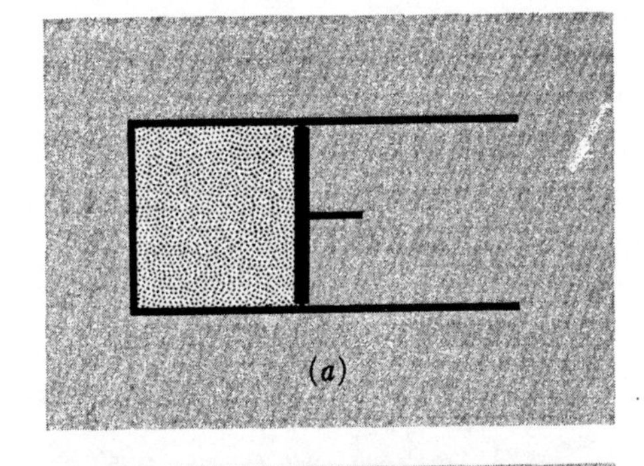

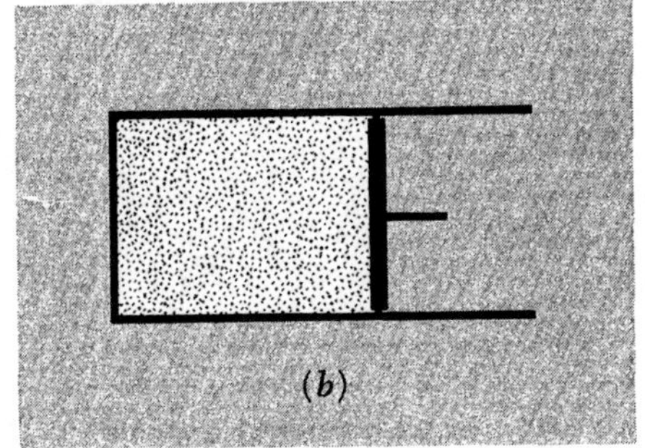

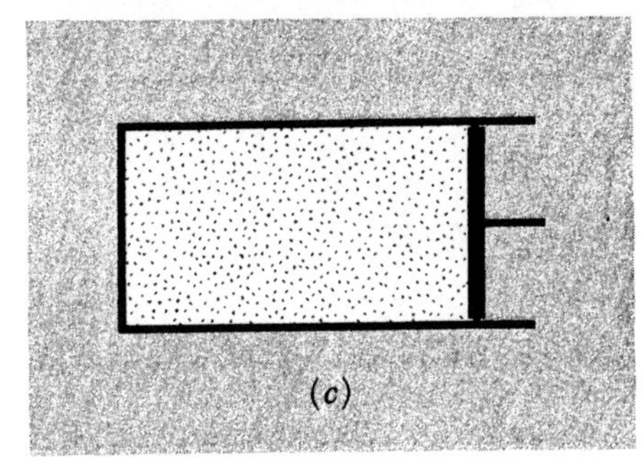

Fig. 3.5 Very slow (quasi-static) expansion of a gas. (*a*) Initial situation. (*b*) Intermediate situation. (*c*) Final situation.

Example (i)

Consider again the system of four spins whose states are listed in Table 3.2, and suppose that the total energy of this system is known to be $-2\mu_0 B$. If the system is in equilibrium, it is then equally likely to be in each one of the four states

$$\{+++-\}, \quad \{++-+\},$$
$$\{+-++\}, \quad \{-+++\}.$$

Focus attention on any one of these spins, say the first. What is the probability P_+ that its magnetic moment points up? Since it points up in three of the four equally likely states accessible to the entire system, this probability is simply

$$P_+ = \tfrac{3}{4}.$$

What is the mean magnetic moment of this spin in the direction of the applied field **B**? Since this moment is μ_0 in three

---

† If the possible values of the parameter are continuous rather than discrete, we can always proceed as in Sec. 2.6 and subdivide the domain of the possible values of y into very small intervals of fixed size δy. Since these intervals can be enumerated and labeled by some index i, we may then simply denote by y_i the value of the parameter when it lies anywhere in the ith such interval. The problem is then reduced to one where the possible values of y are discrete and countable.

‡ The result (20) is so simple because our fundamental postulate asserts that the system has equal probability of being found in each one of its accessible states. In an ensemble of $\mathcal{N}$ systems, the number $\mathcal{N}_i$ of systems with $y = y_i$ is therefore simply proportional to the number Ω_i of states accessible to a system with $y = y_i$. Thus $P_i \equiv \mathcal{N}_i/\mathcal{N} = \Omega_i/\Omega$.

of the accessible states and $-\mu_0$ in one of the accessible states, this mean value is

$$\bar{M} = \frac{3\mu_0 + (-\mu_0)}{4} = \frac{1}{2}\mu_0.$$

Note, incidentally, that one given spin in this system is not equally likely to have a moment which points up or down; i.e., it is not found with equal probability in each one of its two possible states. This result does not, of course, contradict our fundamental statistical postulate since such a single spin is not isolated, but is part of a larger system where it is free to interact and exchange energy with other spins.

Example (ii)

Consider the system of spins discussed in Table 3.3. The total energy of this system is known to be $-3\mu_0 B$ and the system is assumed to be in equilibrium. The system is then equally likely to be found in each one of its 5 accessible states. Let us, for example, focus attention on the subsystem A consisting of 3 spins and denote by M its total magnetic moment in the direction of the field $\mathbf{B}$. It is seen that M can assume the two possible values $3\mu_0$ or $-\mu_0$. The probability $P(M)$ that it assumes any of these values can then be read off directly from the table; thus

$$P(3\mu_0) = \tfrac{2}{5},$$

and

$$P(-\mu_0) = \tfrac{3}{5}.$$

The mean value of M is then given by:

$$\bar{M} = \frac{(2)(3\mu_0) + (3)(-\mu_0)}{5} = \frac{3}{5}\mu_0.$$

The preceding examples are extremely simple since they deal with systems consisting of very few particles. They do, however, illustrate the general procedure used in calculating probabilities and mean values for any system in equilibrium, no matter how complex. The only difference is that the enumeration of accessible states characterized by a particular value of a parameter may become a more difficult task in the case of a macroscopic system consisting of very many particles. The actual computation may thus be correspondingly more complicated.

3.5 *Number of States Accessible to a Macroscopic System*

The preceding four sections contain all the basic concepts needed for a quantitative statistical theory of macroscopic systems in equilibrium and for a qualitative discussion of their approach to equilibrium. We shall use the remainder of this chapter to become familiar with the significance of these concepts and to show how they provide a precise description of some of the qualitative notions introduced in Chap. 1. These preliminaries will prepare us to elaborate these concepts systematically throughout the rest of the book.

As we have seen, the properties of a system in equilibrium can be calculated by counting the number of states accessible to the system under various conditions. Although this counting problem may become difficult, it can ordinarily be circumvented in practice. As is usual in physics, progress may become easy if one strives to gain insight instead of trying to compute blindly. In particular, what is important in the present instance is to recognize some of the general properties exhibited by the number of states accessible to any system consisting of very many particles. Since a qualitative understanding of these properties and some approximate estimates will be quite adequate, fairly crude arguments will suffice to achieve our purpose.

Consider a macroscopic system with given external parameters so that its energy levels are determined. We shall denote the total energy of this system by E. To facilitate the counting of states, we shall group these states according to energy by subdividing the energy scale into equal small intervals of fixed magnitude δE. Thus δE is supposed to be very small on a *macro*scopic scale (i.e., very small compared to the total energy of the system and small compared to the contemplated precision of any macroscopic measurement of its energy). On the other hand, δE is supposed to be large on a *micro*scopic scale (i.e., much larger than the energy of a single particle in the system and thus also much larger than the separation in energy between adjacent energy levels of the system). Any interval δE thus contains very many possible quantum states of the system. We shall introduce the notation

$$\Omega(E) \equiv \text{the number of states with energies lying in the interval between } E \text{ and } E + \delta E. \tag{22}$$

The number of states $\Omega(E)$ depends on the magnitude δE chosen as the subdivision interval in a given discussion. Since δE is macroscopically very small, $\Omega(E)$ must be simply proportional to δE, i.e., we can write†

$$\Omega(E) = \rho(E)\,\delta E \tag{23}$$

where $\rho(E)$ is independent of the size of δE. [The quantity $\rho(E)$ is called the *density of states* because it is equal to the number of states per unit energy range at the given energy E.] Since the interval δE

† The situation here is similar to that encountered in Sec. 2.6 when we discussed continuous probability distributions. The number of states $\Omega(E)$ must vanish when δE approaches zero and must be expressible as a Taylor series in powers of δE. When δE is sufficiently small, this series reduces to (23) since terms involving higher powers of δE are then negligible.

contains very many states, $\Omega(E)$ changes only by a small fraction of itself as one goes from one energy interval to an adjacent one. Hence $\Omega(E)$ can be regarded as a smoothly varying function of the energy E. We shall be interested specifically in examining how sensitively $\Omega(E)$ depends on the energy E of a macroscopic system.

Note, incidentally, that it is possible to obtain $\Omega(E)$ if one knows the quantity

$$\Phi(E) \equiv \text{the total number of states having energies } \textit{less} \text{ than } E. \tag{24}$$

The number $\Omega(E)$ of states having an energy lying between E and $E + \delta E$ is then simply given by

$$\Omega(E) = \Phi(E + \delta E) - \Phi(E) = \frac{d\Phi}{dE}\,\delta E. \tag{25}$$

Before discussing the general properties of $\Omega(E)$ in the case of a macroscopic system, it will be instructive to illustrate how the number of states $\Omega(E)$ can be counted in the case of some very simple systems consisting of a single particle.

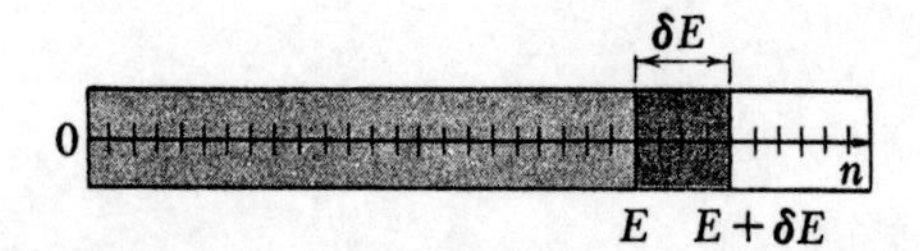

Fig. 3.6 The points on the line indicate the possible values $n = 1, 2, 3, 4, \ldots$ of the quantum number n specifying the state of a single particle in one dimension. The values of n corresponding to the energy values E and $E + \delta E$ are indicated by the vertical lines. The region in light gray includes all the values of n for which the energy of the particle is less than E. The region in dark gray includes all the values of n for which the energy lies between E and $E + \delta E$.

Example (i) Single particle in a one-dimensional box

Consider a single particle of mass m free to move in one dimension in a box of length L. The possible energy levels of this system are then, by (8),

$$E = \frac{\hbar^2}{2m}\frac{\pi^2}{L^2}n^2 \tag{26}$$

where $n = 1, 2, 3, 4 \ldots$. The coefficient of n^2 is very small if L is of macroscopic size. The quantum number n is thus very large for energies of ordinary interest.† By (26), the value of n for a given energy E is

$$n = \frac{L}{\pi\hbar}(2mE)^{1/2}. \tag{27}$$

Since successive quantum states correspond to values of n differing by unity, the total number $\Phi(E)$ of quantum states having an energy less than E, or a quantum number less than n, is then simply equal to $(n/1) = n$. Thus

$$\Phi(E) = n = \frac{L}{\pi\hbar}(2mE)^{1/2}. \tag{28}$$

Accordingly, (25) yields‡

$$\Omega(E) = \frac{L}{2\pi\hbar}(2m)^{1/2}E^{-1/2}\,\delta E. \tag{29}$$

† For example, if $L = 1$ cm and $m \sim 5 \times 10^{-23}$ gm, the mass of a nitrogen molecule given by (1.29), this coefficient is about 10^{-32} erg. But the average energy of such a molecule at room temperature is, by (1.28), of the order of 10^{-14} erg. Hence (26) yields for n a typical value of the order of 10^9.

‡ Since n is very large, a change in n by unity produces only a negligible fractional change in n or E. The fact that n or E can assume only discrete values is then of negligible importance so that these variables can be treated as though they were continuous. In differentiating, it is merely necessary to consider any contemplated change in n to be always large compared to unity so that $dn > 1$, but to be very small in the sense that $dn \ll n$.

Example (ii) Single particle in a three-dimensional box

Consider a single particle of mass m free to move inside a three-dimensional box. For the sake of simplicity, we assume this box to be cubic and of edge length L. The possible energy levels of this system are then given by (15) with $L_x = L_y = L_z = L$; thus

$$E = \frac{\hbar^2}{2m}\frac{\pi^2}{L^2}(n_x^2 + n_y^2 + n_z^2) \quad (30)$$

where $n_x, n_y, n_z = 1, 2, 3, \ldots$. In the "number space" defined by three perpendicular axes labeled by n_x, n_y, n_z, the possible values of these three quantum numbers lie thus geometrically at the centers of cubes of unit edge length, as indicated in Fig. 3.7. As in the preceding example, these quantum numbers are again ordinarily very large for a molecule in a macroscopic box. By (30) it follows that

$$n_x^2 + n_y^2 + n_z^2 = \left(\frac{L}{\pi\hbar}\right)^2(2mE) \equiv R^2.$$

For a given value of E, the values of n_x, n_y, n_z which satisfy this equation lie on a sphere of radius R in Fig. 3.7. Here

$$R = \frac{L}{\pi\hbar}(2mE)^{1/2}.$$

The number $\Phi(E)$ of states with energy less than E is then equal to the number of unit cubes lying within this sphere and having positive values of n_x, n_y, and n_z; i.e., it is simply equal to the volume of one octant of the sphere of radius R. Thus

$$\Phi(E) = \frac{1}{8}\left(\frac{4}{3}\pi R^3\right) = \frac{\pi}{6}\left(\frac{L}{\pi\hbar}\right)^3(2mE)^{3/2}. \quad (31)$$

By (25), the number of states with energy between E and $E + \delta E$ is then

$$\Omega(E) = \frac{V}{4\pi^2\hbar^3}(2m)^{3/2}E^{1/2}\,\delta E \quad (32)$$

where $V = L^3$ is the volume of the box.

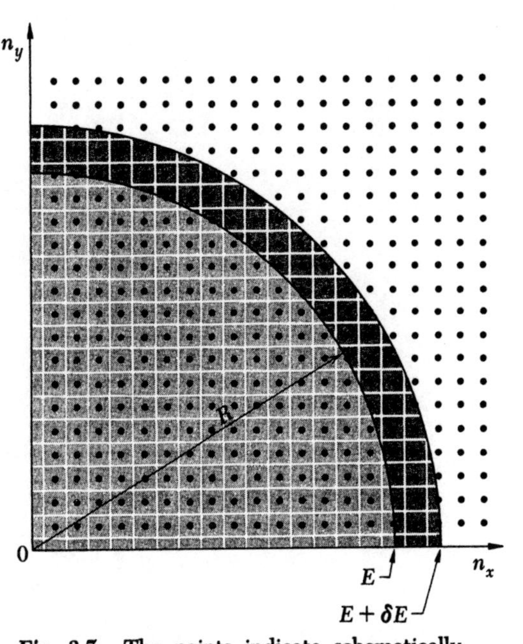

Fig. 3.7 The points indicate schematically in two dimensions the possible values of $n_x, n_y, n_z = 1, 2, 3, 4, \ldots$ of the quantum numbers specifying the state of a single particle in three dimensions. (The n_z axis points out of the paper.) The values of n_x, n_y, and n_z corresponding to the energy values E and $E + \delta E$ lie on the indicated spherical surfaces. The region in light gray includes all values of n for which the energy of the particle is less than E. The region in dark gray includes all the values of n for which the energy lies between E and $E + \delta E$.

Let us now make a crude order-of-magnitude estimate to find roughly how the number of states $\Omega(E)$ or, equivalently $\Phi(E)$, depends on the energy E of a macroscopic system of particles. Any such system can be described by some set of f quantum numbers where f, the *number of degrees of freedom* of the system, is of the order of Avogadro's number. Associated with each quantum number there is a certain contribution ϵ to the total energy E of the system. We shall denote by $\varphi(E)$ the total number of possible values of this quantum number when its associated energy is *less* than ϵ. The number φ is then unity (or of the order of unity) when ϵ has its lowest possible value ϵ_0, and must clearly increase as ϵ increases (although it may approach a constant value in an exceptional case).† Ordinarily we expect φ to increase roughly proportionately to the energy interval $(\epsilon - \epsilon_0)$. Hence we can write the approximate relation that

† This exceptional case arises if a system has only a *finite* total number of possible states and thus also an upper bound to its possible energy. (This can occur only if one ignores the kinetic energies of the particles in a system and considers only their spins.) Therefore, after at first increasing as a function of $(\epsilon - \epsilon_0)$, the quantity φ must in this case ultimately become constant.

ordinarily, $\qquad \varphi(\epsilon) \propto (\epsilon - \epsilon_0)^\alpha \qquad$ where $\alpha \sim 1$, $\qquad$ (33)

i.e., where α is some number of the order of unity.†

Consider now the entire system with f degrees of freedom. Its total energy E (the sum of the kinetic and potential energies of all the particles contained within the system) is the sum of the energies associated with all its degrees of freedom. Hence the value of the energy E (in excess of its lowest possible value E_0) should be roughly f times as large as the average energy ϵ per degree of freedom (in excess of its lowest possible value ϵ_0). Thus

$$E - E_0 \sim f(\epsilon - \epsilon_0). \tag{34}$$

Corresponding to a total energy of the system of E or less, there are then approximately $\varphi(\epsilon)$ possible values which can be assumed by the 1st of its quantum numbers, $\varphi(\epsilon)$ possible values which can be assumed by the 2nd of its quantum numbers, . . . , and $\varphi(\epsilon)$ possible values which can be assumed by the fth of its quantum numbers. The total number of possible combinations of these quantum numbers, i.e., the total number $\Phi(E)$ of states with total energy less than E, is then obtained by multiplying {the number of possible values of the 1st quantum number} by {the number of possible values of the 2nd quantum number}, then multiplying by {the number of possible values of the 3rd quantum number}, and finally multiplying by {the number of possible values of the fth quantum number}. Thus

$$\Phi(E) \sim [\varphi(\epsilon)]^f \tag{35}$$

where ϵ is related to E by (34). The number $\Omega(E)$ of states with energy between E and $E + \delta E$ is then given by (25). Thus

$$\Omega(E) = \frac{d\Phi}{dE}\,\delta E \sim f\varphi^{f-1}\,\frac{d\varphi}{dE}\,\delta E = \varphi^{f-1}\frac{d\varphi}{d\epsilon}\,\delta E \tag{36}$$

since $\qquad d\varphi/dE = f^{-1}\,(d\varphi/d\epsilon) \qquad$ by (34).

The preceding approximate considerations are quite sufficient to lead to some remarkable conclusions based on the fact that f is a very large number. Indeed, since we are dealing with a macroscopic system, f is of the order of Avogadro's number, i.e., $f \sim 10^{24}$. Numbers of this magnitude are so fantastically large that their peculiar properties are unlikely to be familiar from everyday experience.

As the energy E of the system increases, so does the energy ϵ per degree of freedom [see Eq. (34)]. Correspondingly, the number of

† For example, in the case of the quantum number describing the motion of a single particle in one dimension, $\varphi \propto \epsilon^{1/2}$ by virtue of (28). (The lowest possible energy ϵ_0 there is negligible compared to ϵ and thus is essentially zero.)

states $\varphi(\epsilon)$ per degree of freedom increases relatively slowly. But, since the exponents in (35) or (36) are of the order of f and thus extremely large, the number of states $\Phi(E)$, or $\Omega(E)$, of the system with f degrees of freedom increases at a fantastic rate. Hence we arrive at the following conclusion:

The number of states $\Omega(E)$ accessible to any ordinary macroscopic system is an extremely rapidly increasing function of its energy E.	(37)

Indeed, combining (36) with (33) and (34), we obtain for the dependence of Ω on E the approximate relations

$$\Omega(E) \propto (\epsilon - \epsilon_0)^{\alpha f - 1} \propto \left(\frac{E - E_0}{f}\right)^{\alpha f - 1}.$$

Thus we can assert that,

for any ordinary system, $\Omega(E) \propto (E - E_0)^f$ approximately.†	(38)

Here we have neglected 1 compared to f and have simply put $\alpha = 1$ since (38) is intended to indicate only very crudely the dependence of Ω on E. It matters little, for that purpose, whether the exponent in (38) is f, $\frac{1}{2}f$, or any other number of the order of f.

We can also make some statements about the magnitude of $\ln \Omega$. By (36) it follows that

$$\ln \Omega(E) = (f - 1) \ln \varphi + \ln \left(\frac{d\varphi}{d\epsilon}\,\delta E\right). \tag{39}$$

Let us now make the general observation that, if one deals with any large number such as f, its logarithm is always roughly of the order of unity and thus utterly negligible compared to the number itself. For example, if $f = 10^{24}$, $\ln f = 55$; thus $\ln f \lll f$. Consider then the terms on the right side of (39). The first term is of the order of f.‡

---

† The qualifying phrase *for any ordinary system* is intended to exclude the exceptional case (mentioned in the footnote on page 121) where the kinetic energy of the particles in a system is ignored and where their spins have sufficiently high magnetic energy. (Focusing attention on the spins alone may be a justifiable approximation if the translational motion of the particles has only small effect on the orientations of their spins. The spin orientations and translational motion of the particles can then be treated separately.)

‡ This is always true unless the energy E of the system is very close to its ground state energy E_0 when $\varphi \sim 1$ for all degrees of freedom. Indeed, we know from our general comments at the beginning of Sec. 3.1 that, when a system approaches its ground state energy, the number of quantum states accessible to it is of order unity so that $\Omega \sim 1$.

The quantity $(d\varphi/d\epsilon)\,\delta E$ (where δE is any energy interval which is large compared to the spacing between the energy levels of the system) represents the number of possible values of a single quantum number in the interval δE and depends on the size of δE. But irrespective of the size chosen for δE, even the most conservative estimate would lead us to expect that this quantity $(d\varphi/d\epsilon)\,\delta E$ lies somewhere between 1 and 10^{100}. Its logarithm, however, then lies between 0 and 230 and is thus utterly negligible compared to f which is of the order of 10^{24}. Hence the second term on the right of (39) is utterly negligible compared to the first. We thus arrive at the following conclusions:

> In the case of a macroscopic system, the number of states $\Omega(E)$ having an energy between E and $E + \delta E$ is to excellent approximation such that

$$\text{for } E \not\approx E_0, \qquad \ln \Omega(E) \text{ is independent of } \delta E; \tag{40}$$

$$\ln \Omega(E) \sim f. \tag{41}$$

In words, (40) and (41) assert that, as long as the energy E is not very close to its ground state value, $\ln \Omega$ is independent of the magnitude of the chosen subdivision interval δE and is of the order of the number of degrees of freedom of the system.

3.6 *Constraints, Equilibrium, and Irreversibility*

Let us now summarize the general point of view which we have developed and which we shall apply repeatedly to discuss any situation involving macroscopic systems. The starting point for all our considerations is an isolated system.† Such a system is known to satisfy certain conditions which can be described on a macroscopic scale by specifying the value of some macroscopic parameter y of the system (or the values of several such parameters). These conditions act as constraints restricting the states in which the system can be found to those compatible with these conditions, i.e., to what we have called the *accessible states* of the system. The number Ω of these accessible states depends thus on the constraints to which the system is known to be subject so that $\Omega = \Omega(y)$ is some function of the specified macroscopic parameter of the system.

† Any system which is not isolated can always be treated as part of a larger system which *is* isolated.

Example

To keep in mind a specific illustration, consider the familiar system shown in Fig. 3.8*a*. This system consists of an ideal gas contained within the volume V_i constituting the left part of a box. The right part of the box is empty. Here the partition subdividing the box into the two parts acts as a constraint which restricts the accessible states of the gas to those where all its molecules are located in the left part of the box. Hence the number Ω of accessible states of the gas depends on the volume V_i of this left part, i.e., $\Omega = \Omega(V_i)$.

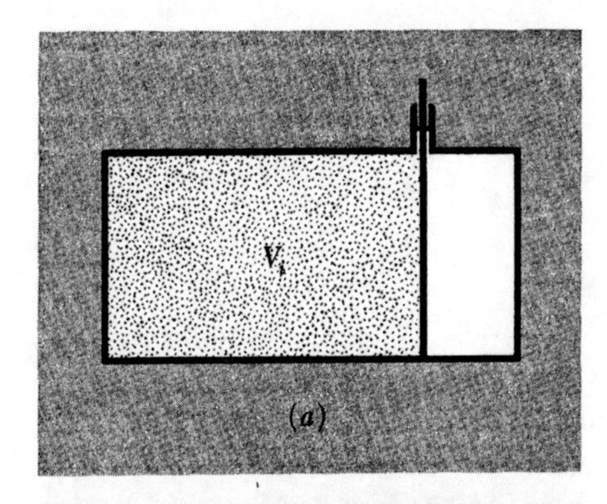

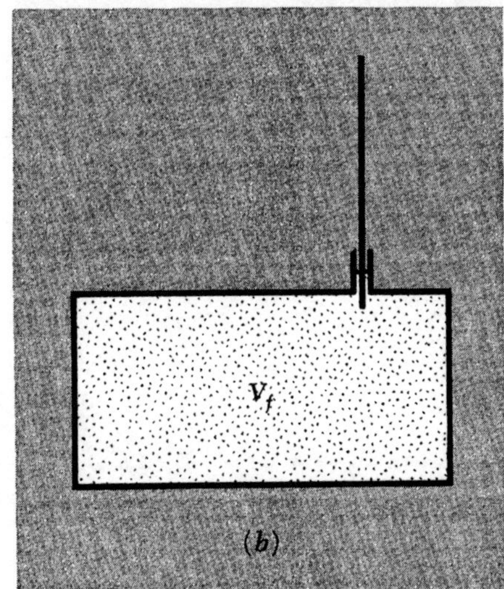

Fig. 3.8 An ideal gas confined within a box. (*a*) Initial situation where the gas is confined by a partition within the left volume V_i of the box. (*b*) Final situation, a long time after the partition is removed, when the gas is confined merely within the entire volume V_f of the box.

The statistical description of the system involves probability statements about an *ensemble* of such systems all of which are subject to the same specified constraints. If the system is in equilibrium, it is found with equal probability in each one of its Ω accessible states, and conversely. If the system is not found with equal probability in each one of its Ω accessible states, then the statistical situation is not independent of time.† The system then tends to change until it ultimately attains the equilibrium situation where it is found with equal probability in each one of its Ω accessible states. The previous statements represent, of course, merely the content of our fundamental postulates (18) and (19).

Consider an isolated system initially in equilibrium under conditions where Ω_i states are accessible to it. The system is then found with equal probability in each one of these states. Suppose that some of the original constraints on the system are now removed. (For instance, in the example of Fig. 3.8, suppose that the partition is now removed.) Since the system is then subject to fewer restrictions than before, the number of states accessible to the system afterward certainly cannot be less than before; indeed, it will ordinarily be very much greater. Denoting by Ω_f this final number of states in the presence of the new constraints, we can thus write

$$\Omega_f \geq \Omega_i. \tag{42}$$

Immediately after the original constraints are removed, the probability of finding the system in any one of its states will be the same as before. Since the system had initially equal probability of being in each one of its Ω_i accessible states, it will thus, immediately after removal of the original constraints, still be found with equal probability in each one of these Ω_i states. Two cases can then arise:

† In other words, in the ensemble of systems, the probability of finding the system in a given state changes with time for at least some of the states.

(i) The special case where $\Omega_f = \Omega_i$
The system is then found with equal probability in each one of the $\Omega_f = \Omega_i$ states which are accessible to it *after* the removal of the original constraints. The equilibrium situation of the system thus remains undisturbed by the removal of the constraints.

(ii) The usual case where $\Omega_f > \Omega_i$
Immediately after removal of the original constraints, the system has equal probability of being found in each one of its original Ω_i states, but has zero probability of being found in any one of the additional $(\Omega_f - \Omega_i)$ states which have now also become accessible to it. This nonuniform probability distribution does *not* correspond to an equilibrium situation. The system tends, therefore, to change in time until it ultimately attains the equilibrium situation where it is found with equal probability in each one of the Ω_f states now accessible to it.

Example

Suppose that the partition in Fig. 3.8 is removed. This leaves the total energy of the gas unchanged, but eliminates the former constraint that prevented molecules from occupying the right part of the box. In the special case where the total volume V_f of the box is such that $V_f = V_i$ (i.e., when the partition is coincident with the right wall of the box), the removal of the partition leaves the volume accessible to the gas unchanged. Thus nothing really happens and the equilibrium of the gas remains undisturbed. In the general case where $V_f > V_i$, removal of the partition causes the volume accessible to each molecule of the ideal gas to be increased by the factor V_f/V_i. Since the number of states accessible to each molecule is, by (32), proportional to the volume of space accessible to it, the number of states accessible to *each* molecule is thus also increased by the factor V_f/V_i. Hence the number of states accessible to *all* the N molecules of the ideal gas is increased by the factor

$$\underbrace{\left(\frac{V_f}{V_i}\right)\left(\frac{V_f}{V_i}\right)\cdots\left(\frac{V_f}{V_i}\right)}_{N \text{ terms}} = \left(\frac{V_f}{V_i}\right)^N.$$

The final number Ω_f of states accessible to the gas after removal of the partition is thus related to the initial number $\Omega_i = \Omega(V_i)$ by

$$\Omega_f = \left(\frac{V_f}{V_i}\right)^N \Omega_i. \tag{43}$$

Even if V_f is barely greater than V_i, $\Omega_f \gg \Omega_i$ if N is a large number of the order of Avogadro's number. *Immediately* after the partition is removed, all the molecules are still in the left part of the box. But this is no longer an equilibrium situation in the absence of the partition. Hence the situation changes with time until the final equilibrium situation is reached where the gas is found with equal probability in each of its Ω_f final accessible states, i.e., where each molecule is equally likely to be found anywhere within the box of volume V_f.

Suppose that the final equilibrium situation has been attained. If $\Omega_f > \Omega_i$, the final distribution of systems in the ensemble is distinctly different from its initial one. Note, in particular, that the initial situation of the ensemble of systems can*not* be restored merely by reimposing the former constraint while keeping the system isolated (i.e., while preventing the system from interacting with any other system with which it can exchange energy).

Of course, if we focus attention on a *single* system of the ensemble, its initial situation might be restored if we waited long enough for the occurrence of the right kind of spontaneous fluctuation. If this fluctuation is such that the system at a particular instant of time is found only among the Ω_i states originally accessible to it, then we could at that instant reimpose the original constraint and thus restore the system to its initial situation. But the probability of occurrence of such a fluctuation is ordinarily extremely small. Indeed, if we consider an ensemble of systems in equilibrium after the constraint has been removed, the probability P_i of encountering in the ensemble a system which is found among only Ω_i of its states is

$$P_i = \frac{\Omega_i}{\Omega_f} \tag{44}$$

since there are Ω_f states accessible to the system. The probability of encountering, among repeated observations of a *single* system, an observation corresponding to the desired fluctuation is then also given by (44). But in the usual case where $\Omega_f \ggg \Omega_i$ [so that the final ensemble of systems differs very substantially from the initial one], (44) shows that spontaneous fluctuations allowing one to restore the initial situation of a single system occur exceedingly rarely.

We shall say that a process is *irreversible* if the initial situation of an *ensemble* of *isolated* systems having undergone this process cannot be restored by simply imposing a constraint. In accordance with this definition, the process in which an isolated system attains a new equilibrium situation after one of its constraints has been removed (thus changing the number of states accessible to it from Ω_i to Ω_f) is irreversible if $\Omega_f > \Omega_i$. Equivalently, our definition implies that a process is irreversible if an isolated system after this process has probability less than unity of being found in its original macrostate; indeed, in the usual case (where $\Omega_f \ggg \Omega_i$) this probability is utterly negligible. Our present definition of irreversibility is thus merely a more precise formulation of that given in Sec. 1.2 in terms of the fluctuations in time of a single isolated system.

We can now also make more quantitative the notion of randomness introduced in Chap. 1. As a statistical measure of the degree of randomness of a system we can take the number of its accessible states actually occupied in an ensemble of such systems. The process of attaining a new equilibrium situation after removing a constraint of an isolated system thus results in increasing the randomness of the system if $\Omega_f > \Omega_i$; the process is correspondingly irreversible.

Example

Once the gas in our previous example has attained its final equilibrium situation so that the molecules are essentially uniformly distributed throughout the entire box, the simple act of replacing the partition does not restore the original situation of an ensemble of such boxes. The molecules now located in the right part of the box still remain there. The process ensuing when the partition is removed is thus irreversible.

To examine possible fluctuations that might restore the original situation in a particular box of gas in the ensemble, let us examine the probability P_i that this box is found with all its molecules on its left side after the final equilibrium situation has been attained. By virtue of (43) and (44), this probability is given by

$$P_i = \frac{\Omega_i}{\Omega_f} = \left(\frac{V_i}{V_f}\right)^N \tag{45}$$

and is thus fantastically small if $V_f > V_i$ and if N is at all large.† The removal of the partition thus illustrates a typical irreversible process in which the randomness of the gas is increased.

The general point of view of this section can be illuminated usefully by two further examples which are prototypes illustrating the interaction between macroscopic systems.

Example (i)

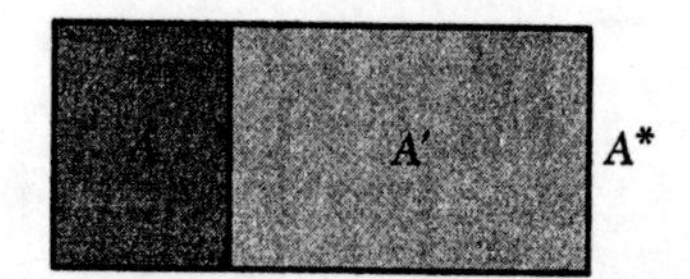

Fig. 3.9 Two systems A and A' with fixed external parameters and free to exchange energy. The combined system A^*, consisting of A and A', is isolated.

Consider an isolated system A^* which consists of two subsystems A and A' with fixed external parameters. (For instance, A and A' might be a piece of copper and a block of ice, respectively.) Suppose that A and A' are separated from each other in space so that they cannot exchange energy with each other. The system A^* is then subject to a constraint requiring that the energy E of A and the energy E' of A' must each separately remain constant. Accordingly, the states accessible to the total system A^* are only those consistent with the conditions that A has some constant specified energy E_i and that A' has some constant specified energy E'_i. If there are Ω^* such states accessible to A^* and if A^* is in equilibrium, then A^* is found with equal probability in each one of these states.

Imagine that the systems A and A' are now placed in contact with each other so that they are free to exchange energy. Then the former constraint has been removed since the energies of A or A' need no longer be separately constant; only the sum $(E + E')$ of their energies, i.e., the *total* energy of the combined system A^*, must remain constant. As a result of removing the constraint, the number of states made accessible to A^* ordinarily becomes much greater, say, equal to Ω_f^*. Therefore, unless $\Omega_f^* = \Omega_i^*$, the system A^* is not in equilibrium immediately after A and A' are placed in contact with each other. The energies of the systems A and A' will, therefore, change (energy in the form of heat passing from one to the

† Note that in the special case where the box is originally divided by the partition into two halves, $V_f = 2V_i$ so that $P_i = 2^{-N}$. This is merely the result (1.1) obtained in Chap. 1 by primitive reasoning.

other) until A^* attains its final equilibrium situation where it is found with equal probability in each one of the Ω_f^* states now accessible to it.

Suppose that the systems A and A' are now separated again from each other so that they can no longer exchange energy. Although the former constraint has thus been restored, the initial situation of A^* has not been restored (unless $\Omega_f^* = \Omega_i^*$). In particular, the mean energies of A and A' in the ensemble are now different from their initial values E_i and E_i'. The preceding process of heat transfer between the systems is thus irreversible.

Example (ii)

Consider an isolated system A^* consisting of two gases A and A' separated by a piston that is clamped in position. The piston then acts as a constraint requiring that the states accessible to A^* be only those consistent with the conditions that the molecules of gas A lie within some fixed volume V_i and that the molecules of gas A' lie within some fixed volume V_i'. If there are Ω_i^* such states accessible to A^* and if A^* is in equilibrium, then A^* is found with equal probability in each one of these states.

Imagine that the piston is now unclamped so that it is free to move. The individual volumes of the gases A and A' then no longer need be separately constant. Consequently, the number of states accessible to A^* becomes ordinarily much greater, say, equal to Ω_f^*. Unless $\Omega_f^* = \Omega_i^*$, the system A^* is thus not in equilibrium after the piston is unclamped. Hence the piston will tend to move and the volumes of A and A' will correspondingly change until A^* attains its final equilibrium situation where it is found with equal probability in each one of the Ω_f^* states now accessible to it. As would be expected (and as will be shown explicitly in Chap. 6), the final volumes of the gases A and A' are then such that their mean pressures are equal, thus guaranteeing that the piston is indeed in mechanical equilibrium when it is in its final position.

The preceding process is again clearly irreversible if $\Omega_f^* > \Omega_i^*$. Leaving A^* isolated while simply clamping the piston again, so that it is not free to move, does not restore the initial volumes of the gases.

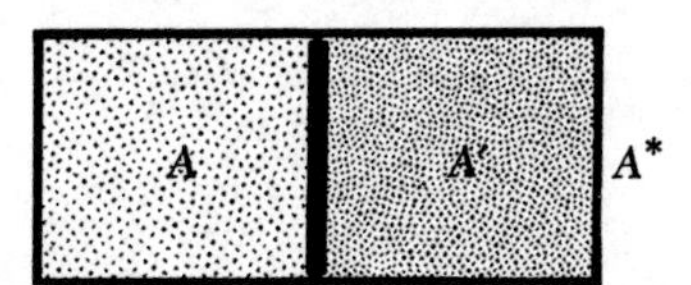

Fig. 3.10 Two gases A and A' separated by a movable piston. The combined system A^*, consisting of A and A', is isolated.

3.7 *Interaction between Systems*

The preceding two examples illustrate specific instances where macroscopic systems interact with each other. Since the study of such interactions is of central importance,† we shall end this chapter by examining explicitly the distinct ways in which macroscopic systems can interact.

Consider two macroscopic systems A and A' which can interact and thus exchange energy with each other. The combined system A^*, consisting of A and A', is then an isolated system whose total energy must remain constant. To describe the interaction between A and A' in statistical terms, we consider an ensemble composed of a very large

† Indeed, the whole discipline of *thermodynamics*, as the name indicates, deals with the macroscopic analysis of thermal and mechanical interactions and the macroscopic consequences that can be derived therefrom.

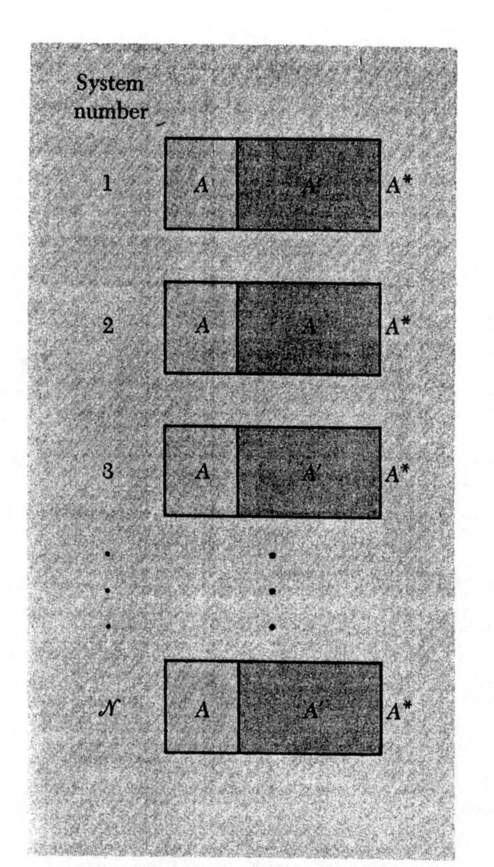

Fig. 3.11 An ensemble of systems A^*, each consisting of two systems A and A' capable of interacting with each other.

number of systems similar to A^*, each consisting of a pair of interacting systems A and A'. The process of interaction between A and A' does not ordinarily result in precisely the same energy transfer between A and A' for each such pair of systems in the ensemble. We can, however, meaningfully discuss the probability that an energy transfer of any specified magnitude results from the interaction process; or, more simply, we can merely ask for the *mean* energy transfer resulting from the interaction process. Let us then, in the ensemble of systems under consideration, denote the initial mean energies of A and A' before the interaction process by $\bar{E}_i$ and $\bar{E}'_i$, respectively; and let us denote the final mean energies of A and A' after the interaction process by $\bar{E}_f$ and $\bar{E}'_f$, respectively. Since the total energy of the isolated system A^* consisting of A and A' remains constant, it follows that

$$\bar{E}_f + \bar{E}'_f = \bar{E}_i + \bar{E}'_i; \tag{46}$$

i.e., the conservation of energy implies simply that

$$\Delta\bar{E} + \Delta\bar{E}' = 0 \tag{47}$$

where $\quad \Delta\bar{E} \equiv \bar{E}_f - \bar{E}_i \quad$ and $\quad \Delta\bar{E}' \equiv \bar{E}'_f - \bar{E}'_i \qquad (48)$

denote the changes of mean energy of each of the two systems A and A'.

We can now refine our discussion of Sec. 1.5 by examining systematically the various ways in which two macroscopic systems A and A' can interact. For this purpose, we shall inquire what happens to the external parameters of the systems during the interaction process.†

Thermal interaction

The interaction between the systems is particularly simple if all their external parameters are kept fixed so that their energy levels remain unchanged. We shall call such an interaction process *thermal interaction.* [Example (i) at the end of the preceding section provides a specific illustration.] The resulting increase (positive or negative) of the mean energy of a system is called the *heat absorbed* by the system and is conventionally denoted by the letter Q. Correspondingly, the decrease (positive or negative) of the mean energy of a system is called

† As mentioned at the beginning of Sec. 3.2, an external parameter of a system is a macroscopic parameter (such as the applied magnetic field **B** or the volume V) which affects the motion of the particles in the system and thus also the energy levels of this system. The energy E_r of each quantum state r of a system thus depends ordinarily on all the external parameters of the system.

the *heat given off* by the system and is given by $-Q$. Thus we can write

$$Q \equiv \Delta \bar{E} \qquad \text{and} \qquad Q' \equiv \Delta \bar{E}' \tag{49}$$

for the heat Q absorbed by system A and the heat Q' absorbed by system A', respectively.† The conservation of energy (47) implies then that

$$Q + Q' = 0, \tag{50}$$

or

$$Q = -Q'.$$

This last relation asserts merely that the heat absorbed by A must be equal to the heat given off by A'. In accord with the definitions already introduced in connection with Eq. (1.15), the system absorbing a positive amount of heat is said to be the *colder* system, while the system absorbing a negative amount of heat (or giving off a positive amount of heat) is said to be the *warmer* or *hotter* system.

The characteristic feature of thermal interaction, where all external parameters are kept fixed, is that the energy levels of the systems remain unchanged while energy on the atomic scale is transferred from one system to the other. The mean energy of one of the systems increases then at the expense of the other *not* because the energies of its possible quantum states have changed, but because the system after the interaction is more likely to be found in those of its states which have higher energy.

Thermal insulation (or adiabatic isolation)

Thermal interaction between two systems can be prevented if they are properly separated from each other. Two systems are said to be *thermally insulated* or *adiabatically isolated* from each other if they cannot exchange energy as long as their external parameters are kept fixed.‡ Thermal insulation can be achieved by separating the systems sufficiently in space, or can be approximated by separating them from each other by a sufficiently thick partition made of a suitable material (such as asbestos or fiberglass). Such a partition is said to be *thermally insulating* or *adiabatic* if any two systems separated by it are found

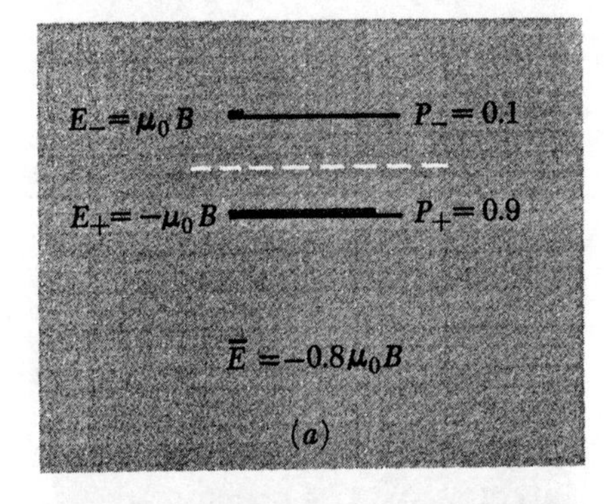

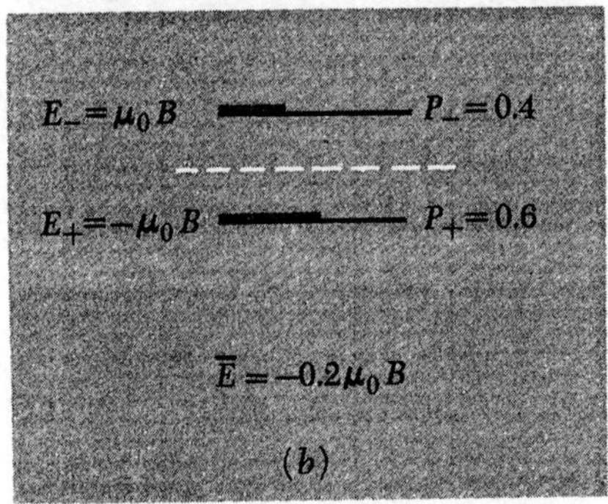

Fig. 3.12 The effect of thermal interaction on a very simple system A consisting of a single spin $\frac{1}{2}$ having a magnetic moment μ_0 and located in a magnetic field B. The diagram shows the two possible energy levels of A. These two quantum states are labeled by $+$ and $-$ and their corresponding energies are denoted by E_+ and E_-. The probability of finding A in a given state is denoted by P_+ and P_-, respectively, and its magnitude is indicated graphically by the length of the gray bar. The energy levels remain unchanged since the applied magnetic field (the one external parameter of this system) is supposed to be fixed. The initial equilibrium situation (a) is one where the spin is embedded in some solid. The solid, with its included spin, is then immersed in some liquid and one waits until the final equilibrium situation (b) is attained. In this process the spin system A absorbs heat from the system A' consisting of the solid and liquid. If the probabilities change as indicated in the diagram, the heat Q absorbed by A is equal to $Q = 0.6\mu_0 B$.

† Note that, since our present discussion has been more careful in its use of statistical concepts than that of Sec. 1.5, we have now defined heat in terms of the change of *mean* energy of a system.

‡ The word *adiabatic*, meaning "heat is not able to go through," comes from the Greek word *adiabatikos* (*a*, not + *dia*, through + *bainein*, to go). We shall always use the word in this sense, although it is sometimes also used in physics to denote a different concept.

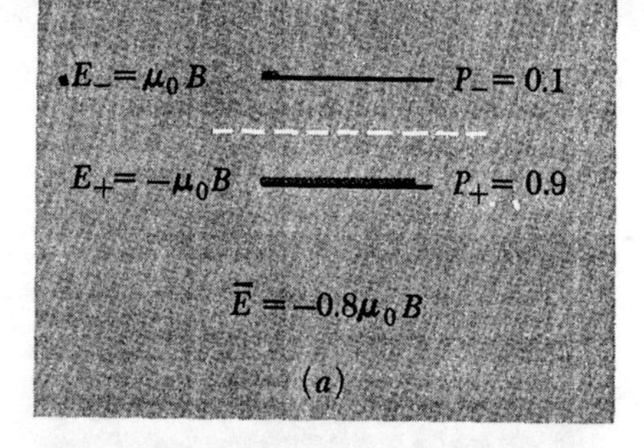

Fig. 3.13 The effect of adiabatic interaction on a very simple system A consisting of a single spin $\frac{1}{2}$ having a magnetic moment μ_0 and located in a magnetic field B. The initial situation and notation is the same as in Fig. 3.12, but the spin is then adiabatically isolated. Suppose that the magnetic field B is changed by means of an electromagnet. The amount of work done depends in general on just how the process is carried out. Panel (*b*) shows the final equilibrium situation if the magnetic field is changed from B to B_1 very slowly; the work W done on A is then $W = -0.8\mu_0(B_1 - B)$. Panel (*c*) shows a final equilibrium situation which might result if the magnetic field is changed from B to B_1 in some arbitrary way; the work done in the particular case shown here is $W = -0.4\mu_0 B_1 + 0.8\mu_0 B$.

to be thermally insulated from each other; i.e., if any such systems, initially in equilibrium, are found to remain in equilibrium as long as their external parameters are kept fixed.† A process which occurs while a system is thermally insulated from all other systems is called an *adiabatic process* of this system.

Adiabatic interaction

When two systems A and A' are thermally insulated from each other, they can still interact and thus exchange energy with each other, provided that at least some of their external parameters change in the process. We shall call such an interaction process *adiabatic interaction*. [Example (ii) at the end of Sec. 3.6 provides a specific illustration, if the piston is made of thermally insulating material.] The increase (positive or negative) of the mean energy of an adiabatically isolated system is called the *macroscopic work done on the system*‡ and will be denoted by W. Correspondingly, the decrease (positive or negative) of the mean energy of a system is called the *macroscopic work done by the system* and is given by $-W$. Hence we can write

$$W = \Delta\bar{E} \quad \text{and} \quad W' = \Delta\bar{E}' \tag{51}$$

for the work done on system A and the work done on system A', respectively. If the total system $A + A'$ is isolated, the conservation of energy (47) implies then that

$$W + W' = 0 \tag{52}$$

or

$$W = -W'.$$

This last relation asserts merely that the work done on one system must be equal to the work done by the other system.

Since adiabatic interaction involves changes in some of the external parameters of the systems, at least some of the energy levels of these systems are changed in the process. The mean energy of one of these interacting systems is thus ordinarily changed both because the energy of each of its states changes and because the probability that the system is found in any such state also changes.§

---

† If a partition is *not* thermally insulating, it is said to be *thermally conducting*.

‡ The macroscopic work, defined in terms of a *mean* energy difference, is a statistical quantity equal to the mean value of the work done on each system in the ensemble. Where no confusion is likely to arise, we shall henceforth simply use the term *work* to denote the macroscopic work thus defined.

§ It is worth noting a special case. If a system is in an exact quantum state whose energy depends on an external parameter, then the system will simply remain in this state and have its energy changed accordingly if the external parameter is changed *sufficiently slowly*.

General interaction

In the most general case, interacting systems are neither adiabatically isolated, nor are their external parameters kept fixed. It is then useful to write the total mean energy change of an interacting system, such as A, in the additive form

$$\boxed{\Delta\bar{E} = W + Q} \tag{53}$$

where W represents the change of mean energy of A caused by changes of external parameters and Q represents the change of mean energy which is not caused by changes of external parameters. The decomposition (53) of $\Delta\bar{E}$ into work W done on the system and heat Q absorbed by the system is meaningful when these contributions can be experimentally separated. Thus suppose that the system A interacts simultaneously with one system A'_1 which is separated from A by a thermally insulating partition, and another system A'_2 whose external parameters are kept fixed. Then the work W in (53) must simply be equal to the work done by (or decrease in mean energy of) the system A'_1 which is adiabatically isolated; similarly, the heat Q in (53) must simply be equal to the heat given off by (or decrease in mean energy of) the system A'_2 whose external parameters are kept fixed.

The general relation (53) is, for historical reasons, called the *first law of thermodynamics*. It explicitly recognizes work and heat as forms of energy transferred in different ways. Since both work and heat represent energies, these quantities are, of course, measured in units of energy, i.e., typically in terms of ergs or joules.†

Infinitesimal general interaction

An interaction process is particularly simple if it is infinitesimal in the sense that it takes a system from an initial macrostate to a final macrostate which differs from the initial macrostate only infinitesimally. The energy and external parameters of the system in the final macrostate differ then only very slightly from their values in the initial macrostate. Correspondingly, the infinitesimal increase in mean energy of the system can be written as the differential $d\bar{E}$. In addition, we shall use the symbol $đW$, instead of W, to denote the infinitesimal amount of work done on the system in the process; this is merely a convenient notation designed to emphasize that the work is infinitesimally small. It is important to point out that $đW$ does *not* designate any difference

† In the older literature of physics, and much of the present-day literature of chemistry, one still finds heat measured in terms of an old unit, the *calorie*, introduced in the eighteenth century before it was realized that heat is a form of energy. The calorie is now *defined* so that 1 calorie $\equiv$ 4.184 joules *exactly*.

Fig. 3.14 Count Rumford (born Benjamin Thompson, 1753–1814). Born in Massachusetts and very much of an adventurer, he had Royalist sympathies during the American Revolution and left America to serve for a time as minister of war to the Elector of Bavaria. There his attention was attracted by the temperature rise produced by the boring of cannons, an observation which led him to suggest in 1798 that heat is merely a form of motion of the particles in a body. Although this was a valuable insight, it was too qualitative to have much impact on contemporary thinking which conceived of heat as a substance ("phlogiston") that is conserved. (*From a portrait in the Fogg Art Museum, painted in 1783 by T. Gainsborough, reproduced by courtesy of Harvard University.*)

between works. Indeed, that would be a meaningless statement. The work done is a quantity referring to the interaction *process* itself; one *cannot* talk of the work in the system before and after the process, or of the difference between these. Similar comments apply to $đQ$ which also denotes merely the infinitesimal amount of heat absorbed in a process, *not* any meaningless difference betwen heats. With this notation, the relation (53) for an infinitesimal process can be written in the form

$$d\bar{E} = đW + đQ. \tag{54}$$

Remark

An infinitesimal process can be discussed particularly simply in statistical terms if it is carried out quasi-statically, i.e., so slowly that the system remains always very close to equilibrium. Let P_r denote the probability that the system A is in a state r having an energy E_r. The mean energy of this system is then, by definition,

$$\bar{E} = \sum_r P_r E_r \tag{55}$$

where the sum extends over all possible states r of the system. In an infinitesimal process the energies E_r change only by small amounts as a result of changes in the external parameters; furthermore, if the process is carried out very slowly, the probabilities P_r also change at most by small amounts. Hence the mean energy change in the process can be written as the differential

$$d\bar{E} = \sum_r (P_r\, dE_r + E_r\, dP_r). \tag{56}$$

The heat absorbed corresponds to the increase of mean energy resulting when the external parameters are kept fixed, i.e., when the energy levels E_r are kept fixed so that $dE_r = 0$. Thus one can write

$$đQ = \sum_r E_r\, dP_r. \tag{57}$$

Hence the work done on the system is given by

$$đW = d\bar{E} - đQ = \sum_r P_r\, dE_r. \tag{58}$$

This infinitesimal work is thus merely the mean energy change resulting from the shift of the energy levels produced by the infinitesimal change of external parameters, the probabilities P_r retaining their initial values appropriate to the equilibrium situation.

Summary of Definitions

(Some of these definitions are more precise versions of definitions already encountered in previous chapters.)

microstate (or simply state) A particular quantum state of a system. It corresponds to the most detailed possible specification of a system described by quantum mechanics.

macrostate (or macroscopic state) A complete specification of a system in terms of macroscopically measurable parameters.

accessible state Any microstate in which a system can be found without contradicting the macroscopic information available about the system.

number of degrees of freedom The number of distinct quantum numbers necessary to describe completely the microstate of a system. It is equal to the number of independent coordinates (including spin coordinates) of all the particles in the system.

external parameter A macroscopically measurable parameter whose value affects the motion of the particles in a system and thus the energies of the possible quantum states of the system.

isolated system A system which does not interact with any other system so as to exchange energy with it.

total energy of a system The sum of the kinetic and potential energies of all the particles in the system.

internal energy of a system The total energy of a system measured in the frame of reference where its center of mass is at rest.

equilibrium An isolated system is said to be in equilibrium if the probability of finding the system in any one of its accessible states is independent of time. (The mean values of all macroscopic parameters of the system are then independent of time.)

constraint A macroscopic condition to which a system is known to be subject.

irreversible process A process which is such that the initial situation of an *ensemble* of *isolated* systems subjected to this process *cannot* be restored by simply imposing a constraint.

reversible process A process which is such that the initial situation of an *ensemble* of *isolated* systems subjected to this process *can* be restored by simply imposing a constraint.

thermal interaction Interaction in which the external parameters (and thus also the energy levels) of the interacting systems remain unchanged.

adiabatic isolation (or thermal insulation) A system is said to be adiabatically isolated (or thermally insulated) if it cannot interact thermally with any other system.

adiabatic interaction Interaction in which the interacting systems are adiabatically isolated. In this case the interaction process involves changes in some of the external parameters of the systems.

heat absorbed by a system Increase in mean energy of a system whose external parameters are kept fixed.

work done on a system Increase in mean energy of a system which is adiabatically isolated.

cold A comparative term applied to the system which absorbs positive heat as a result of thermal interaction with another system.

warm (or hot) A comparative term applied to the system which gives off positive heat as a result of thermal interaction with another system.

Fig. 3.15 Julius Robert Mayer (1814–1878). A German physician, Mayer suggested in 1842 the equivalence and conservation of all forms of energy, including heat. Although he made some quantitative estimates, his writing was too philosophical to be convincing and his work remained unrecognized for about twenty years. *(From G. Holton and D. Roller, "Foundations of Modern Physical Science," Addison-Wesley Publishing Co., Inc., Cambridge, Mass., 1958. By permission of the publishers.)*

Important Relations

Relation between mean energy, work, and heat:

$$\Delta \bar{E} = W + Q \tag{i}$$

Suggestions for Supplementary Reading

Completely macroscopic discussions of heat, work, and energy:

M. W. Zemansky, *Heat and Thermodynamics*, 4th ed., secs. 3.1–3.5, 4.1–4.6 (McGraw-Hill Book Company, New York, 1957).

H. B. Callen, *Thermodynamics*, secs. 1.1–1.7 (John Wiley & Sons, Inc., New York, 1960).

Historical and biographical accounts:

G. Holton and D. Roller, *Foundations of Modern Physical Science* (Addison-Wesley Publishing Company, Inc., Reading, Mass., 1958). Chapters 19 and 20 contain a historical account of the development of the ideas leading to the recognition of heat as a form of energy.

S. G. Brush, *Kinetic Theory*, vol. I (Pergamon Press, Oxford, 1965). The historical introduction by the author and the reprints of original papers by Mayer and Joule may be of particular interest.

S. B. Brown, *Count Rumford, Physicist Extraordinary* (Anchor Books, Doubleday & Company, Inc., Garden City, N.Y., 1962). A short biography of Count Rumford.

Problems

3.1 *Simple example of thermal interaction*

Consider the system of spins described in Table 3.3. Suppose that, when the systems A and A' are initially separated from each other, measurements show the total magnetic moment of A to be $-3\mu_0$ and the total magnetic moment of A' to be $+4\mu_0$. The systems are now placed in thermal contact with each other and are allowed to exchange energy until the final equilibrium situation has been reached. Under these conditions calculate:

(*a*) The probability $P(M)$ that the total magnetic moment of A assumes any one of its possible values M.

(*b*) The mean value $\bar{M}$ of the total magnetic moment of A.

(*c*) Suppose that the systems are now again separated so that they are no longer free to exchange energy with each other. What are the values of $P(M)$ and $\bar{M}$ of the system A after this separation?

3.2 *One spin in thermal contact with a small spin system*

Consider a system A consisting of a spin $\frac{1}{2}$ having magnetic moment μ_0, and another system A' consisting of 3 spins $\frac{1}{2}$ each having magnetic moment μ_0. Both systems are located in the same magnetic field **B**. The systems are placed in contact with each other so that they are free to exchange energy. Suppose that, when the moment of A points up (i.e., when A is in its $+$ state), two of the moments of A' point up and one of them points down. Count the total number of states accessible to the combined system $A + A'$ when the moment of A points up, and when it points down. Hence calculate the ratio P_-/P_+, where P_- is the probability that the moment of A points down and P_+ is the probability that it points up. Assume that the total system $A + A'$ is isolated.

3.3 *One spin in thermal contact with a large spin system*

Generalize the preceding problem by considering the case where the system A' consists of some arbitrarily large number N of spins $\frac{1}{2}$, each having magnetic moment μ_0. The system A consists again of a single spin $\frac{1}{2}$ with magnetic moment μ_0. Both A and A' are located in the same magnetic field **B** and are placed in contact with each other so that they are free to exchange energy. When the moment of A points up, n of the moments of A' point up and the remaining $n' = N - n$ of the moments of A' point down.

(*a*) When the moment of A points up, find the number of states accessible to the combined system $A + A'$. This is, of course, just the number of ways in which the N spins of A' can be arranged so that n of them point up and n' of them point down.

(*b*) Suppose now that the moment of A points down. The total energy of the combined system $A + A'$ must, of course, remain unchanged. How many of the moments of A' now point up, and how many of them point down? Correspondingly, find the number of states accessible to the combined system $A + A'$.

(*c*) Calculate the ratio P_-/P_+, where P_- is the probability that the moment of A points down and P_+ is the probability that it points up. Simplify your result by using the fact that $n \gg 1$ and $n' \gg 1$. Is the ratio P_-/P_+ larger or smaller than unity if $n > n'$?

3.4 *Generalization of the preceding problem*

Suppose that in the preceding problem the magnetic moment of A had the value $2\mu_0$. Calculate again the ratio P_-/P_+ of the probabilities that this moment points down or up.

3.5 *Arbitrary system in thermal contact with a large spin system*

The considerations of the preceding problems can readily be extended to treat the following general case. Consider any system A whatsoever, which might be a single atom or a macroscopic system. Suppose that this system A is placed in thermal contact with a system A' with which it is free to exchange energy. The system A' is supposed to be located in a magnetic field $\mathbf{B}$ and to consist of N spins $\frac{1}{2}$, each having a magnetic moment μ_0. The number N is assumed to be very large compared to the number of degrees of freedom of the relatively much smaller system A. When the system A is in its lowest state of energy E_0, we suppose that n of the moments of A' point up and the remaining $n' = N - n$ moments of A' point down. Here $n \gg 1$ and $n' \gg 1$, since all numbers are very large.

(*a*) When the system A is in its lowest possible state of energy E_0, find the total number of states accessible to the combined system $A + A'$.

(*b*) Suppose now that the system A is in some other state, call it r, where it has an energy E_r higher than E_0. In order to conserve the total energy of the combined system $A + A'$, there must then be $(n + \Delta n)$ moments of A' which point up, and $(n - \Delta n)$ moments of A' which point down. Express Δn in terms of the energy difference $(E_r - E_0)$. You can assume that $(E_r - E_0) \gg \mu_0 B$.

(*c*) When the system A is in the state r with energy E_r, find the total number of states accessible to the combined system $A + A'$.

(*d*) Let P_0 denote the probability that the system A is in the state of energy E_0, and P_r the probability that it is in the state r of energy E_r. Find the ratio P_r/P_0. Use the approximation that $\Delta n \ll n$ and $\Delta n \ll n'$.

(*e*) Using the results just derived, show that the probability P_r of finding the system A in any state r having an energy E_r is of the form

$$P_r = Ce^{-\beta E_r}$$

where C is a constant of proportionality. Express β in terms of $\mu_0 B$ and the ratio n/n'.

(*f*) If $n > n'$, is β positive or negative? Suppose that the system A is such that its states, labeled by a quantum number r, are equally separated in energy by an amount b. (For example, A might be a simple harmonic oscillator.) Thus $\epsilon_r = a + br$, where $r = 0, 1, 2, 3 \ldots$ and a is some constant. Compare the probability of finding the system A in any one of these states with the probability that it is found in its lowest state $r = 0$.

3.6 *Pressure exerted by an ideal gas (quantum mechanical calculation)*

Consider a single particle, of mass m, confined within a box of edge lengths L_x, L_y, L_z. Suppose that this particle is in a particular quantum state r specified by particular values of the three quantum numbers n_x, n_y, n_z. The energy E_r of this state is then given by (15).

When the particle is in the particular state r, it exerts on the right wall of the box (i.e., the wall $x = L_x$) some force F_r in the x direction. This wall must then exert on the particle a force $-F_r$ (i.e., in the $-x$ direction). If the right wall of the box is slowly moved to the right by an amount dL_x, the work performed on the particle in this state is thus $-F_r\,dL_x$ and must be equal to the increase in energy dE_r of the particle in this state. Thus one has

$$dE_r = -F_r\,dL_x. \tag{i}$$

The force F_r exerted by a particle in the state r is thus related to the energy E_r of the particle in this state by

$$F_r = -\frac{\partial E_r}{\partial L_x}. \tag{ii}$$

Here we have written a partial derivative since the dimensions L_y and L_z are supposed to remain constant in deriving the expression (ii).

(*a*) Using (ii) and the expression (15) for the energy, calculate the force F exerted by the particle on the right wall when the particle is in a state specified by given values of n_x, n_y, and n_z.

(*b*) Suppose that the particle is not isolated, but is one of the many particles which constitute a gas confined within the container. The particle, being able to interact weakly with the other particles, can then be in any one of many possible states characterized by different values of n_x, n_y, and n_z. Express the mean force $\bar{F}$ exerted by the particle in terms of $\overline{n_x^2}$. For simplicity, assume that the box is cubic so that $L_x = L_y = L_z = L$; the symmetry of the situation then implies that $\overline{n_x^2} = \overline{n_y^2} = \overline{n_z^2}$. Use this result to relate $\bar{F}$ to the mean energy $\bar{E}$ of the particle.

(*c*) If there are N similar particles in the gas, the mean force exerted by all of them is simply $N\bar{F}$. Hence show that the mean pressure $\bar{p}$ of the gas (i.e., the mean force exerted by the gas per unit area of the wall) is simply given by

$$\bar{p} = \frac{2}{3}\frac{N}{V}\bar{E} \tag{iii}$$

where $\bar{E}$ is the mean energy of *one* particle in the gas.

(*d*) Note that the result (iii) agrees with that derived in (1.21) on the basis of approximate arguments using classical mechanics.

3.7 *Typical number of states accessible to a gas molecule*

The result (iii) of the preceding problem, or Eq. (1.21), permits one to estimate the mean energy of a gas molecule, such as nitrogen (N_2), at room temperature. Using the density and the known pressure exerted by such a gas, the mean energy $\bar{E}$ of such a molecule was found in (1.28) to be about 6×10^{-14} erg.

(*a*) Use (31) to calculate numerically the number of states $\Phi(\bar{E})$, with energy less than $\bar{E}$, accessible to such a molecule enclosed in a box having a volume of one liter (10^3 cm^3).

(*b*) Consider a small energy interval $\delta E = 10^{-24}$ erg, which is very much smaller than $\bar{E}$ itself. Calculate the number of states $\Omega(\bar{E})$ accessible to the molecule in the range between $\bar{E}$ and $\bar{E} + \delta E$.

(*c*) Show that the preceding number of states is very large, despite the small magnitude of the energy interval δE.

3.8 ***Number of states of an ideal gas***

Consider an ideal gas consisting of N particles confined within a box with edge lengths L_x, L_y, and L_z. Here N is supposed to be of the order of Avogadro's number. By considering the energy contribution from each quantum number separately and using approximate arguments similar to those used in Sec. 3.5, show that the number of states $\Omega(E)$ in a given energy interval between E and $E + \delta E$ is given by

$$\Omega(E) = CV^N E^{(3/2)N}\,\delta E$$

where C is a constant of proportionality and $V = L_x L_y L_z$ is the volume of the box.

*** 3.9** ***Number of states of a spin system***

A system consists of N spins $\frac{1}{2}$, each having magnetic moment μ_0, and is located in an applied magnetic field **B**. The system is of macroscopic size so that N is of the order of Avogadro's number. The energy of the system is then equal to

$$E = -(n - n')\mu_0 B$$

if n denotes the number of its magnetic moments which point up, and $n' = N - n$ the number which point down.

(*a*) Calculate for this spin system the number of states $\Omega(E)$ which lie in a small energy interval between E and $E + \delta E$. Here δE is understood to be large compared to individual spin energies, i.e., $\delta E \gg \mu_0 B$.

(*b*) Find an explicit expression for $\ln \Omega$ as a function of E. Since both n and n' are very large, apply the result $\ln n! \approx n \ln n - n$ derived in (M.10) to calculate both $n!$ and $n'!$. Show thus that, to excellent approximation,

$$\ln \Omega(E) = N \ln (2N) - \tfrac{1}{2}(N - E') \ln (N - E') - \tfrac{1}{2}(N + E') \ln (N + E')$$

where
$$E' \equiv \frac{E}{\mu_0 B}.$$

(*c*) Make a rough sketch showing the behavior of $\ln \Omega$ as a function of E. Note that $\Omega(E)$ does not always increase as a function of E. The reason is that a system of spins is anomalous in that it has not only a lowest possible energy $E = -N\mu_0 B$, but also a highest possible energy $E = N\mu_0 B$. On the other hand, in all ordinary systems where one does not ignore the kinetic energy of the particles (as we did in discussing the spins), there is no upper bound on the magnitude of the kinetic energy of the system.

Chapter 4

Thermal Interaction

Chapter 4 *Thermal Interaction*

The preceding chapter has provided us with all the postulates and essential theoretical framework necessary for a quantitative discussion of macroscopic systems. We are, therefore, ready to explore the power of our concepts by applying them to some problems of major physical importance.

We shall begin by examining in detail the thermal interaction between systems. The analysis of this situation is particularly simple since the external parameters, and hence also energy levels, of the systems remain fixed. In addition, thermal interaction is one of the processes occurring most commonly in the world around us. The particular questions which we shall want to investigate are the following: What conditions must be satisfied so that two systems in thermal interaction with each other are in equilibrium? What happens when these conditions are not satisfied? What probability statements can be made? We shall see that these questions can be answered quite simply to yield results of remarkable generality and great utility. Indeed, in this chapter we shall attain a good understanding of the concept of temperature and a precise definition of the "absolute temperature." Furthermore, we shall obtain some eminently practical methods for calculating the properties of any macroscopic system in equilibrium on the basis of our knowledge about the atoms or molecules of which it is composed. Finally, we shall apply these methods to deduce explicitly the macroscopic properties of some specific systems.

4.1 Distribution of Energy between Macroscopic Systems

Consider two macroscopic systems A and A'. We shall denote the energies of these systems by E and E', respectively. To facilitate the counting of states, we proceed as in Sec. 3.5 and imagine the energy scales to be subdivided into very small equal intervals of fixed size δE. (The magnitude of δE is, however, supposed to be large enough to contain many states.) We shall then denote by $\Omega(E)$ the number of states accessible to A when its energy lies between E and $E + \delta E$, and by $\Omega'(E')$ the number of states accessible to A' when its energy lies between E' and $E' + \delta E$. The counting problem can be simplified since we can, to excellent approximation, treat all energies as if they could assume only discrete values separated by the small amount δE. In particular, we can lump together all states of A which have an energy in the small interval between E and $E + \delta E$ and treat them as if they simply had an energy equal to E; there are thus $\Omega(E)$ such states. Similarly, we can lump together all states of A' which have an energy in the small interval between E' and $E' + \delta E$ and treat them as if they simply had an energy equal to E'; there are thus $\Omega'(E')$ such states. If we adopt this procedure, the statement that A has an energy E means physically that A has an energy anywhere between E and $E + \delta E$. Similarly, the statement that A' has an energy E' means physically that A' has an energy anywhere between E' and $E' + \delta E$.

The systems A and A' have fixed external parameters, but are assumed to be free to exchange energy. (Any energy transfer between them is thus, by definition, in the form of heat.) Although the energy of each system separately is then not constant, the combined system $A^* \equiv A + A'$ is isolated so that its total energy E^* must remain unchanged. Hence†

$$E + E' = E^* = \text{constant.} \tag{1}$$

When the energy of A is known to be equal to E, the energy of A' is then determined to be

$$E' = E^* - E. \tag{2}$$

Now consider the situation where A and A' are in equilibrium with each other, i.e., where the combined system A^* is in equilibrium. The energy of A can then assume many possible values. The question of interest is, however, the following: What is the probability $P(E)$ that the energy of A is equal to E (i.e., that it lies in the interval between E and $E + \delta E$) where E has any specified value? [The energy of A', of course, then has a corresponding value E' given by (2).] The answer to this question is readily obtained by focusing attention on the combined *isolated* system A^*, since the basic postulate (3.19) asserts that such a system is equally likely to be found in each one of its accessible states. We have merely to pose the following question: Among the total number Ω^*_{tot} of states accessible to A^*, what is the number $\Omega^*(E)$ of states of A^* which are such that the subsystem A has an energy

† In our discussion E denotes the energy of A irrespective of A', and E' denotes the energy of A' irrespective of A. The total energy E^* written as the simple sum (1) neglects, therefore, any interaction energy E_i which depends on *both* A and A', i.e., any work necessary to bring the systems together. By definition, thermal interaction is supposed to be sufficiently weak so that E_i is negligible, i.e., so that $E_i \ll E$ and $E_i \ll E'$.

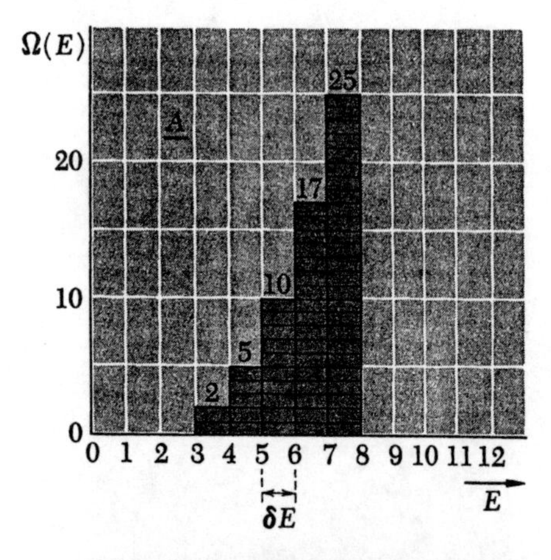

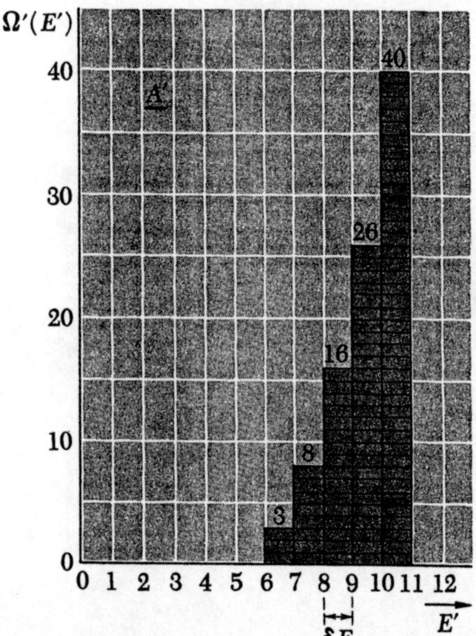

Fig. 4.1 Graphs showing, in the case of two special and very small systems A and A', the number of states $\Omega(E)$ accessible to A and the number of states $\Omega'(E')$ accessible to A' as a function of their respective energies E and E'. The energies are measured in terms of an arbitrary unit; only a few values of $\Omega(E)$ and $\Omega'(E')$ are shown.

equal to E? By the general argument leading to (3.20), the desired probability $P(E)$ then is simply given by

$$P(E) = \frac{\Omega^*(E)}{\Omega^*_{tot}} = C\,\Omega^*(E) \tag{3}$$

where $C = (\Omega^*_{tot})^{-1}$ is merely some constant independent of E.

The number $\Omega^*(E)$ can be readily expressed in terms of the numbers of states accessible to A and A', respectively. When A has an energy E, it can be in any one of its $\Omega(E)$ possible states. The system A' then must have an energy E' given by (2) by virtue of the conservation of energy. Hence A' can be in any one of the $\Omega'(E') = \Omega'(E^* - E)$ states accessible to it under these conditions. Since every possible state of A can be combined with every possible state of A' to give a different possible state of the total system A^*, it follows that the number of distinct states accessible to A^* when A has an energy E is simply given by the product

$$\Omega^*(E) = \Omega(E)\,\Omega'(E^* - E). \tag{4}$$

Correspondingly the probability (3) that the system A has an energy E is simply given by

$$\boxed{P(E) = C\,\Omega(E)\,\Omega'(E^* - E).} \tag{5}$$

Example

The following simple example uses very small numbers unrepresentative of real macroscopic systems, but serves to illustrate the essential ideas of the preceding paragraphs. Consider two special systems A and A' for which $\Omega(E)$ and $\Omega'(E')$ depend on their respective energies E and E' in the manner shown in Fig. 4.1. Here the energies E and E' are measured in terms of some arbitrary unit and are subdivided into unit energy intervals. Suppose that the combined energy E^* of both systems is equal to 13 units. A possible situation would be one where $E = 3$; it then follows that $E' = 10$. In this case A can be in either one of its 2 possible states, and A' in any one of its 40 possible states. There are then a total of $\Omega^* = 2 \times 40 = 80$ different possible states accessible to the combined system A^*. Table 4.1 enumerates systematically the possible situations compatible with the specified total energy E^*. Note that it would be most probable in a statistical ensemble of such systems to find the combined system A^* in a state where $E = 5$ and $E' = 8$. This situation would be likely to occur twice as frequently as the situation where $E = 3$ and $E' = 10$.

Let us now investigate the dependence of $P(E)$ on the energy E. Since A and A' are both systems with very many degrees of freedom, we know by (3.37) that $\Omega(E)$ and $\Omega'(E)$ are extremely rapidly increasing functions of E and E', respectively. Considering the expression (5) as a function of increasing energy E, it thus follows that the factor $\Omega(E)$ *increases* extremely rapidly while the factor $\Omega'(E^* - E)$ *decreases* extremely rapidly. The result is that the product of these two factors, i.e., the probability $P(E)$, exhibits a very sharp maximum† for some particular value $\tilde{E}$ of the energy E. Thus the dependence of $P(E)$ on E must show the general behavior illustrated in Fig. 4.2 where the width ΔE of the region where $P(E)$ has appreciable magnitude is such that $\Delta E \lll \tilde{E}$.

It is actually more convenient to investigate the behavior of $\ln P(E)$ rather than that of $P(E)$ itself since the logarithm is a much more slowly varying function of E. In addition, it follows by (5) that this logarithm involves the numbers Ω and Ω' as a simple sum rather than as a product, i.e.,

$$\ln P(E) = \ln C + \ln \Omega(E) + \ln \Omega'(E') \tag{6}$$

where $E' = E^* - E$. The value $E = \tilde{E}$ which corresponds to the maximum of $\ln P(E)$ is determined by the condition‡

$$\frac{\partial \ln P}{\partial E} = \frac{1}{P}\frac{\partial P}{\partial E} = 0 \tag{7}$$

and thus corresponds also to the maximum of $P(E)$ itself. By using (6) and (2), the condition (7) becomes simply

$$\frac{\partial \ln \Omega(E)}{\partial E} + \frac{\partial \ln \Omega'(E')}{\partial E'}(-1) = 0$$

or

$$\boxed{\beta(E) = \beta'(E')} \tag{8}$$

where we have introduced the definition

$$\boxed{\beta(E) \equiv \frac{\partial \ln \Omega}{\partial E} = \frac{1}{\Omega}\frac{\partial \Omega}{\partial E}} \tag{9}$$

and the corresponding definition for $\beta'(E')$. The relation (8) is thus the fundamental condition which determines the particular value $\tilde{E}$ of the energy of A (and corresponding value $\tilde{E}' \equiv E^* - \tilde{E}$ of the energy of A') which occurs with the greatest probability $P(E)$.

† Note that the behavior of $P(E)$ is analogous to that of the preceding simple example, except for the fact that the maximum of $P(E)$ is *enormously* sharper for macroscopic systems where $\Omega(E)$ and $\Omega'(E')$ are such rapidly varying functions.

‡ We write this as a partial derivative to emphasize that all external parameters of the system are supposed to remain fixed throughout the discussion.

E	E'	$\Omega(E)$	$\Omega'(E')$	$\Omega^*(E)$
3	10	2	40	80
4	9	5	26	130
5	8	10	16	160
6	7	17	8	136
7	6	25	3	75

Table 4.1 Enumeration of the possible numbers of states compatible with a specified total energy $E^* = 13$ of the systems A and A' described in Fig. 4.1.

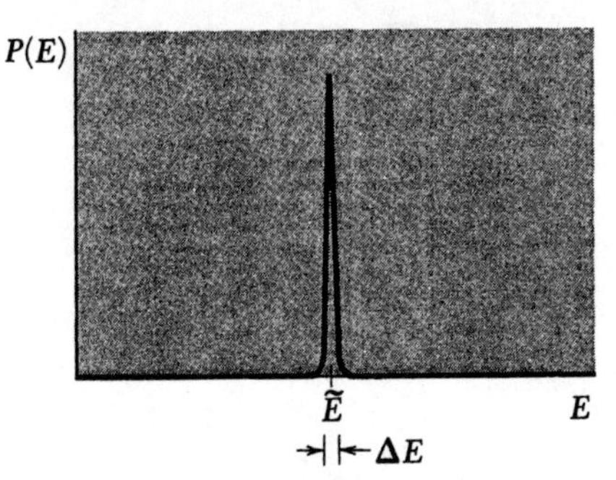

Fig. 4.2 Schematic illustration showing the dependence of the probability $P(E)$ on the energy E.

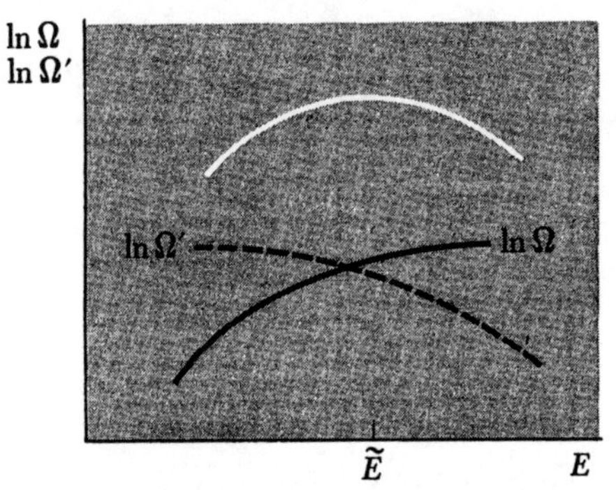

Fig. 4.3 Schematic sketch of the dependence of $\ln \Omega(E)$ and of $\ln \Omega'(E') \equiv \ln \Omega'(E^* - E)$ on the energy E. By virtue of (3.38), the energy dependence of $\ln \Omega$ is roughly of the form $\ln \Omega(E) \sim f \ln (E - E_0) +$ constant. Since the curves are thus concave downward, their sum (shown in white) exhibits a unique maximum at some value $\tilde{E}$. This gentle maximum of the slowly varying logarithm of $P(E)$, given by (6), corresponds then to an extremely sharp maximum of $P(E)$ itself.

Sharpness of the maximum of $P(E)$

By examining the behavior of $\ln P(E)$ near its maximum, one can readily estimate how rapidly $P(E)$ decreases when E becomes different from $\tilde{E}$. Indeed, it is shown in Appendix A.3 that $P(E)$ becomes negligibly small compared to its maximum value when E differs from $\tilde{E}$ by appreciably more than an amount ΔE whose magnitude is roughly of the order

$$\Delta E \sim \frac{\tilde{E}}{\sqrt{f}}. \tag{10}$$

Here f is the number of degrees of freedom of the smaller of the two interacting systems and $\tilde{E}$ is assumed to be much greater than the lowest (or ground state) energy of A. For a typical system consisting of a mole of atoms, f is of the order of Avogadro's number so that $f \sim 10^{24}$ and

$$\Delta E \sim 10^{-12}\tilde{E}. \tag{11}$$

Hence the probability $P(E)$ has ordinarily an exceedingly sharp maximum at the value $\tilde{E}$, becoming negligibly small when E differs from $\tilde{E}$ by as little as a few parts in 10^{12}. Thus the energy of A practically never differs appreciably from $\tilde{E}$; in particular, the mean value $\bar{E}$ of A must then also be equal to $\tilde{E}$, i.e., $\bar{E} = \tilde{E}$. Here we encounter another instance where the relative magnitude of the fluctuations of a quantity about its mean value is exceedingly small when one is dealing with a system consisting of very many particles.

Some conventional definitions

The preceding discussion shows that the quantities $\ln \Omega$ and β of the system A (and the corresponding quantities of A') are of crucial importance in the study of thermal interaction. It is, therefore, convenient to introduce some alternative symbols and conventional names for these quantities.

Note first that the parameter β has, according to its definition (9), the dimensions of a reciprocal energy. It is thus often useful to express β^{-1} as a multiple of some positive constant k which has the dimensions of energy and thus can be expressed in ergs. (This constant is called *Boltzmann's constant* and its magnitude can be chosen once and for all in some convenient way.) The parameter β^{-1} can then be written in the form

$$\boxed{\frac{1}{\beta} \equiv kT} \tag{12}$$

where the quantity T thus defined provides a measure of energy in units of the quantity k. This new parameter T is called the *absolute temperature* of the system under consideration and its magnitude is commonly expressed in terms of *degrees.*† The physical reason for the name "temperature" will become more apparent in Sec. 4.3.

† For example, an absolute temperature of 5 *degrees* corresponds merely to an energy of $5k$.

By virtue of (9), the definition of T in terms of $\ln \Omega$ can now be written in the form

$$\frac{1}{T} \equiv \frac{\partial S}{\partial E} \tag{13}$$

where we have introduced the quantity S defined by

$$\boxed{S \equiv k \ln \Omega.} \tag{14}$$

This quantity S is called the *entropy* of the system under consideration. It has the dimensions of energy because its definition involves the constant k. According to its definition (14), the entropy of a system is merely a logarithmic measure of the number of states accessible to the system. In accordance with the comments at the end of Sec. 3.6, the entropy thus provides a quantitative measure of the degree of randomness of the system.†

In terms of the previous definitions, the condition that the probability $P(E)$ be maximum is, by virtue of (3), equivalent to the statement that the entropy $S^* \equiv k \ln \Omega^*$ of the total system is maximum with respect to the energy E of the subsystem A. Using (6), the condition of maximum probability is thus equivalent to the statement that

$$S^* = S + S' = \text{maximum}. \tag{15}$$

This condition is fulfilled if (8) is satisfied, i.e., if

$$T = T'. \tag{16}$$

Our discussion thus shows that the energy E of A adjusts itself in such a way that the entropy of the total isolated system A^* is as large as possible. The system A^* is then distributed over the largest possible number of states, i.e., it is in its most random macrostate.

4.2 *The Approach to Thermal Equilibrium*

As we have seen, the probability $P(E)$ has an exceedingly sharp maximum at the energy $E = \tilde{E}$. Therefore, in the equilibrium situation where A and A' are in thermal contact, the system A almost always has an energy E extremely close to $\tilde{E}$, while the system A' correspondingly has an energy E' extremely close to $\tilde{E}' = E^* - \tilde{E}$. To excellent ap-

† Note that the entropy defined by (14) has a definite value which is, by virtue of (3.40), essentially independent of the magnitude of the energy subdivision interval δE used in our discussion. Furthermore, since δE is some fixed interval independent of E, the derivative (9) defining β or T is certainly also independent of δE.

proximation, the respective mean energies of the systems are then also equal to these energies, i.e.,

$$\bar{E} = \tilde{E} \qquad \text{and} \qquad \bar{E}' = \tilde{E}'. \tag{17}$$

Now consider a situation where the systems A and A' are initially isolated from each other and separately in equilibrium, their mean energies being $\bar{E}_i$ and $\bar{E}_i'$, respectively. Suppose that A and A' are then placed in thermal contact so that they are free to exchange energy with each other. The resulting situation is ordinarily an extremely improbable one, except in the special case where the respective energies of the systems were initially very close to $\tilde{E}$ and $\tilde{E}'$. In accordance with our postulate (3.18), the systems will then tend to exchange energy until they attain the ultimate equilibrium situation discussed in the preceding section. The final mean energies $\bar{E}_f$ and $\bar{E}_f$ of the systems will then, by (17), be equal to

$$\bar{E}_f = \tilde{E} \qquad \text{and} \qquad \bar{E}_f' = \tilde{E}' \tag{18}$$

so that the probability $P(E)$ becomes maximum. The β parameters of the systems will then correspondingly be equal, i.e.,

$$\beta_f = \beta_f' \tag{19}$$

where $\qquad \beta_f \equiv \beta(\bar{E}_f) \qquad \text{and} \qquad \beta_f' \equiv \beta(\bar{E}_f').$

The conclusion that the systems exchange energy until they reach the situation of maximum probability $P(E)$ is, by virtue of (6) and the definition (14), equivalent to the statement that they exchange energy until their total entropy becomes maximum. Thus the final probability (or entropy) can never be less than the original one, i.e.,

$$S(\bar{E}_f) + S'(\bar{E}_f') \geq S(\bar{E}_i) + S'(\bar{E}_i')$$

or

$$\boxed{\Delta S + \Delta S' \geq 0,} \tag{20}$$

where $\qquad \Delta S \equiv S(\bar{E}_f) - S(\bar{E}_i)$

and $\qquad \Delta S' \equiv S'(\bar{E}_f') - S'(\bar{E}_i')$

denote the entropy changes of A and A', respectively.

In the process of exchanging energy, the *total* energy of the systems is, of course, conserved. In accordance with (3.49) and (3.50), it thus follows that

$$\boxed{Q + Q' = 0} \tag{21}$$

where Q and Q' denote the heats absorbed by A and A' respectively.

The relations (20) and (21) summarize completely the conditions which must be satisfied in any process of thermal interaction.

Our discussion shows that there are two possible cases which may arise:

(i) The initial energies of the systems are such that $\beta_i = \beta_i'$, where $\beta_i \equiv \beta(\bar{E}_i)$ and $\beta_i' \equiv \beta(\bar{E}_i')$. The systems are then already in their most probable situation, i.e., their total entropy is already maximum. Hence the systems remain in equilibrium and no net exchange of heat occurs between them.

(ii) More generally, the initial energies of the systems are such that $\beta_i \neq \beta_i'$. The systems are then in a rather improbable situation where their total entropy is not maximum. Hence the situation will change in time while energy, in the form of heat, is exchanged between the systems until the final equilibrium is reached where the total entropy is maximum and $\beta_f = \beta_f'$.

4.3 *Temperature*

In the last section, we noted that the parameter β [or equivalently the parameter $T = (k\beta)^{-1}$] has the following two properties:

(i) If two systems separately in equilibrium are characterized by the *same* value of the parameter, then equilibrium will be preserved and *no* heat transfer will occur when the systems are brought into thermal contact with each other.

(ii) If the systems are characterized by *different* values of the parameter, then equilibrium will not be preserved and heat transfer *will* occur when the systems are brought into thermal contact with each other.

These statements allow us to derive several important consequences. In particular, they permit us to formulate in a precise and quantitative way the qualitative notions explored in Sec. 1.5.

Imagine, for example, that there are three systems A, B, and C which are separately in equilibrium. Suppose that no heat transfer occurs when C is placed in thermal contact with A, and that no heat transfer occurs when C is placed in thermal contact with B. Then we know that $\beta_C = \beta_A$ and that $\beta_C = \beta_B$ (where β_A, β_B, and β_C denote the β parameters of systems A, B, and C, respectively). But from these two equalities we can conclude that $\beta_A = \beta_B$, and hence that no heat transfer will occur if systems A and B are placed in thermal contact with each other.

Thus we are led to the following general conclusion:

If two systems are in thermal equilibrium with a third system, then they must be in thermal equilibrium with each other.	(22)

The statement (22) is called the *zeroth law of thermodynamics*. Its validity makes possible the use of test systems, called *thermometers*, which allow measurements to decide whether any two systems will, or will not, exchange heat when brought into thermal contact with each other. Such a thermometer can be *any* macroscopic system M chosen in accordance with the following two specifications:

(*a*) Among the many macroscopic parameters characterizing the system M, select one (call it θ) which varies by appreciable amounts when M gains or loses energy by thermal interaction. All the other macroscopic parameters of M are held fixed. The parameter θ which is allowed to vary is called the *thermometric parameter* of M.

(*b*) The system M is ordinarily chosen to be much smaller (i.e., to have many fewer degrees of freedom) than the systems which it is designed to test. This is desirable in order to minimize any energy transfer to these systems and thus to reduce to a minimum their disturbance by the testing process.

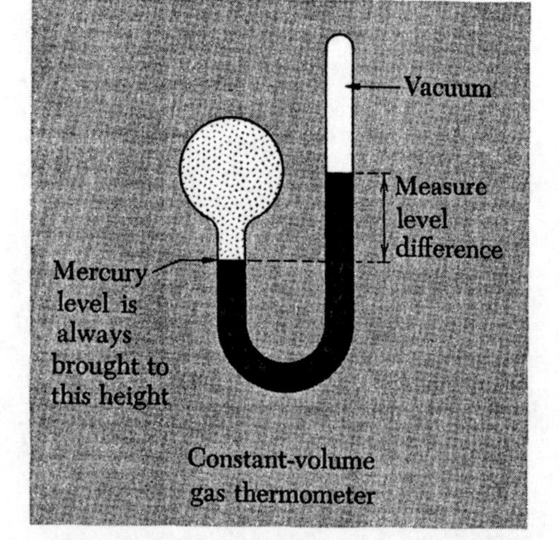

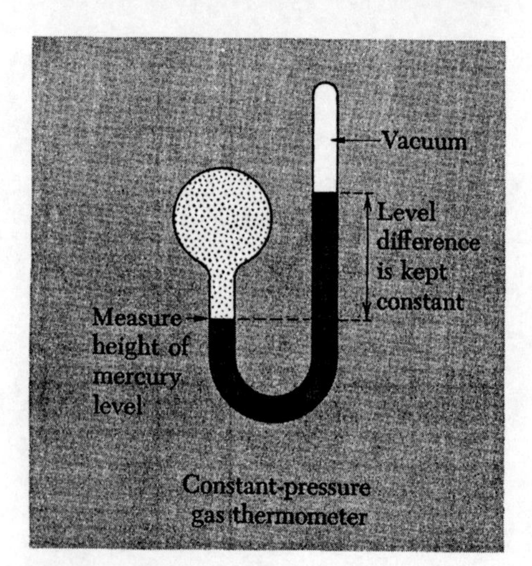

Fig. 4.4 Constant-volume and constant-pressure gas thermometers.

Examples of thermometers

There are many possible systems which can serve as thermometers. We mention only a few of those which are commonly used.

(i) A liquid, such as mercury or alcohol, confined within a glass tube of small diameter. This is the familiar type of thermometer which we have already described in Sec. 1.5. Here the thermometric parameter θ is the height of the liquid in the tube.

(ii) A gas confined within a bulb in such a way that its volume is kept constant. This is called a *constant-volume gas thermometer*. Here the thermometric parameter θ is the pressure exerted by the gas.

(iii) A gas confined within a bulb in such a way that its pressure is kept constant. This is called a *constant-pressure gas thermometer*. Here the thermometric parameter θ is the volume occupied by the gas.

(iv) An electrical conductor (e.g., a coil of platinum wire) maintained at constant pressure and carrying a small current. This is called a *resistance thermometer*. Here the thermometric parameter θ is the electrical resistance of the conductor.

(v) A paramagnetic sample maintained at constant pressure. Here the thermometric parameter θ is the magnetic susceptibility of the sample (i.e., the ratio of its mean magnetic moment per unit volume to the applied magnetic field.) This quantity can be determined, for example, by measuring the self-inductance of a coil wound around the sample.

A thermometer M is used in the following way. It is successively placed in thermal contact with the systems under test, call them A and B, and is allowed to come to equilibrium with each.

(i) Suppose that the thermometric parameter θ of M (e.g., the length of the liquid column of a mercury-in-glass thermometer) assumes the same value in both cases. This implies that, after M has come to equilibrium with A, it remains in equilibrium after being placed in thermal contact with B. Hence the zeroth law allows one to conclude that A and B will remain in equilibrium if they are brought into thermal contact with each other.

(ii) Suppose that the thermometric parameter θ of M does *not* assume the same value in both cases. Then one can conclude that A and B will *not* remain in equilibrium if they are brought into thermal contact with each other. To make the argument explicit, suppose that A and B *did* remain in equilibrium. After M attains thermal equilibrium with A, it would then, according to the zeroth law, have to remain in equilibrium when brought into thermal contact with B. But then the parameter θ cannot change if M is brought into thermal contact with B, contrary to hypothesis.†

Consider *any* thermometer M with *any one* parameter θ chosen as its thermometric parameter. The value assumed by θ when the thermometer M has come to thermal equilibrium with some system A will, by definition, be called the *temperature of A with respect to the particular thermometric parameter θ of the particular thermometer M.* According to this definition the temperature can be a length, a pressure, or any other quantity. Note that, even if two different thermometers have parameters of the same kind, they do not ordinarily yield the same value of the temperature for the same body.‡ Furthermore, if a body C has a temperature halfway between the temperatures of bodies A and B when measured by one thermometer, this statement is not necessarily true with respect to the temperatures measured by some other thermometer. Nevertheless, our discussion shows that the tem-

---

† All the preceding measurements could be performed with any other thermometer M' having a thermometric parameter θ'. There exists ordinarily a one-to-one relation between any value of θ and the corresponding value of θ'. In exceptional cases a particular thermometer M might, however, be multiple-valued so that a given value of θ corresponds to more than one value of θ' for almost every other thermometer M'. Peculiar thermometers which are multiple-valued in the experimental range of interest are rarely useful and will be excluded from our discussion. (See Prob. 4.1.)

‡ For example, the two thermometers might both consist of glass tubes filled with liquid, so that the length of the liquid column is the thermometric parameter in both cases. The liquid might, however, be mercury in one thermometer and alcohol in the other.

perature concept which we have defined has the following useful property:

> Two systems will remain in equilibrium when placed in thermal contact with each other if, and only if, they have the same temperature with respect to the same thermometer. (23)

The temperature concept which we have introduced is important and useful, but is rather arbitrary in the sense that the temperature assigned to a system depends in an essential way on the peculiar properties of the particular system M used as the thermometer. On the other hand, we can exploit the properties of the parameter β to obtain a much more useful temperature. Indeed, suppose that we have a thermometer M whose parameter β is known as a function of its thermometric parameter θ. If this thermometer is placed in thermal contact with some system A, then we know that in equilibrium $\beta = \beta_A$. Thus the thermometer measures, by virtue of (9), a fundamental property of the system A, namely the fractional increase of the number of its states with energy. Furthermore, suppose that we take *any other* thermometer M' whose parameter β' is also known as a function of its thermometric parameter θ'. If this thermometer is placed in thermal contact with the system A, then we know that in equilibrium $\beta' = \beta_A$. Thus $\beta' = \beta$ and we arrive at the following conclusion:

> If the parameter β is used as the thermometric parameter of a thermometer, then *every* such thermometer yields the *same* temperature reading when used to measure the temperature of a particular system. Furthermore, this temperature measures a fundamental property of the number of states of the system under test. (24)

The parameter β is, therefore, a particularly useful and fundamental temperature parameter. This is the reason for the name of *absolute temperature* given to the corresponding temperature parameter $T \equiv (k\beta)^{-1}$ defined in terms of β. We shall postpone until the next chapter a discussion of the following two points: (i) practical procedures for finding numerical values of β or T by appropriate measurements, and (ii) the international convention adopted to assign a particular numerical value to k.

Properties of the absolute temperature

By virtue of its definition (9), the absolute temperature is given by

$$\frac{1}{kT} \equiv \beta \equiv \frac{\partial \ln \Omega}{\partial E}. \tag{25}$$

We saw in (3.37) that $\Omega(E)$ is, for any ordinary system, an extremely rapidly increasing function of its energy E. Hence (25) implies that,

$$\text{for any ordinary system,} \qquad \beta > 0 \quad \text{or} \quad T > 0. \tag{26}$$

In other words,

> the absolute temperature of any ordinary system is positive.† (27)

We can also readily estimate the order of magnitude of the absolute temperature of a system. The approximate functional dependence of $\Omega(E)$ is ordinarily of the form given by (3.38),

$$\Omega(E) \propto (E - E_0)^f \tag{28}$$

where f is the number of degrees of freedom of the system and E is the energy of the system whose ground state energy is E_0. Thus

$$\ln \Omega \sim f \ln (E - E_0) + \text{constant}$$

so that

$$\beta = \frac{\partial \ln \Omega}{\partial E} \sim \frac{f}{E - E_0}. \tag{29}$$

The magnitude of T can then be estimated by putting $E = \bar{E}$, the mean energy of the system. Hence we can conclude that,

$$\text{for an ordinary system,} \qquad kT = \frac{1}{\beta} \sim \frac{\bar{E} - E_0}{f}. \tag{30}$$

In other words,

> for any ordinary system at the absolute temperature T, the quantity kT is roughly equal to the mean energy (above the ground state) per degree of freedom of the system. (31)

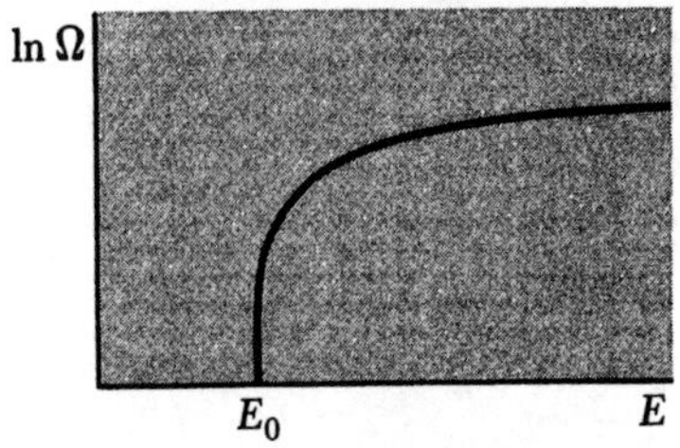

Fig. 4.5 Schematic sketch showing the behavior of $\ln \Omega$ as a function of the energy E. The slope of the curve is the absolute temperature parameter β.

† As pointed out in connection with (3.38), the qualifying phrase *for any ordinary system* is meant to exclude specifically the exceptional case of a system where the kinetic energy of the particles is ignored and where their spins have a magnetic energy which is sufficiently high.

The condition of equilibrium (8) between two systems in thermal contact asserts that their respective absolute temperatures must be equal. By virtue of (31) we see that this condition is roughly equivalent to the statement that the total energy of the interacting systems is shared between them in such a way that the mean energy per degree of freedom is the same for both systems. This last statement was essentially the one which we used in our qualitative discussion of Sec. 1.5.

How does the parameter β, or T, vary with the energy E of a system? The quantity β measures the slope of the curve of $\ln \Omega$ versus E. We already noted in the caption of Fig. 4.3 that this curve must be concave downward so as to guarantee the physical requirement that a situation with a unique maximum probability results when two systems are placed in thermal contact with each other. Hence it follows that the slope of the curve must decrease as E increases; i.e.,

for *any* system,
$$\frac{\partial \beta}{\partial E} < 0. \tag{32}$$

In the case of an ordinary system, this result follows also from the approximate functional form (28) since differentiation of (29) yields explicitly

$$\frac{\partial \beta}{\partial E} \sim -\frac{f}{(E - E_0)^2} < 0. \tag{33}$$

Since we have just shown that β decreases as E increases and since, by its definition $T \equiv (k\beta)^{-1}$, the absolute temperature T increases as β decreases, we can then conclude from (32) that

the absolute temperature of any system is an increasing function of its energy.	(34)

In more mathematical terms,

$$\frac{\partial T}{\partial E} = \frac{\partial}{\partial E}\left(\frac{1}{k\beta}\right) = -\frac{1}{k\beta^2}\frac{\partial \beta}{\partial E}$$

so that (32) implies

$$\frac{\partial T}{\partial E} > 0. \tag{35}$$

This last relation permits us to establish a general connection between absolute temperatures and the direction of heat flow. Consider two systems A and A' which are initially in equilibrium at different absolute temperatures T_i and T_i', and which are then brought into thermal contact with each other. One system then absorbs heat, and the other gives off heat, until the systems attain a final equilibrium at some common absolute temperature T_f. Suppose that system A is the one which absorbs heat and thus gains energy; then it follows by (34) that $T_f > T_i$. Correspondingly system A' must give off heat and thus lose energy; then it follows by (34) that $T_f < T_i'$. Hence the initial and final temperatures are such that

$$T_i < T_f < T_i'.$$

This means that the system A which absorbs heat has an initial absolute temperature T_i which is less than the initial absolute temperature T_i' of the system A' which gives off heat. In short,

> when any two ordinary systems are placed in thermal contact with each other, heat is given off by the system with the larger absolute temperature and is absorbed by the system with the smaller absolute temperature.† (36)

Since we defined the *warmer* system as the one which gives off heat and the *colder* system as the one which absorbs heat, (36) is equivalent to the statement that *a warmer system has a higher absolute temperature than a colder one.*

4.4 *Small Heat Transfer*

The preceding sections complete our general discussion of the thermal interaction between macroscopic systems. We shall now turn our attention to some special simple cases which are of particular importance.

Suppose that, when a system A is placed in thermal contact with some other system, it absorbs an amount of heat Q which is so small that

$$|Q| \ll \bar{E} - E_0, \tag{37}$$

i.e., so that the resulting mean energy change $\Delta\bar{E} = Q$ of A is small compared to the mean energy $\bar{E}$ of A above its ground state. The

† In the exceptional case of a system of spins, this statement may need to be qualified because $T \to \pm\infty$ when $\beta \to 0$. See Prob. 4.30.

absolute temperature of the system A changes then by a negligible amount. Indeed, putting $E = \bar{E}$, (29) and (33) yield the estimates

$$\Delta\beta = \frac{\partial\beta}{\partial E} Q \sim -\frac{f}{(\bar{E} - E_0)^2} Q \sim -\frac{\beta}{\bar{E} - E_0} Q.$$

Hence (37) implies that

$$|\Delta\beta| = \left|\frac{\partial\beta}{\partial E} Q\right| \ll \beta. \tag{38}$$

Since $T = (k\beta)^{-1}$ or $\ln T = -\ln\beta - \ln k$, it follows correspondingly that $(\Delta T/T) = -(\Delta\beta/\beta)$ so that (38) is also equivalent to

$$|\Delta T| \ll T. \tag{39}$$

We shall say that the heat Q absorbed by a system is *small* whenever (38) is valid, i.e., whenever Q is sufficiently small so that the absolute temperature of the system remains essentially unchanged.

Suppose that the system A absorbs such a small amount of heat Q. Its initial and final energies are then, with overwhelming probability, equal to their respective mean values $\bar{E}$ and $\bar{E} + Q$. In the process of absorbing this heat, the number of states $\Omega(E)$ accessible to A also changes. By expanding in a Taylor series, we find

$$\begin{aligned}\ln\Omega(\bar{E} + Q) - \ln\Omega(\bar{E}) &= \left(\frac{\partial\ln\Omega}{\partial E}\right) Q + \frac{1}{2}\left(\frac{\partial^2\ln\Omega}{\partial E^2}\right) Q^2 + \cdots \\ &= \beta Q + \frac{1}{2}\frac{\partial\beta}{\partial E} Q^2 + \cdots.\end{aligned}$$

But since the absorbed heat Q is supposed to be small, the absolute temperature of A remains essentially unchanged. Hence the term involving $\partial\beta/\partial E$ can be neglected, in accordance with (38). The change in the quantity $\ln\Omega$ then becomes simply

$$\Delta(\ln\Omega) = \frac{\partial\ln\Omega}{\partial E} Q = \beta Q. \tag{40}$$

In the process of absorbing an amount of heat Q, the entropy $S \equiv k\ln\Omega$ of a system at the absolute temperature $T = (k\beta)^{-1}$ changes thus by an amount ΔS which is,

if Q is small,

$$\Delta S = \frac{Q}{T}. \tag{41}$$

We emphasize that, even if the heat Q is large in absolute magnitude, it may yet be relatively small in the comparative sense of (37) or (39) so that the relation (41) remains valid. If the absorbed heat is actually infinitesimal in magnitude, we can denote it by $đQ$; the corresponding infinitesimal entropy change is then equal to

$$\boxed{dS = \frac{đQ}{T}}. \tag{42}$$

Note that the heat $đQ$ is merely an infinitesimal quantity. The quantity dS is, however, an actual differential, i.e., an infinitesimal *difference* between the entropies of A in its final and initial macrostates.

The heat Q absorbed by a system A will always be very small in the comparative sense of (37) or (39) when A is placed in thermal contact with any other system B sufficiently smaller than A itself. Indeed, any amount of heat Q which A can absorb from B then is at most of the order of the total energy of B (above its ground state) and is thus much smaller than the energy $\bar{E} - E_0$ of A itself. A system A is said to act as a *heat reservoir* (or *heat bath*) with respect to some other set of systems if it is sufficiently large so that its temperature remains essentially unchanged in any thermal interaction with those systems. Equation (41) thus is always valid in relating the entropy change ΔS of a heat reservoir to the heat Q absorbed by it.

4.5 *System in Contact with a Heat Reservoir*

Most systems which one encounters in practice are not isolated, but are free to exchange heat with their environment. Since such a system is usually small compared to its environment, it constitutes a relatively small system in thermal contact with a heat reservoir provided by other systems in its environment. (For example, any object in a room, say a table, is in thermal contact with the heat reservoir consisting of the room itself with its floor, walls, other furniture, and enclosed air.) In this section we shall, therefore, consider any relatively small system A in contact with a heat reservoir A', and ask the following detailed question about the small system A: Under conditions of equilibrium, what is the probability P_r of finding the system A in any *one* particular state r of energy E_r?

This is a rather important and very general kind of question. Note that in the present context the system A can be any system which has many fewer degrees of freedom than the heat reservoir A'. Thus A

might be any relatively small *macro*scopic system. (For example, it might be a piece of copper immersed in the water of a lake, the latter acting as a heat reservoir.) Alternatively A might also be a distinguishable *micro*scopic system which can be clearly identified.† (For example, it might be an atom located at a particular lattice site in a solid, the latter acting as a heat reservoir.)

To facilitate counting the states of the reservoir A', we again imagine its energy scale to be subdivided into small fixed intervals of magnitude δE and denote by $\Omega'(E')$ the number of states accessible to A' when its energy is equal to E' (i.e., when it lies between E' and $E' + \delta E$). (Here δE is assumed to be very small compared to the separation between the energy levels of A, but to be large enough to contain many possible states of the reservoir A'.) It is then quite easy to use reasoning similar to that of Sec. 4.1 to find the desired probability P_r that the system A is in its state r. Although the reservoir can have any energy E', the conservation of energy applied to the isolated system A^* consisting of both A and A' requires that the energy of A^* have some constant value, say E^*. When the system A is in its state r of energy E_r, the reservoir A' must then have an energy

$$E' = E^* - E_r. \tag{43}$$

But, when A is in this *one* definite state r, the number of states accessible to the combined system A^* is simply the number of states $\Omega'(E^* - E_r)$ which are accessible to A'. Our fundamental statistical postulate asserts, however, that the isolated system A^* is equally likely to be found in each one of its accessible states. Hence the probability of occurrence of a situation where A is in the state r is simply proportional to the number of states accessible to A^* when A is known to be in the state r, i.e.,

$$\boxed{P_r \propto \Omega'(E^* - E_r).} \tag{44}$$

Up to this point our discussion has been completely general. Let us now make use of the fact that A is very much smaller than the reservoir A' in the sense that any energy E_r of interest to us satisfies the relation

$$E_r \lll E^*. \tag{45}$$

---

† This qualifying remark is pertinent since it may not always be possible, in a quantum-mechanical description, to identify a particular particle among particles which are basically indistinguishable.

We can then find an excellent approximation for (44) by expanding the slowly varying logarithm of $\Omega'(E')$ about the value $E' = E^*$. Analogously to (40), we thus get for the heat reservoir A'

$$\ln \Omega'(E^* - E_r) = \ln \Omega'(E^*) - \left[\frac{\partial \ln \Omega'}{\partial E'}\right] E_r$$
$$= \ln \Omega'(E^*) - \beta E_r. \tag{46}$$

Here we have written

$$\beta \equiv \left[\frac{\partial \ln \Omega'}{\partial E'}\right] \tag{47}$$

for the derivative evaluated at the fixed energy $E' = E^*$. Thus $\beta = (kT)^{-1}$ is simply the constant temperature parameter of the *heat reservoir* A'.† Hence (46) yields the result

$$\Omega'(E^* - E_r) = \Omega'(E^*)\, e^{-\beta E_r}. \tag{48}$$

Since $\Omega'(E^*)$ is merely a constant independent of r, (44) then becomes simply

$$\boxed{P_r = C\, e^{-\beta E_r}} \tag{49}$$

where C is a constant of proportionality independent of r.

Let us examine the physical content of the results (44) or (49). If A is known to be in a definite state r, the reservoir A' can be in any one of the large number $\Omega'(E^* - E_r)$ of states accessible to it under these conditions. But the number of states $\Omega'(E')$ accessible to the reservoir is ordinarily a rapidly increasing function of its energy E' [i.e., β in (47) is ordinarily positive]. Suppose then that we compare the probabilities of finding the system A in *any* two of its states which have different energies. If A is in the state where its energy is higher, the conservation of energy for the total system then implies that the energy of the reservoir is correspondingly lower; hence the number of states accessible to the reservoir is markedly reduced. Accordingly, the probability of encountering this situation is very much less. The exponential dependence of P_r on E_r in (49) merely expresses the result of this argument in mathematical terms.

---

† For the sake of simplicity, we shall not embellish the symbol β with a prime.

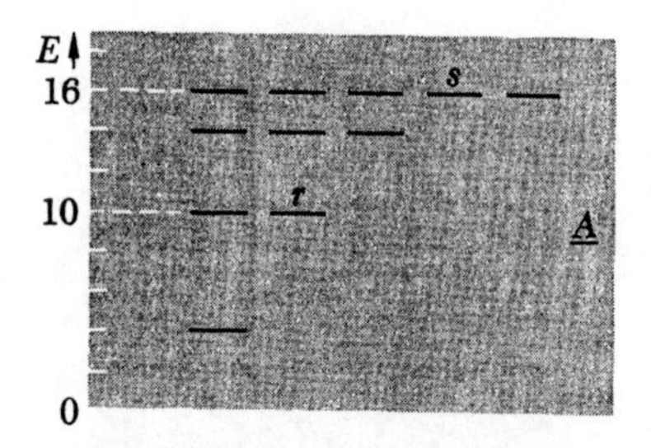

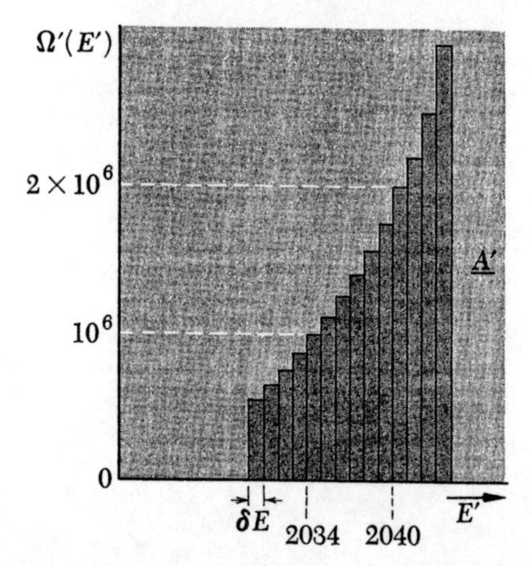

Fig. 4.6 Schematic illustration showing the states accessible to a particular system A and to a special (rather small) heat reservoir A'. The top diagram shows the energy levels corresponding to a few distinct states of A. The bottom diagram shows, for a few values of E', the number of states $\Omega'(E')$ accessible to A' as a function of its energy E'. The energy is measured in terms of an arbitrary unit.

Example

Let us illustrate the preceding comments with a simple example. Consider a system A which has energy levels some of which are shown in the top part of Fig. 4.6. Consider also the much larger system A' whose energy scale is subdivided into intervals of size $\delta E = 1$ unit and whose number of states $\Omega'(E')$ is shown as a function of its energy E' in the bottom part of Fig. 4.6. We assume that A is in thermal equilibrium with the heat reservoir A' and that the total energy E^* of the combined system A^* has some value $E^* = 2050$ units. Suppose that A is in the particular state r where its energy $E_r = 10$ units. The energy of the reservoir A' must then be $E' = 2040$; accordingly, A' can be in any one of 2×10^6 possible states. In an ensemble containing many isolated systems A^* consisting of A and A', the number of instances where A is found in the state r will then be proportional to 2×10^6. On the other hand, suppose that A is in the particular state s where its energy $E_s =$ 16 units. The energy of the reservoir A' must then be $E' = 2034$; accordingly, A' can now be in any one of only 10^6 possible states. In the ensemble of systems, the number of instances where A is found in the state s will thus be proportional to 10^6 and will consequently be only half as large as the number of instances where A is in the state r of lower energy.

The probability (49) is a very general result of fundamental importance in statistical mechanics. The exponential factor $e^{-\beta E_r}$ is called the *Boltzmann factor;* the corresponding probability distribution (49) is known as the *canonical distribution.* An ensemble of systems all of which are in contact with a heat reservoir of known temperature T [i.e., all of which are distributed over their states in accordance with (49)] is called a *canonical ensemble.*

The constant of proportionality C in (49) can be readily determined by the normalization condition that the system must have probability unity of being in some one of its states; i.e.,

$$\sum_r P_r = 1 \tag{50}$$

where the sum extends over all possible states of A irrespective of energy. By virtue of (49) this condition determines C to be such that

$$C \sum_r e^{-\beta E_r} = 1.$$

Hence (49) can be written in the explicit form

$$P_r = \frac{e^{-\beta E_r}}{\sum_r e^{-\beta E_r}}. \tag{51}$$

The probability distribution (49) allows us to calculate very simply the mean values of various parameters characterizing the system A in contact with a heat reservoir at the absolute temperature $T = (k\beta)^{-1}$. For example, let y be any quantity assuming the value y_r in state r of the system A. Then the mean value of y is given by

$$\bar{y} \equiv \sum_r P_r\, y_r = \frac{\sum_r e^{-\beta E_r} y_r}{\sum_r e^{-\beta E_r}} \tag{52}$$

where the summation is over all states r of the system A.

Remarks pertinent when A is a macroscopic system

The fundamental result (49) gives the probability P_r of finding A in any *one* particular state r of energy E_r. The probability $P(E)$ that A has an energy in a small *range*, say between E and $E + \delta E$, is then simply obtained by adding the probabilities for all states r whose energy E_r lies in the range $E < E_r < E + \delta E$; i.e.,

$$P(E) = \sum_r{}' P_r$$

where the prime on the summation symbol indicates that the sum extends only over the states having nearly identical energies lying in this small range. But the probability P_r is then, by (49), essentially the same for all these states and proportional to $e^{-\beta E}$. Hence the probability $P(E)$ of interest is obtained simply by multiplying {the probability of finding A in any one of these states} by {the number $\Omega(E)$ of its states in this energy range}; i.e.,

$$P(E) = C\,\Omega(E)\,e^{-\beta E}. \tag{53}$$

To the extent that A itself is a large system (although very much smaller than A'), $\Omega(E)$ is a rapidly *in*creasing function of E. The presence of the rapidly *de*creasing factor $e^{-\beta E}$ in (53) results then in a maximum of the product $\Omega(E)\,e^{-\beta E}$. This maximum in $P(E)$ is sharper the larger A is, i.e., the more rapidly $\Omega(E)$ increases with E. Thus we arrive again at the conclusions obtained in Sec. 4.1 for a macroscopic system.

When a system in contact with a heat reservoir is itself of macroscopic size, the relative magnitude of the fluctuations of its energy E is so exceedingly small that its energy is practically always equal to its mean value $\bar{E}$. On the other hand, if the system were removed from contact with the heat reservoir and were thermally isolated, its energy could not fluctuate at all. The distinction between this situation and the previous one, however, is so small that it is practically irrelevant; in particular, the mean values of all physical parameters of the system (such as its mean pressure or mean magnetic moment) remain quite unaffected. Hence it makes no difference whether these mean values are calculated by considering the macroscopic system to be an isolated system with a fixed energy in some small range between E and $E + \delta E$, or by considering this system to be in thermal contact with a heat reservoir of such a temperature that the mean energy $\bar{E}$ of the system is equal to E. The latter point of view, however, makes calculations much easier. The reason is that use of the canonical distribution reduces the computation of mean value to the evaluation of the sums (52) over all states without restriction; thus it does *not* require the much more difficult task of counting the number Ω of states of a particular type lying in a *specified* small energy range.

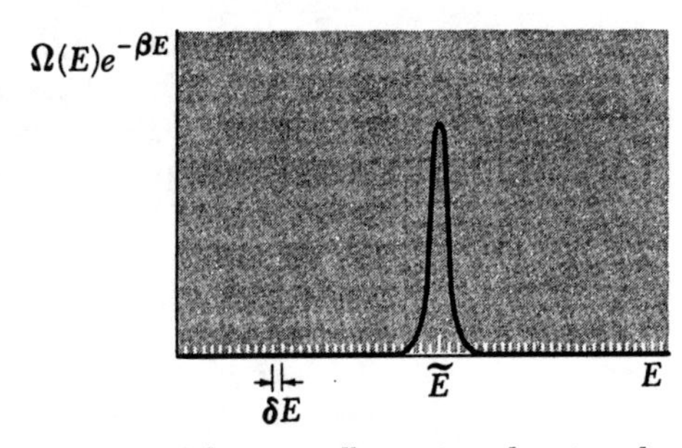

Fig. 4.7 Schematic illustration showing the dependence of the function $\Omega(E)e^{-\beta E}$ on the energy E of a macroscopic system in contact with a heat reservoir.

Fig. 4.8 Ludwig Boltzmann (1844–1906). This Austrian physicist, who did pioneering work in statistical physics, contributed greatly to the development of the atomic theory of gases which he put in its modern quantitative form. His work in 1872 led to fundamental insights into the microscopic explanation of irreversibility. He also laid the foundation for statistical mechanics and introduced the basic relation $S = k \ln \Omega$ connecting the entropy with the number of accessible states. Boltzmann's work came under heavy attack by a whole school of thought represented by such men as Ernst Mach (1838–1916) and Wilhelm Ostwald (1853–1932), who argued that physical theory should deal only with macroscopically observable quantities and should reject purely hypothetical concepts such as atoms. Discouraged, Boltzmann wrote in 1898, "I am conscious of being only an individual struggling weakly against the stream of time." Increasingly subject to depressions, Boltzmann committed suicide in 1906 only shortly before Perrin's experiment on Brownian motion (1908) and Millikan's oil-drop experiment (1909) provided very direct evidence for the discrete atomic structure of matter. (*Photograph by courtesy of Professor W. Thirring, University of Vienna.*)

Fig. 4.9 Josiah Willard Gibbs (1839–1903). The first American theoretical physicist of note, he was born and died in New Haven where he spent all his working life as professor at Yale University. In the 1870s he made fundamental contributions to *thermodynamics*, couching its purely macroscopic reasoning in powerful analytical form and using it to treat many important problems of physics and chemistry. Around 1900 he developed a very general formulation of statistical mechanics in terms of the concept of ensembles. Despite the modifications brought about by quantum mechanics, the basic framework of his formulation still stands and is essentially the one upon which we have built our systematic discussion beginning with Chap. 3. The name *canonical ensemble* originated with Gibbs. (*Photograph by courtesy of the Beinecke Rare Book and Manuscript Library, Yale University.*)

4.6 *Paramagnetism*

The canonical distribution can be used to discuss a large number of situations of great physical interest. As a first application we shall investigate the magnetic properties of a substance which contains N_0 magnetic atoms per unit volume and which is placed in an external magnetic field **B**. We consider the particularly simple case where each magnetic atom has spin $\frac{1}{2}$ and an associated magnetic moment μ_0. In a quantum-mechanical description the magnetic moment of each atom can then point either "up" (i.e., parallel to the external field) or "down" (i.e., antiparallel to the field). The substance is said to be *paramagnetic* since its magnetic properties are due to the orientation of individual magnetic moments. Suppose that the substance is at an absolute temperature T. What then is $\bar{\mu}$, the *mean* component of the magnetic moment of one of its atoms along the direction of the magnetic field **B**?

We assume that each magnetic atom interacts only weakly with all the other atoms of the substance. In particular, we assume that the magnetic atoms are sufficiently far separated from each other so that we can neglect the magnetic field produced at the position of one magnetic atom by a neighboring magnetic atom. It is then permissible to focus attention on a single magnetic atom as the small system under consideration and to regard all the other atoms of the substance as constituting a heat reservoir at the absolute temperature T of interest.†

Each atom can be in two possible states: the state (+) where its magnetic moment points up, and the state (−) where its magnetic moment points down. Let us discuss these states in turn.

In the state (+), the atomic magnetic moment is parallel to the field so that $\mu = \mu_0$. The corresponding magnetic energy of the atom is then $\epsilon_+ = -\mu_0 B$. The canoncial distribution (49) thus yields for the probability P_+ of finding the atom in this state the result

$$P_+ = Ce^{-\beta\epsilon_+} = Ce^{\beta\mu_0 B} \tag{54}$$

where C is a constant of proportionality and $\beta = (kT)^{-1}$. This is the state of lower energy and thus is the state in which the atom is more likely to be found.

† This assumes that it is possible to identify a single atom unambiguously, an assumption which is justified if the atoms are localized at definite lattice sites of a solid or if they form a dilute gas where the atoms are widely separated. In a sufficiently concentrated gas of identical atoms the assumption breaks down since the atoms are then indistinguishable in a quantum-mechanical description. It would then be necessary to adopt a point of view (which is always permissible, although more complicated) which considers the *entire* gas of atoms as a small macroscopic system in contact with some heat reservoir.

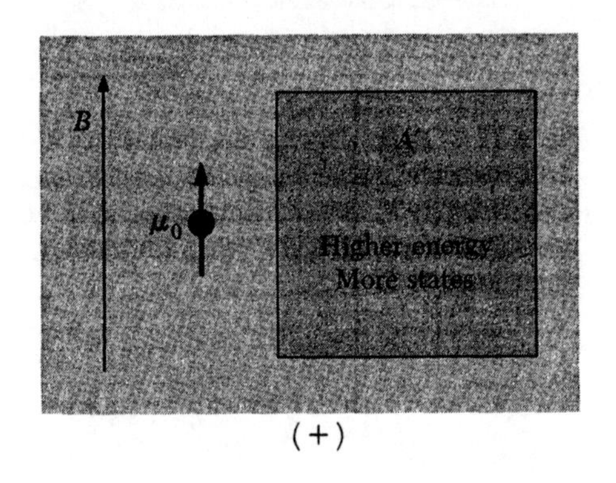

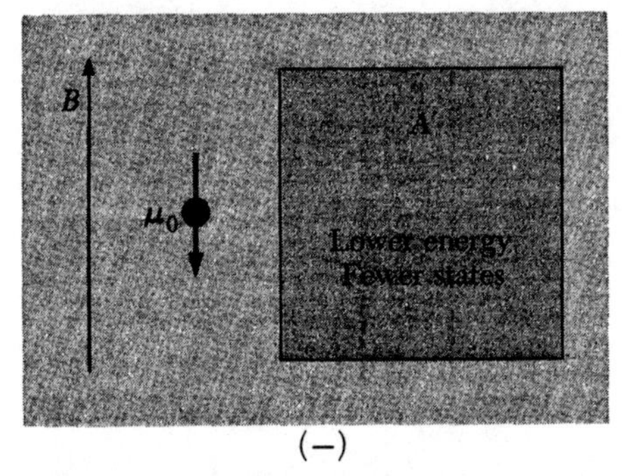

Fig. 4.10 An atom with spin $\frac{1}{2}$ shown in thermal contact with a heat reservoir A'. When the magnetic moment of the atom points up, its energy is less by an amount $2\mu_0 B$ than when it points down; accordingly the energy of the reservoir is greater by an amount $2\mu_0 B$ so that the reservoir can be in many more states. Hence the situation in which the moment points up occurs with greater probability than that in which it points down.

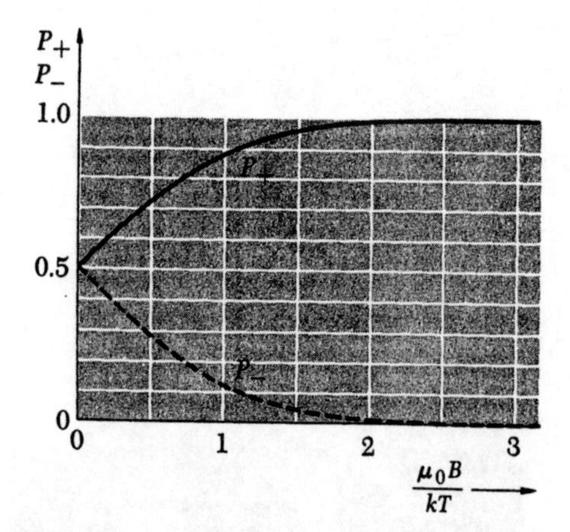

Fig. 4.11 Graph showing the probability P_+ that a magnetic moment μ_0 points parallel (and the probability P_- that it points antiparallel) to an applied magnetic field B when the absolute temperature is T.

In the state $(-)$ the atomic magnetic moment is antiparallel to the field so that $\mu = -\mu_0$. The corresponding energy of the atom is then $\epsilon_- = +\mu_0 B$. The probability P_- of finding the atom in this state is thus

$$P_- = Ce^{-\beta\epsilon_-} = Ce^{-\beta\mu_0 B}. \tag{55}$$

This is the state of higher energy and thus is the state in which the atom is less likely to be found.

The constant C is immediately determined by the normalization requirement that the probability of finding the atom in either one of these states must be unity. Thus

$$P_+ + P_- = C(e^{\beta\mu_0 B} + e^{-\beta\mu_0 B}) = 1$$

or

$$C = \frac{1}{e^{\beta\mu_0 B} + e^{-\beta\mu_0 B}}. \tag{56}$$

Since the atom is more likely to be in the state $(+)$ where its magnetic moment is parallel to the field $\mathbf{B}$, the *mean* magnetic moment $\bar{\mu}$ must point in the direction of the field $\mathbf{B}$. By virtue of (54) and (55), the significant parameter characterizing the orientation of the magnetic moment is the quantity

$$w \equiv \beta\mu_0 B = \frac{\mu_0 B}{kT} \tag{57}$$

which measures the ratio of the magnetic energy $\mu_0 B$ to the characteristic thermal energy kT. It is apparent that if T is very large (i.e., if $w \ll 1$), the probability that the magnetic moment is parallel to the field is almost the same as that it is antiparallel. In this case the magnetic moment is almost completely randomly oriented so that $\bar{\mu} \approx 0$. On the other hand, if T is very small (i.e., if $w \gg 1$), it is much more probable that the magnetic moment is parallel to the field than antiparallel to it. In this case $\bar{\mu} \approx \mu_0$.

All these qualitative conclusions can readily be made quantitative by actually calculating the mean value $\bar{\mu}$. Thus we find

$$\bar{\mu} \equiv P_+(\mu_0) + P_-(-\mu_0) = \mu_0 \frac{e^{\beta\mu_0 B} - e^{-\beta\mu_0 B}}{e^{\beta\mu_0 B} + e^{-\beta\mu_0 B}}. \tag{58}$$

Alternatively, this result can be written in the form

$$\boxed{\bar{\mu} = \mu_0 \tanh\left(\frac{\mu_0 B}{kT}\right)} \tag{59}$$

where we have used the definition of the hyperbolic tangent

$$\tanh w \equiv \frac{e^{w} - e^{-w}}{e^{w} + e^{-w}}. \tag{60}$$

The mean magnetic moment per unit volume of the substance (i.e., its *magnetization*) points then in the direction of the magnetic field. Its magnitude $\bar{M}_0$ is simply given by

$$\bar{M}_0 = N_0\bar{\mu} \tag{61}$$

if there are N_0 magnetic atoms per unit volume.

We can easily verify that $\bar{\mu}$ exhibits the behavior discussed previously in qualitative terms. If $w \ll 1$, then $e^{w} = 1 + w + \cdots$ and $e^{-w} = 1 - w + \cdots$. Hence,

$$\text{for } w \ll 1, \qquad \tanh w = \frac{(1 + w + \cdots) - (1 - w + \cdots)}{2} = w.$$

On the other hand, if $w \gg 1$, then $e^{w} \gg e^{-w}$. Hence,

$$\text{for } w \gg 1, \qquad \tanh w = 1.$$

The relation (59) predicts thus the following limiting behavior:

$$\text{for } \mu_0 B \ll kT, \qquad \bar{\mu} = \mu_0\left(\frac{\mu_0 B}{kT}\right) = \frac{\mu_0^2 B}{kT}; \tag{62}$$

$$\text{for } \mu_0 B \gg kT, \qquad \bar{\mu} = \mu_0. \tag{63}$$

When $\mu_0 B \ll kT$, the value of $\bar{\mu}$ is rather small. By (62), $\bar{\mu}$ is then less than its maximum possible value μ_0 by the ratio $(\mu_0 B/kT)$. Note that $\bar{\mu}$ in this limit is simply proportional to the magnetic field B and inversely proportional to the absolute temperature T. Using (61) and (62), the magnetization then becomes,

$$\text{for } \mu_0 B \ll kT, \qquad \bar{M}_0 = N_0\bar{\mu} = \frac{N_0\mu_0^2 B}{kT} \equiv \chi B \tag{64}$$

where χ is a constant of proportionality independent of B. This parameter χ is called the *magnetic susceptibility* of the substance.†

† The magnetic susceptibility is conventionally defined in terms of the magnetic field H so that $\chi \equiv M_0/H$. But since the concentration N_0 of magnetic atoms is supposed to be small, $H = B$ to excellent approximation.

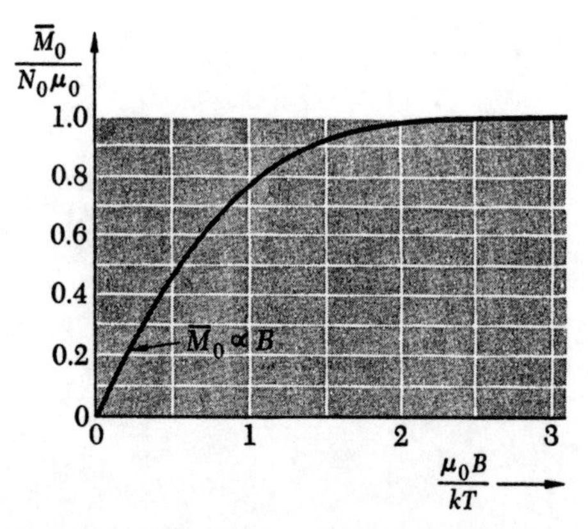

Fig. 4.12 Dependence of the magnetization $\bar{M}_0$ on the magnetic field B and temperature T in the case of weakly interacting magnetic atoms of spin $\frac{1}{2}$ and magnetic moment μ_0.

Equation (64) thus provides the following explicit expression for χ in terms of microscopic quantities:

$$\boxed{\chi = \frac{N_0\mu_0^2}{kT}.} \tag{65}$$

The fact that χ is inversely proportional to the absolute temperature is known as *Curie's law.*

When $\mu_0 B \gg kT$, the mean magnetic moment $\bar{\mu}$ attains its maximum possible value μ_0. Correspondingly the magnetization becomes

$$\text{for } \mu_0 B \gg kT, \qquad \bar{M}_0 = N_0\mu_0 \tag{66}$$

which is its maximum possible (or *saturation*) value and is thus independent of B or T. The complete dependence of the magnetization $\bar{M}_0$ on the absolute temperature T and magnetic field B is shown in Fig. 4.12.

4.7 *Mean Energy of an Ideal Gas*

Consider a gas of N identical molecules, each of mass m, confined within a box of edge lengths L_x, L_y, and L_z. Suppose that the gas is sufficiently dilute, i.e., that the number N of molecules in the given volume $V = L_xL_yL_z$ is sufficiently small so that the mean separation between the molecules is correspondingly large. The following two simplifying conditions are then satisfied:

(i) The mean potential energy of mutual interaction between the molecules is very small compared to their mean kinetic energy. (The gas is then said to be *ideal.*)

(ii) It is permissible to focus attention on any one particular molecule as an entity which can be identified despite the essential indistinguishability of the molecules. (The gas is then said to be *nondegenerate.*)†

We shall assume that the gas is sufficiently dilute so that both of these conditions are satisfied.‡

† This last conclusion is based on the fact that the mean separation between the molecules is then large compared to the typical de Broglie wavelength of a molecule. If this is not the case, quantum-mechanical limitations make it impossible to focus attention unambiguously on a particular molecule and a rigorous quantum-mechanical treatment of a system of indistinguishable particles becomes essential. (The gas is then said to be *degenerate* and is described by either the so-called Bose-Einstein or Fermi-Dirac statistics.)

‡ Condition (ii) is satisfied by practically all ordinary gases; the extent of its validity will be examined more quantitatively at the end of Sec. 6.3. When a gas is made less dilute, condition (i) usually breaks down long before condition (ii) is violated. If the interaction between molecules is very small, however, a gas may satisfy condition (i) so that it is ideal, but violate condition (ii).

Suppose that the gas is in equilibrium at an absolute temperature T. By virtue of condition (ii), we can focus attention on a particular molecule of the gas and regard it as a small system in thermal contact with a heat reservoir (at temperature T) consisting of all the other molecules of the gas. The probability of finding the molecule in any one of its quantum states r, where its energy is ϵ_r, is then simply given by the canonical distribution (49) or (51); i.e.,

$$P_r = \frac{e^{-\beta\epsilon_r}}{\sum_r e^{-\beta\epsilon_r}}, \qquad \text{where } \beta \equiv \frac{1}{kT}. \tag{67}$$

In calculating the energy ϵ_r of the molecule, the condition (i) allows us to neglect any energy of interaction of the molecule with other molecules.

Consider, for instance, the particularly simple case of a *monatomic* gas [such as helium (He) or argon (Ar)] in which each molecule consists of a single atom only. The energy of such a molecule is then simply its kinetic energy. Each possible quantum state r of the molecule is accordingly labeled by some particular values of the three quantum numbers $\{n_x,n_y,n_z\}$ and has an energy given by Eq. (3.15). Thus

$$\epsilon_r = \frac{\pi^2\hbar^2}{2m}\left(\frac{n_x^2}{L_x^2} + \frac{n_y^2}{L_y^2} + \frac{n_z^2}{L_z^2}\right). \tag{68}$$

The probability of finding the molecule in any such state is then given by (67).

On the other hand, consider the case of a *polyatomic* gas [such as oxygen (O_2), nitrogen (N_2), or methane (CH_4)] in which each molecule consists of two or more atoms. Then the energy ϵ of each molecule is given by

$$\epsilon = \epsilon^{(k)} + \epsilon^{(i)}. \tag{69}$$

Here $\epsilon^{(k)}$ is the kinetic energy of translational motion of the center of mass of the molecule, while $\epsilon^{(i)}$ is the intramolecular energy associated with the rotation and vibration of the atoms with respect to the center of mass. Since the center of mass moves like a simple particle having the mass of the molecule, the state of translational motion of the molecule is again specified by the set of three quantum numbers $\{n_x,n_y,n_z\}$ and its translational kinetic energy $\epsilon^{(k)}$ is again given by (68). The state of intramolecular motion is specified by one or more other quantum numbers (denote them collectively simply by n_i) describing the state of rotational and vibrational motion of the atoms in the molecule; the energy $\epsilon^{(i)}$ depends then on n_i. A particular state r of the molecule is then specified by specific values of the quantum numbers $\{n_x,n_y,n_z;n_i\}$ and has a corresponding energy ϵ_r given by

$$\epsilon_r = \epsilon^{(k)}(n_x,n_y,n_z) + \epsilon^{(i)}(n_i). \tag{70}$$

Note that the translational motion of the molecule is affected by the presence of the confining walls of the container; hence $\epsilon^{(k)}$ depends on the dimensions L_x, L_y, L_z of the container, as seen explicitly in (68). On the other hand, the intramolecular motion of the atoms with respect to the molecular center of mass does *not* involve the dimensions of the container; hence $\epsilon^{(i)}$ is independent of the dimensions of the container, i.e.,

$$\epsilon^{(i)} \text{ is independent of } L_x,\ L_y,\ L_z. \tag{71}$$

Calculation of the mean energy

If a molecule is found with probability P_r in a state r of energy ϵ_r, then its mean energy is simply given by

$$\bar{\epsilon} = \sum_r P_r \epsilon_r = \frac{\sum_r e^{-\beta\epsilon_r}\epsilon_r}{\sum_r e^{-\beta\epsilon_r}} \tag{72}$$

where we have used (67) and where the sums are over all possible states r of the molecule. The relation (72) can be considerably simplified by noting that the sum in the numerator can be readily expressed in terms of the sum in the denominator. Thus we can write

$$\sum_r e^{-\beta\epsilon_r}\epsilon_r = -\sum_r \frac{\partial}{\partial\beta}(e^{-\beta\epsilon_r}) = -\frac{\partial}{\partial\beta}\left(\sum_r e^{-\beta\epsilon_r}\right)$$

where we have used the fact that the derivative of a sum of terms is equal to the sum of the derivatives of these terms. Introducing for the denominator in (72) the convenient abbreviation

$$Z \equiv \sum_r e^{-\beta\epsilon_r}, \tag{73}$$

the relation (72) becomes then

$$\bar{\epsilon} = \frac{-\dfrac{\partial Z}{\partial\beta}}{Z} = -\frac{1}{Z}\frac{\partial Z}{\partial\beta}$$

or

$$\boxed{\bar{\epsilon} = -\frac{\partial \ln Z}{\partial\beta}.} \tag{74}$$

The calculation of the mean energy $\bar{\epsilon}$ thus requires only the evaluation of the single sum Z of (73). (The sum Z over all states of the molecule is called the *partition function* of the molecule.)

In the case of a monatomic gas, the energy levels are given by (68) and the sum Z of (73) becomes accordingly†

$$Z = \sum_{n_x} \sum_{n_y} \sum_{n_z} \exp\left[-\frac{\beta\pi^2\hbar^2}{2m}\left(\frac{n_x^2}{L_x^2} + \frac{n_y^2}{L_y^2} + \frac{n_z^2}{L_z^2}\right)\right] \tag{75}$$

where the triple sum is over all possible values of n_x, n_y, and n_z [each ranging over all integral values from 1 to ∞ in accordance with (3.14)]. But the exponential function factors immediately into a simple product of exponentials, i.e.,

$$\exp\left[-\frac{\beta\pi^2\hbar^2}{2m}\left(\frac{n_x^2}{L_x^2} + \frac{n_y^2}{L_y^2} + \frac{n_z^2}{L_z^2}\right)\right]$$

$$= \exp\left[-\frac{\beta\pi^2\hbar^2}{2m}\frac{n_x^2}{L_x^2}\right]\exp\left[-\frac{\beta\pi^2\hbar^2}{2m}\frac{n_y^2}{L_y^2}\right]\exp\left[-\frac{\beta\pi^2\hbar^2}{2m}\frac{n_z^2}{L_z^2}\right].$$

Here n_x appears only in the first factor, n_y only in the second, and n_z only in the third. Hence the sum (75) breaks up into a simple product, i.e.,

$$Z = Z_x Z_y Z_z \tag{76}$$

where

$$Z_x \equiv \sum_{n_x=1}^{\infty} \exp\left[-\frac{\beta\pi^2\hbar^2}{2m}\frac{n_x^2}{L_x^2}\right], \tag{77a}$$

$$Z_y \equiv \sum_{n_y=1}^{\infty} \exp\left[-\frac{\beta\pi^2\hbar^2}{2m}\frac{n_y^2}{L_y^2}\right], \tag{77b}$$

$$Z_z \equiv \sum_{n_z=1}^{\infty} \exp\left[-\frac{\beta\pi^2\hbar^2}{2m}\frac{n_z^2}{L_z^2}\right]. \tag{77c}$$

It only remains to evaluate a typical sum such as Z_x. This is easily done if we note that, for any container where L_x is of macroscopic size, the coefficient of n_x^2 in (77*a*) is very small unless β is extremely large (i.e., unless T is extremely small). Since successive terms in the sum differ thus only very slightly in magnitude, it is an excellent approximation to replace this sum by an integral. By regarding a term in the

† To prevent the long expression in the exponent from overwhelming the poor letter e, we here use the standard alternative notation $\exp u \equiv e^u$ to denote the exponential function.

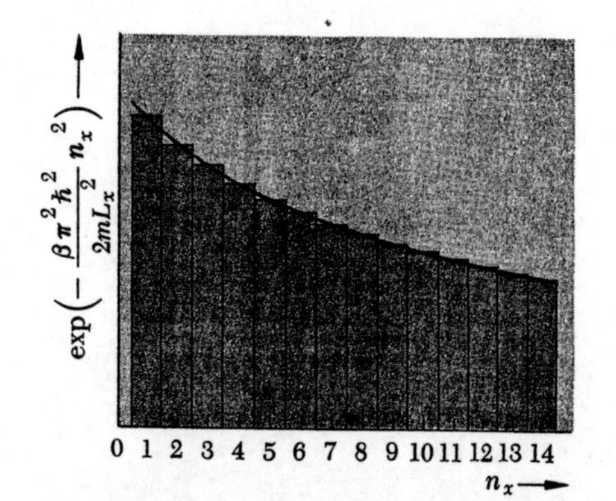

Fig. 4.13 Schematic diagram illustrating the replacement of a sum over integral values of n_x (the sum being equal to the area beneath the rectangles) by an integral over continuous values of n_x (the integral being equal to the area beneath the curve).

sum as a function of n_x (considered as a continuous variable defined also for nonintegral values), the sum Z_x thus becomes

$$Z_x = \int_{1/2}^{\infty} \exp\left[-\frac{\beta\pi^2\hbar^2}{2m}\frac{n_x^2}{L_x^2}\right] dn_x$$

$$= \left(\frac{2m}{\beta}\right)^{1/2}\left(\frac{L_x}{\pi\hbar}\right)\int_0^{\infty} \exp\left[-u^2\right] du, \tag{78}$$

where

$$u \equiv \left(\frac{\beta}{2m}\right)^{1/2}\left(\frac{\pi\hbar}{L_x}\right) n_x \tag{79}$$

or

$$n_x = \left(\frac{2m}{\beta}\right)^{1/2}\left(\frac{L_x}{\pi\hbar}\right) u.$$

The lower limit in the last integral of (78) has been put equal to zero without noticeable error since the coefficient of n_x in (79) is extremely small. The last definite integral in (78) is merely equal to some constant so that (78) is of the form

$$Z_x = b\frac{L_x}{\beta^{1/2}} \tag{80}$$

where b is some constant involving the mass of the molecule.† The corresponding expressions for Z_y and Z_z are, of course, analogous to (80). Hence (76) yields the result

$$Z = \left(b\frac{L_x}{\beta^{1/2}}\right)\left(b\frac{L_y}{\beta^{1/2}}\right)\left(b\frac{L_z}{\beta^{1/2}}\right)$$

or

$$Z = b^3\frac{V}{\beta^{3/2}}, \qquad \text{where } V \equiv L_xL_yL_z \tag{81}$$

is the volume of the box. Correspondingly, we obtain

$$\boxed{\ln Z = \ln V - \tfrac{3}{2}\ln\beta + 3\ln b.} \tag{82}$$

Our calculation is now essentially completed. Indeed, (74) yields for the mean energy $\bar{\epsilon}$ of a molecule the result

$$\bar{\epsilon} = -\frac{\partial \ln Z}{\partial\beta} = -\left(-\frac{3}{2}\frac{1}{\beta}\right) = \frac{3}{2}\left(\frac{1}{\beta}\right).$$

We have thus arrived at the important conclusion that

$$\boxed{\begin{gathered}\text{for a monatomic molecule,}\\ \bar{\epsilon} = \tfrac{3}{2}kT.\end{gathered}} \tag{83}$$

† Although this will be of no particular importance to us, we note that the last integral in (78) has, by (M.21), the value $\sqrt{\pi}/2$; thus $b = (m/2\pi)^{1/2}\hbar^{-1}$.

The mean kinetic energy of a molecule is thus independent of the size of the container and simply proportional to the absolute temperature T of the gas.

If the molecules of the gas are not monatomic, the additive expression (69) yields for the mean energy of a molecule the result

$$\bar{\epsilon} = \bar{\epsilon}^{(k)} + \bar{\epsilon}^{(i)} = \tfrac{3}{2}kT + \overline{\epsilon^{(i)}}(T) \tag{84}$$

since the mean kinetic energy of translation $\overline{\epsilon^{(k)}}$ of the center of mass is again given by (83). The mean intramolecular energy $\overline{\epsilon^{(i)}}$ is, by (71), independent of the dimensions of the container and thus can only be a function of the absolute temperature T.

Since the gas is ideal so that the interaction between the molecules is negligible, the total mean energy $\bar{E}$ of the gas is simply equal to the sum of the mean energies of the N individual molecules. Thus we have

$$\bar{E} = N\bar{\epsilon}. \tag{85}$$

Even in the most general case, the mean energy of an ideal gas is independent of the dimensions of the container and a function of the temperature only; i.e.,

> for an ideal gas,
>
> $$\bar{E} = \bar{E}(T) \tag{86}$$
>
> independent of the dimensions of the container.

This result is physically plausible. The translational kinetic energy and intramolecular energy of a molecule do not depend on the separations between the molecules. Hence a change of the container dimensions (at the fixed temperature T) leaves these energies unchanged so that $\bar{E}$ remains unaffected. If the gas were not ideal, this conclusion would no longer be true. Indeed, if the gas is sufficiently dense, the mean separation between molecules is small enough so that their mean potential energy of mutual interaction *is* appreciable. A change of the container dimensions (at the fixed temperature T) then results in a change in the mean separation of the molecules; correspondingly, it affects their mean intermolecular *potential* energy which is included in the total mean energy $\bar{E}$ of the gas.

4.8 Mean Pressure of an Ideal Gas

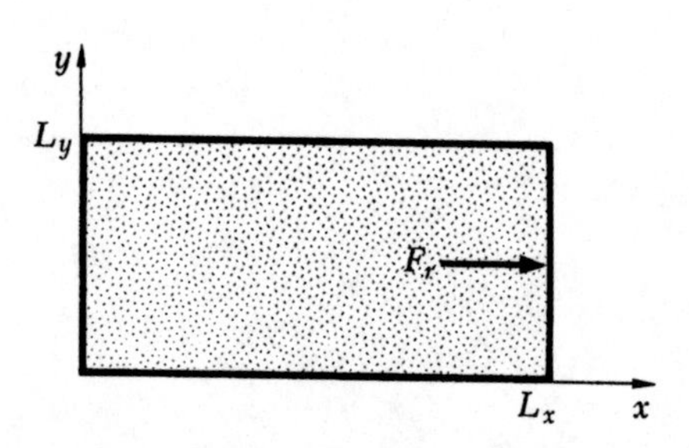

Fig. 4.14 A box containing an ideal gas. A molecule in a given state r exerts on the right wall of the box a force F_r in the x direction.

It is experimentally quite easy to measure the mean pressure (i.e., the mean force per unit area) exerted by a gas on the walls of the container within which it is confined. Hence it is of particular interest to calculate the mean pressure exerted by an ideal gas. Let us denote by F the force in the x direction exerted by a single molecule on the right wall (i.e., the wall $x = L_x$) of the box containing the gas. Let us denote by F_r the value of this force when the molecule is in a particular quantum state r where its energy is ϵ_r. The force F_r can be readily related to the energy ϵ_r. Indeed, suppose that the right wall of the box were displaced very slowly to the right by an amount dL_x. In this process the molecule would do on the wall an amount of work $F_r\,dL_x$ which must be equal to the decrease $-d\epsilon_r$ of the energy of the molecule. Hence

$$F_r\,dL_x = -d\epsilon_r$$

or

$$F_r = -\frac{\partial \epsilon_r}{\partial L_x}. \tag{87}$$

Here we have written a partial derivative to indicate that the other dimensions L_y and L_z were supposed to remain constant in the argument leading to (87).

The *mean* force $\bar{F}$ exerted by a molecule on the wall is then given by averaging the force F_r over all possible states r of the molecule. Thus

$$\bar{F} = \sum_r P_r F_r = \frac{\sum_r e^{-\beta\epsilon_r}\left(-\frac{\partial \epsilon_r}{\partial L_x}\right)}{\sum_r e^{-\beta\epsilon_r}} \tag{88}$$

where we have used the expression (67) for the probability P_r of finding the molecule in any state r. The relation (88) can be simplified since the sum in the numerator can again be expressed in terms of the sum in the denominator. Thus the numerator can be written

$$-\sum_r e^{-\beta\epsilon_r}\frac{\partial \epsilon_r}{\partial L_x} = -\sum_r\left(-\frac{1}{\beta}\right)\frac{\partial}{\partial L_x}(e^{-\beta\epsilon_r})$$
$$= \frac{1}{\beta}\frac{\partial}{\partial L_x}\left(\sum_r e^{-\beta\epsilon_r}\right).$$

Using again the abbreviation Z of (73), the expression (88) becomes

$$\bar{F} = \frac{\frac{1}{\beta}\frac{\partial Z}{\partial L_x}}{Z} = \frac{1}{\beta}\frac{1}{Z}\frac{\partial Z}{\partial L_x}$$

or

$$\boxed{\bar{F} = \frac{1}{\beta}\frac{\partial \ln Z}{\partial L_x}.} \tag{89}$$

This general relation can now be applied to the result (82) already obtained for ln Z in the case of a monatomic molecule. Remembering that $V = L_xL_yL_z$, the partial differentiation yields immediately

$$\bar{F} = \frac{1}{\beta}\frac{\partial \ln Z}{\partial L_x} = \frac{1}{\beta}\frac{\partial \ln V}{\partial L_x} = \frac{1}{\beta L_x}$$

or

$$\boxed{\bar{F} = \frac{kT}{L_x}.} \tag{90}$$

If the molecule is not monatomic, the expression (87) for the force F_r becomes by (70)

$$F_r = -\frac{\partial}{\partial L_x}[\epsilon_r^{(k)} + \epsilon_r^{(i)}] = -\frac{\partial \epsilon_r^{(k)}}{\partial L_x}.$$

Here we have used the fact, noted in (71), that the intramolecular energy $\epsilon^{(i)}$ does not depend on the dimension L_x of the box. Hence the force F_r is calculable from the center-of-mass translational energy alone. The preceding computation, which was based on this translational energy, remains therefore equally valid for a polyatomic molecule. The expression (90) for $\bar{F}$ is accordingly a completely general result.

Since the gas is ideal, the molecules move about without influencing each other appreciably. Hence the *total* mean normal force (i.e., force in the x direction) exerted on the right wall by all the molecules of the gas is simply obtained by multiplying {the mean force $\bar{F}$ exerted by one molecule} by {the total number N of molecules in the gas}. Dividing this result by the area L_yL_z of the wall gives then the mean pressure $\bar{p}$ exerted by the gas on this wall. The relation (90) thus leads to the result

$$\bar{p} = \frac{N\bar{F}}{L_yL_z} = \frac{N}{L_yL_z}\frac{kT}{L_x} = \frac{N}{V}kT.$$

Hence

$$\boxed{\bar{p}V = NkT} \tag{91}$$

or

$$\boxed{\bar{p} = nkT} \tag{92}$$

where $V = L_xL_yL_z$ is the volume of the container and where $n \equiv N/V$ is the number of molecules per unit volume. Note that (92) makes no specific reference to the particular wall which was considered in the calculation. The calculation would thus yield properly the same result for the mean pressure $\bar{p}$ exerted on *any* wall.† ‡

Discussion

The important relations (91) and (92) can be expressed in another equivalent form. Thus the total number N of molecules is ordinarily deduced from macroscopic measurements of the number ν of moles of gas present in the container. Since the number of molecules per mole is, by definition, equal to Avogadro's number N_a, it follows that $N = \nu N_a$. Hence (91) can also be written in the form

$$\bar{p}V = \nu RT \tag{93}$$

where we have introduced a new constant R, called the *gas constant*, defined by the relation

$$R \equiv N_a k. \tag{94}$$

The equation relating the pressure, the volume, and the absolute temperature of a substance in equilibrium is called the *equation of state* of that substance. Thus the equations (91) through (93) are dif-

† Indeed, an elementary (and completely macroscopic) analysis of forces in a fluid in mechanical equilibrium shows that the pressure on any element of area must be the same anywhere in the fluid (if gravity is neglected) and must be independent of the spatial orientation of this element of area.

‡ Note concerning Secs. 4.7 and 4.8:
Although our calculations of the mean energy and pressure have been carried through for a gas confined within a container in the simple shape of a rectangular parallelepiped, all our results are actually quite general and independent of the shape of the container. The physical reason is the following. At any ordinary temperature the momentum of a molecule is so large that its de Broglie wavelength is negligibly small compared to the dimensions of any macroscopic container. Practically every region within the container is thus many wavelengths away from the container walls. Accordingly, the nature of the possible wave functions which can exist in such a region is utterly insensitive to the detailed boundary conditions imposed at the walls or to details involving the precise shape of these walls.

ferent forms of the equation of state for an ideal gas. This equation of state, which we have derived on the basis of our theory, makes several important predictions:

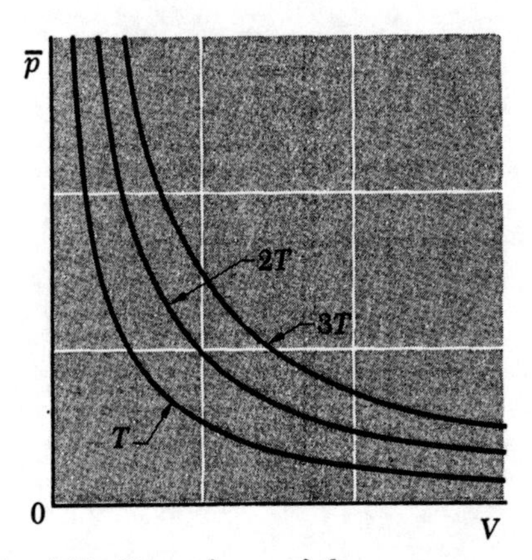

Fig. 4.15 Dependence of the mean pressure $\bar{p}$ of an ideal gas on its volume V at the absolute temperatures T, $2T$, and $3T$.

(i) If a given amount of gas, sufficiently dilute to be ideal, is maintained at a constant temperature, it follows by (91) that

$$\bar{p}V = \text{constant},$$

i.e., that its pressure should be inversely proportional to its volume. This result was discovered experimentally by Boyle in 1662 (long before the advent of the atomic theory of matter) and is accordingly called *Boyle's law*.

(ii) If a given amount of gas, sufficiently dilute to be ideal, is maintained at constant volume, its mean pressure should be proportional to its absolute temperature. This result can be conveniently exploited, as will be shown in the next chapter, as a method for measuring the absolute temperature.

(iii) The equation of state (91) depends only on the number of molecules, but makes no reference to the nature of these molecules. Hence the equation of state should be the same for *any* gas (whether it be He, H_2, N_2, O_2, CH_4, etc.), as long as it is made sufficiently dilute to be ideal. This prediction can be checked experimentally and is well verified.

Summary of Definitions

absolute temperature The absolute temperature T of a macroscopic system [or its related parameter $\beta = (kT)^{-1}$] is defined by

$$\frac{1}{kT} \equiv \beta \equiv \frac{\partial \ln \Omega}{\partial E}$$

where $\Omega(E)$ is the number of states accessible to the system in a small energy interval between E and $E + \delta E$, and where k is a conventionally chosen constant called *Boltzmann's constant.*

entropy The entropy S of a system is defined in terms of the number Ω of states accessible to it by the relation

$$S \equiv k \ln \Omega.$$

The entropy thus provides a logarithmic measure of the degree of randomness of the system.

thermometer A relatively small macroscopic system arranged so that only one of its macroscopic parameters changes when the system gains or loses energy as a result of thermal interaction.

thermometric parameter The variable macroscopic parameter of a thermometer.

temperature of a system with respect to a given thermometer The value of the particular thermometric parameter of this thermometer when the latter is in thermal equilibrium with the system.

heat reservoir A macroscopic system which is sufficiently large with respect to a set of other systems under consideration so that its temperature remains essentially unchanged in any thermal interaction with those systems.

Boltzmann factor The factor $e^{-\beta E}$ where β is related to the absolute temperature T by $\beta \equiv (kT)^{-1}$ and where E denotes an energy.

canonical distribution The probability distribution according to which the probability P_r of finding a system in a state r of energy E_r is given by

$$P_r \propto e^{-\beta E_r}$$

where $\beta = (kT)^{-1}$ is the absolute temperature parameter of the heat reservoir with which the system is in equilibrium.

ideal gas A gas in which the energy of interaction between the molecules is almost negligible compared to their kinetic energy.

nondegenerate gas A gas sufficiently dilute so that the mean separation between molecules is large compared to the mean de Broglie wavelength of a molecule.

equation of state The relation connecting the volume, the mean pressure, and the absolute temperature of a given macroscopic system.

Important Relations

Definition of absolute temperature:

$$\frac{1}{kT} \equiv \beta \equiv \frac{\partial \ln \Omega}{\partial E}. \tag{i}$$

Definition of entropy:

$$S \equiv k \ln \Omega. \tag{ii}$$

Entropy increase of a system at absolute temperature T when it absorbs a small amount of heat $đQ$:

$$dS = \frac{đQ}{T}. \tag{iii}$$

Canonical distribution for a system in thermal equilibrium with a heat reservoir at absolute temperature T:

$$P_r \propto e^{-\beta E_r}. \tag{iv}$$

Equation of state of an ideal nondegenerate gas:

$$\bar{p} = nkT. \tag{v}$$

Suggestions for Supplementary Reading

The following books provide alternative derivations of some of the results of this chapter:

C. W. Sherwin, *Basic Concepts of Physics,* secs. 7.3–7.5 (Holt, Rinehart and Winston, Inc., New York, 1961).

G. S. Rushbrooke, *Introduction to Statistical Mechanics,* chaps. 2 and 3 (Oxford University Press, Oxford, 1949).

F. C. Andrews, *Equilibrium Statistical Mechanics,* secs. 6–8 (John Wiley & Sons, Inc., New York, 1963).

Historical and biographical accounts:

H. Thirring, "*Ludwig Boltzmann,*" *J. of Chemical Education,* p. 298 (June 1952).

E. Broda, *Ludwig Boltzmann; Mensch, Physiker, Philosoph* (Franz Deuticke, Vienna, 1955). In German.

L. Boltzmann, *Lectures on Gas Theory,* translated by S. G. Brush (University of California Press, Berkeley, 1964). The introduction by Brush describes briefly the historical development of the description of matter in terms of atomic concepts.

B. A. Leerburger, *Josiah Willard Gibbs, American Theoretical Physicist* (Franklin Watts, Inc., New York, 1963).

L. P. Wheeler, *Josiah Willard Gibbs, the History of a Great Mind* (Yale University Press, New Haven, 1951; paperback ed., 1962).

M. Rukeyser, *Willard Gibbs* (Doubleday & Company, Inc., Garden City, N.Y., 1942).

Problems

4.1 *Example of a peculiar thermometer*

The density of alcohol, like that of most substances, is found to decrease with increasing absolute temperature. The properties of water are, however, somewhat unusual. As the absolute temperature is increased above the temperature at which water melts (i.e., turns from ice into liquid water), its density at first increases and then, after going through a maximum, decreases.

Suppose that an ordinary thermometer, of the type consisting of a liquid column within a glass tube, is filled with colored water instead of the colored alcohol which is commonly used. The temperature θ indicated by this type of thermometer is, as usual, taken to be the length of the liquid column. Imagine that this thermometer, when placed in contact with two systems A and B, indicates that they have respective temperatures θ_A and θ_B.

(*a*) Suppose that the temperature θ_A of system A is larger (higher) than the temperature θ_B of system B. Can one necessarily conclude that heat will flow from system A to system B when these two systems are placed in thermal contact with each other?

(*b*) Suppose that the temperatures θ_A and θ_B of the two systems are found to be equal. Can one necessarily conclude that no heat will flow from system A to system B when these two systems are placed in thermal contact with each other?

4.2 *Value of kT at room temperature*

One mole of any gas at room temperature and atmospheric pressure (10^6 dynes/cm^2) is found experimentally to occupy a volume of approximately 24 liters (i.e., 24×10^3 cm^3). Use this result to estimate the value of kT at room temperature. Express your answer in ergs and also in electron volts. (1 electron volt $= 1.60 \times 10^{-12}$ erg.)

4.3 *Actual variation of the number of states with energy*

Consider *any* macroscopic system at room temperature.

(*a*) Use the definition of absolute temperature to find the percentage increase in the number of states accessible to such a system when its energy is increased by 10^{-3} electron volt.

(*b*) Suppose that such a system absorbs a single photon of visible light (having a wavelength of 5×10^{-5} cm). By what factor does the number of states accessible to the system increase as a result?

4.4 *Achievement of atomic spin polarization*

Consider a substance which contains magnetic atoms having a spin $\frac{1}{2}$ and a magnetic moment μ_0. Since this moment is due to an unpaired electron, it is of the order of a Bohr magneton, i.e., $\mu_0 \approx 10^{-20}$ erg/gauss. In order to do scattering experiments from atoms whose spins are preferentially polarized in a given direction, one can apply a large magnetic field B and cool the substance to a sufficiently low absolute temperature to achieve appreciable polarization.

The largest magnetic field which can conveniently be produced in the laboratory is about 50,000 gauss. Find the absolute temperature T which must be attained so that the number of atomic moments pointing parallel to the field is at least 3 times as large as the number of moments pointing in the opposite direction. Express your answer in terms of the ratio T/T_R where T_R is room temperature.

4.5 *Proposed method for producing polarized proton targets*

In research in nuclear physics and elementary particles, it is of great interest to do scattering experiments on targets consisting of protons whose spins are preferentially polarized in a given direction. Each proton has a spin $\frac{1}{2}$ and a magnetic moment $\mu_0 = 1.4 \times 10^{-23}$ erg/gauss. Suppose that one tries to apply the method of the preceding problem by taking a sample of paraffin (which contains many protons), applying a magnetic field of 50,000 gauss, and cooling the sample to some very low absolute temperature T. How low would this temperature have to be so that, after equilibrium has been reached, the number of proton moments pointing parallel to the field is at least 3 times as large as the number of proton moments pointing in the opposite direction? Express your answer again in terms of the ratio T/T_R where T_R is room temperature.

4.6 *Nuclear magnetic resonance absorption*

A sample of water is placed in an external magnetic field B. Each proton of the H_2O molecule has a nuclear spin $\frac{1}{2}$ and a small magnetic moment μ_0. Since each proton can point either "up" or "down," it can be in one of two possible states of respective energies $\mp\mu_0 B$. Suppose that one applies a radio-frequency magnetic field of frequency ν which is such that it satisfies the resonance condition $h\nu = 2\mu_0 B$, where $2\mu_0 B$ is the energy difference between these two proton states and h is Planck's constant. Then the radiation field produces transitions between these two states, causing the proton to go from the "up" state to the "down" state, or vice versa, with equal probability. The net power absorbed by the protons from the radiation field is then proportional to the *difference* between the numbers of protons in the two states.

Assume that the protons always remain very close to equilibrium at the absolute temperature T of the water. How does the absorbed power depend on the temperature T? Use the excellent approximation based on the fact that μ_0 is so small that $\mu_0 B \ll kT$.

4.7 *Relative numbers of atoms in an atomic beam experiment*

Precision measurements of the magnetic moment of the electron were instrumental in leading to our modern understanding of the quantum theory of the electromagnetic field. The first such precision experiment, performed by Kusch and Foley [*Physical Review* 74, 250 (1948)], was based on a comparison of the measured total magnetic moments of gallium (Ga) atoms in two different sets of states labeled (in standard spectroscopic notation) as the $^2P_{1/2}$ and $^2P_{3/2}$

states, respectively. The ${}^2P_{1/2}$ states are the states of lowest possible energy of the atom. (There are *two* such states with the same energy; they correspond merely to the two possible spatial orientations of the total angular momentum of the atom in this set of states.) The ${}^2P_{3/2}$ states have an energy higher than that of the ${}^2P_{1/2}$ states by an amount accurately known from spectroscopic measurements and equal to 0.102 electron volt. (There are *four* such states having the same energy; they correspond again merely to the four possible spatial orientations of the total angular momentum of the atom in this set of states.)

To perform the desired comparison, the number of atoms in the ${}^2P_{3/2}$ states must be comparable to the number of atoms in the ${}^2P_{1/2}$ states. The atoms can be produced by heating gallium in an oven to a high absolute temperature T. A small hole in the oven wall then permits a few of the atoms to escape into a surrounding vacuum where they form an *atomic beam* on which the actual measurements are made.

(*a*) Suppose that the absolute temperature T of the oven is $3T_R$, where T_R is room temperature. What is then the proportion of gallium atoms in the beam which are in the ${}^2P_{1/2}$ and ${}^2P_{3/2}$ states, respectively?

(*b*) The highest temperature which can be produced conveniently in such an oven is about $6T_R$. What then is the proportion of gallium atoms in the ${}^2P_{1/2}$ and ${}^2P_{3/2}$ states, respectively? Is this proportion adequate for a successful experiment?

4.8 *Mean energy of a system with two discrete energy levels*

A system consists of N weakly interacting particles, each of which can be in either one of two states with respective energies ϵ_1 and ϵ_2, where $\epsilon_1 < \epsilon_2$.

(*a*) Without explicit calculation, make a qualitative plot of the mean energy $\bar{E}$ of the system as a function of its absolute temperature T. What is $\bar{E}$ in the limit of very low and very high temperatures? Roughly near what temperature does $\bar{E}$ change from its low to its high temperature limiting values?

(*b*) Find an explicit expression for the mean energy of this system. Verify that this expression exhibits the temperature dependence established qualitatively in (*a*).

4.9 *Elastic properties of rubber*

A strip of rubber maintained at an absolute temperature T is fastened at one end to a peg; from its other end hangs a weight w. Assume the following simple microscopic model for the rubber band: It consists of a linked polymer chain of N segments joined end to end; each segment has length a and can be oriented either parallel or antiparallel to the vertical direction. Find an expression for the resultant mean length $\bar{L}$ of the rubber band as a function of w. (Neglect the kinetic energies or weights of the segments themselves, or any interaction between the segments.)

4.10 *Polarization due to impurity atoms in a solid*

The following represents a simple two-dimensional model of a situation of actual physical interest. A solid at absolute temperature T contains N_0 negatively charged impurity ions per unit volume, these ions replacing some of the ordinary atoms of the solid. The solid as a whole is, of course, electrically neutral. This is so because each negative ion with charge $-e$ has in its vicinity one positive ion with charge $+e$. The positive ion is small and thus free to move between lattice sites. In the absence of an external electric field it will, therefore, be found with equal probability in any one of the four equidistant sites surrounding the stationary negative ion. (See Fig. 4.16; the lattice spacing is a.)

If a small electrical field $\mathcal{E}$ is applied along the x direction, calculate the electric polarization, i.e., the mean electric dipole moment per unit volume along the x direction.

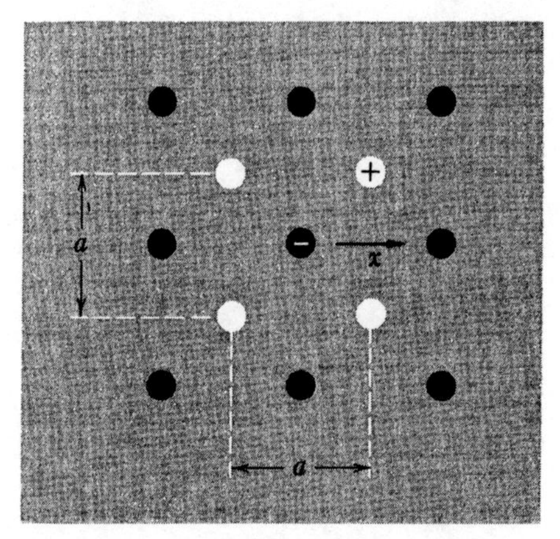

Fig. 4.16 Impurity atoms in the crystal lattice of a solid.

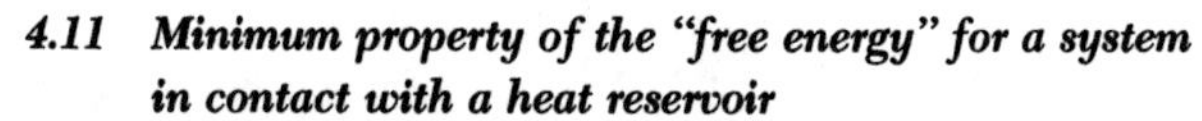

4.11 *Minimum property of the "free energy" for a system in contact with a heat reservoir*

When two systems A and A' are placed in thermal contact, their total entropy tends to increase in accordance with the relation (20), i.e.,

$$\Delta S + \Delta S' \geq 0. \tag{i}$$

The equilibrium situation ultimately attained after system A has absorbed some heat $Q = \Delta \bar{E}$ thus corresponds to that where the total entropy $S + S'$ of the combined isolated system is maximum.

Suppose now that A is small compared to A' so that A' acts as a heat reservoir at some constant absolute temperature T'. The entropy change $\Delta S'$ of A' can then be expressed very simply in terms of $\Delta \bar{E}$ and T'. Show that (i) implies in this case that the quantity $F \equiv \bar{E} - T'S$ tends to *decrease* and becomes a *minimum* in the equilibrium situation. (The function F is called the *Helmholtz free energy* of the system A at the constant temperature T'.)

4.12 *Quasi-static compression of a gas*

Consider a thermally insulated ideal gas of particles confined within a container of volume V. The gas is initially at some absolute temperature T. Assume now that the volume of this container is very slowly reduced by moving a piston to a new position. Give qualitative answers to the following questions:

(*a*) What happens to the energy levels of each particle?

(*b*) Does the mean energy of a particle increase or decrease?

(*c*) Is the work done on the gas in reducing its volume positive or negative?

(*d*) Does the mean energy of a particle, measured above its ground state energy, increase or decrease?

(*e*) Does the absolute temperature of the gas increase or decrease?

4.13 *Quasi-static magnetization of a magnetic substance*

Consider a thermally insulated system consisting of N spins $\frac{1}{2}$, each having a magnetic moment μ_0 and located in an applied magnetic field B. The system is initially at some positive absolute temperature T. Suppose that the magnetic field is now very slowly increased to some new value. Give qualitative answers to the following questions:

(*a*) What happens to the energy levels of each spin?

(*b*) Does the mean energy of each spin increase or decrease?

(*c*) Is the work done on the system in increasing the magnetic field positive or negative?

(*d*) Does the mean energy of a spin, measured above its ground state energy, increase or decrease?

(*e*) Does the absolute temperature of the system increase or decrease?

4.14 *Equation of state of an ideal gas mixture*

Consider a container of volume V which contains a gas consisting of N_1 molecules of one type and N_2 molecules of another type. (For example, these might be O_2 and N_2 molecules.) Assuming that the gas is sufficiently dilute to be ideal, what is the mean pressure $\bar{p}$ of this gas if its absolute temperature is T?

4.15 *Pressure and energy density of an ideal gas*

Use the expressions derived in Secs. 4.7 and 4.8 for the mean pressure $\bar{p}$ and the mean energy $\bar{E}$ of a gas to show that

$$\bar{p} = \tfrac{2}{3}\bar{u} \tag{i}$$

where $\bar{u}$ is the mean *kinetic* energy per unit volume of the gas. Compare the exact result (i) with the expression (1.21) derived in Chap. 1 where approximate reasoning based on classical arguments considered the individual impacts of gas molecules with the container walls.

4.16 *Pressure and energy density of any ideal nonrelativistic gas*

Rederive the result of the preceding problem so as to appreciate its full generality and recognize the origin of the factor $\frac{2}{3}$. Consider thus an ideal gas of N monatomic particles enclosed in a box of edge lengths L_x, L_y, and L_z. If the particle is nonrelativistic, its energy ϵ is related to its momentum $\hbar\mathbf{K}$ by

$$\epsilon = \frac{(\hbar K)^2}{2m} = \frac{\hbar^2}{2m}(K_x^2 + K_y^2 + K_z^2) \tag{i}$$

where the possible values of K_x, K_y, and K_z are given by (3.13).

(*a*) Use this expression to calculate the force F_r exerted by a particle on the right wall of the container when the particle is in a given state r specified by n_x, n_y, n_z.

(*b*) By simply averaging, derive an expression for the mean force $\bar{F}$ in terms of the mean energy $\bar{\epsilon}$ of a particle. Use the symmetry requirement that $\overline{K_x^2} = \overline{K_y^2} = \overline{K_z^2}$ when the gas is in equilibrium.

(*c*) Hence show that the mean pressure $\bar{p}$ exerted by the gas is given by

$$\bar{p} = \tfrac{2}{3}\bar{u} \tag{ii}$$

where $\bar{u}$ is the mean energy per unit volume of the gas.

4.17 ***Pressure and energy density of electromagnetic radiation***

Consider electromagnetic radiation (i.e., a gas of photons) enclosed in a box of edge lengths L_x, L_y, and L_z. Since a photon moves with the speed c of light, it is a *relativistic* particle. Hence its energy ϵ is related to its momentum $\hbar\mathbf{K}$ by

$$\epsilon = c\hbar K = c\hbar(K_x^2 + K_y^2 + K_z^2)^{1/2} \tag{i}$$

where the possible values of K_x, K_y, and K_z are again given by (3.13).

(*a*) Use this expression to calculate the force F_r exerted by a photon on the right wall of the container when the photon is in a given state r specified by n_x, n_y, n_z.

(*b*) By simply averaging, derive an expression for the mean force $\bar{F}$ in terms of the mean energy $\bar{\epsilon}$ of a photon. Use the symmetry argument that $\overline{K_x^2} = \overline{K_y^2} = \overline{K_z^2}$ when the radiation is in equilibrium with the container walls.

(*c*) Hence show that the mean pressure $\bar{p}$ exerted on the walls by the radiation is given by

$$\bar{p} = \tfrac{1}{3}\bar{u} \tag{ii}$$

where $\bar{u}$ is the mean electromagnetic energy per unit volume of the radiation.

(*d*) Why is the constant of proportionality in (ii) equal to $\frac{1}{3}$ instead of being equal to the value $\frac{2}{3}$ derived in the preceding problem for a nonrelativistic gas?

4.18 ***Mean energy expressed in terms of partition function***

Consider any system, no matter how complicated, in thermal equilibrium with a heat reservoir at the absolute temperature $T = (k\beta)^{-1}$. The probability that the system is in any one of its states r of energy E_r is then given by the canonical distribution (49). Obtain an expression for the mean energy $\bar{E}$ of this system. In particular, show that the arguments used in Sec. 4.7 are generally applicable and derive thus the very general result

$$\bar{E} = -\frac{\partial \ln Z}{\partial \beta}. \tag{i}$$

Here

$$Z \equiv \sum_r e^{-\beta E_r} \tag{ii}$$

represents a sum over all possible states of the system and is called the *partition function* of the system.

4.19 ***Mean pressure expressed in terms of partition function***

Consider again the system described in Prob. 4.18. The system is in thermal equilibrium with a heat reservoir at the absolute temperature T, but may be arbitrarily complicated (e.g., it might be a gas, a liquid, or a solid). For the sake of simplicity, assume that the system is confined within a container in the shape of a rectangular parallelepiped with edge lengths L_x, L_y, and L_z. Show that the arguments used in Sec. 4.8 are generally applicable and establish thus the following very general results:

(*a*) Show that the mean force $\bar{F}$ exerted by the system on its right boundary wall can always be expressed in terms of the partition function Z of the system by the relation

$$\bar{F} = \frac{1}{\beta} \frac{\partial \ln Z}{\partial L_x}. \qquad \text{(i)}$$

Here Z is defined by the relation (ii) of the preceding problem.

(*b*) In the case of any isotropic system, the function Z does not depend on the individual dimensions L_x, L_y, and L_z, but is merely a function of the volume $V = L_x L_y L_z$ of the system. Show that (i) implies then that the mean pressure $\bar{p}$ exerted by the system can be expressed in the form

$$\bar{p} = \frac{1}{\beta} \frac{\partial \ln Z}{\partial V}. \qquad \text{(ii)}$$

4.20 ***Partition function of an entire gas***

Consider an ideal gas consisting of N monatomic molecules.

(*a*) Write down the expression for the partition function Z of this entire gas. By exploiting the properties of the exponential function, show that Z can be written in the form

$$Z = Z_0^N \qquad \text{(i)}$$

where Z_0 is the partition function for a single molecule and was already calculated in Sec. 4.7.

(*b*) Use (i) to calculate the mean energy $\bar{E}$ of the gas by means of the general relation derived in Prob. 4.18. Show that the functional form of (i) implies immediately that $\bar{E}$ must be simply N times as large as the mean energy per molecule.

(*c*) Use (i) to calculate the mean pressure $\bar{p}$ of the gas by means of the general relation derived in Prob. 4.19. Show that the functional form of (i) implies again that $\bar{p}$ must be simply N times as large as the mean pressure exerted by a single molecule.

4.21 ***Mean energy of a magnetic moment***

Consider a single spin $\frac{1}{2}$ in contact with a heat reservoir at the absolute temperature T. The spin has a magnetic moment μ_0 and is located in an external magnetic field B.

(*a*) Calculate the partition function Z of this spin.

(*b*) Using your result for Z, apply the general relation (i) of Prob. 4.18 to obtain the mean energy $\bar{E}$ of this spin as a function of T and B.

(*c*) Verify that your expression for $\bar{E}$ satisfies the expression $\bar{E} = -\bar{\mu}B$ where $\bar{\mu}$ is the value of the mean component of magnetic moment previously derived in Eq. (59).

4.22 *Mean energy of a harmonic oscillator*

A harmonic oscillator has a mass and spring constant which are such that its classical angular frequency of oscillation is equal to ω. In a quantum-mechanical description, such an oscillator is characterized by a set of discrete states having energies E_n given by

$$E_n = (n + \tfrac{1}{2})\hbar\omega. \qquad \text{(i)}$$

The quantum number n which labels these states can here assume all the integral values

$$n = 0, 1, 2, 3, \ldots. \qquad \text{(ii)}$$

A particular instance of a harmonic oscillator might, for example, be an atom vibrating about its equilibrium position in a solid.

Suppose that such a harmonic oscillator is in thermal equilibrium with some heat reservoir at the absolute temperature T. To find the mean energy $\bar{E}$ of this oscillator, proceed as follows:

(*a*) First calculate the partition function Z for this oscillator, using the definition (ii) of Prob. 4.18. (To evaluate the sum, note that it is merely a geometric series.)

(*b*) Apply the general relation (i) of Prob. 4.18 to calculate the mean energy of the oscillator.

(*c*) Make a qualitative sketch showing how the mean energy $\bar{E}$ depends on the absolute temperature T.

(*d*) Suppose that the temperature T is very small in the sense that $kT \ll \hbar\omega$. Without any calculation whatever, using only the energy levels of (i), what can you say about the value of $\bar{E}$ in this case? Does the result you obtained in (*b*) properly approach this limiting case?

(*e*) Suppose that the temperature T is very high so that $kT \gg \hbar\omega$. What then is the limiting value of the mean energy $\bar{E}$ obtained in (*b*)? How does it depend on T? How does it depend on ω?

4.23 *Mean rotational energy of a diatomic molecule

The kinetic energy of a diatomic molecule, rotating about an axis perpendicular to the line joining the two atoms, is classically given by

$$E = \frac{\mathbf{J}^2}{2A} = \frac{J^2}{2A}$$

where $\mathbf{J}$ is the angular momentum and A is the moment of inertia of the molecule. In a quantum-mechanical description, this energy can assume the discrete values

$$E_j = \frac{\hbar^2 j(j+1)}{2A} \tag{i}$$

where the quantum number j, which determines the *magnitude* of the angular momentum $\mathbf{J}$, can assume the possible values

$$j = 0, 1, 2, 3, \ldots . \tag{ii}$$

For each value of j, there are $(2j + 1)$ distinct possible quantum states which correspond to the discrete possible spatial orientations of the angular momentum vector $\mathbf{J}$.

Suppose that the diatomic molecule is in a gas in equilibrium at the absolute temperature T. To calculate the mean energy of rotation of this diatomic molecule, proceed as follows:

(*a*) First calculate the partition function Z, using the definition (ii) of Prob. 4.18. (Be careful to remember that this is a sum containing a term for *each* individual state.) Assume that T is sufficiently large so that $kT \gg \hbar^2/2A$, a condition satisfied for most diatomic molecules at room temperature. Show that the sum Z can then be approximated by an integral, using $u = j(j + 1)$ as a continuous variable.

(*b*) Now apply the general relation (i) of Prob. 4.18 to calculate the mean rotational energy $\bar{E}$ of the diatomic molecule in this temperature range.

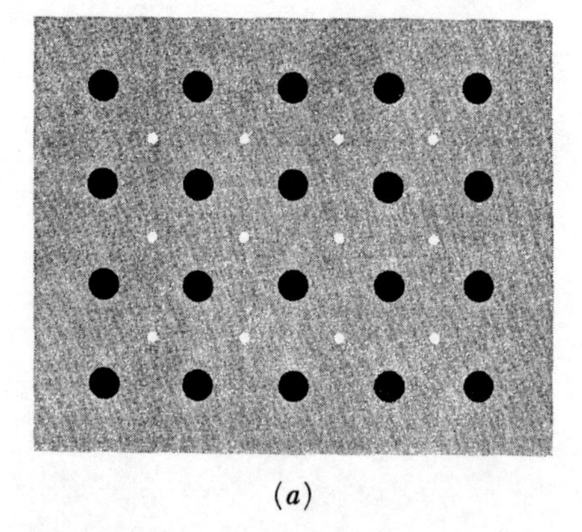

(*a*)

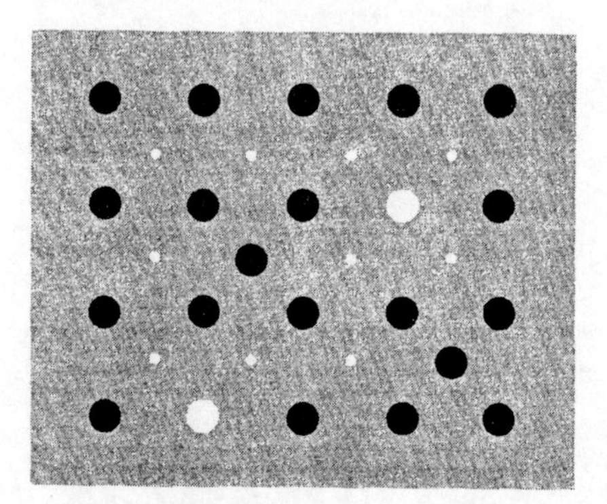

(*b*)

Fig. 4.17 In (*a*) all the atoms of the solid (indicated by black circles) are shown in their normal positions while the possible interstitial positions (indicated by white dots) are unoccupied. At higher temperatures some of the interstitial positions are also occupied, as shown in (*b*).

4.24 *Number of interstitial atoms in a solid (approximate analysis)*

Consider a monatomic crystalline solid consisting of N atoms and maintained at an absolute temperature T. The atoms are ordinarily located at the normal lattice position indicated by the black circles in Fig. 4.17*a*. An atom may, however, also be located at one of the ***interstitial positions*** indicated by the white dots in this figure. If an atom is in such an interstitial position, its energy is larger by an amount ϵ than when it is in a normal position. When the absolute temperature is very low, all atoms will, therefore, be in normal positions. When the absolute temperature T is appreciable, however, this is no longer the case. Suppose that there are N atoms which can be located in N possible normal and N possible interstitial positions. The question of interest is then the following: At any absolute temperature T, what is the mean number $\bar{n}$ of atoms located in interstitial positions? An approximate way of solving the problem is the following:

(*a*) Focus attention first on an individual atom and consider that it can be in either one of only two particular positions—one normal, the other interstitial. This system can then be in only two possible configurations, call them A and B:

(A): atom in normal position,
no atom in interstitial position.
(B): no atom in normal position,
atom in interstitial position.

What is the ratio P_B/P_A of the probabilities P_B and P_A of encountering these two configurations?

(*b*) Focusing attention now on the whole solid, suppose that there are $\bar{n}$ atoms in interstitial positions. Then there must also be $\bar{n}$ atoms absent from normal positions. Since any one of the $\bar{n}$ empty normal positions can be combined with any one of the $\bar{n}$ occupied interstitial positions, a B configuration can arise in any of $\bar{n}^2$ possible ways. According to this argument, the probability P_B of encountering a single atom in the solid in a B configuration should be simply proportional to $\bar{n}^2$ if the empty normal and occupied interstitial positions are assumed to be distributed at random. Thus $P_B \propto \bar{n}^2$. Show by a similar argument that $P_A \propto (N - \bar{n})^2$.

(*c*) Combining the results of parts (*a*) and (*b*) and assuming the usual situation where $\bar{n} \ll N$, show that

$$\frac{\bar{n}}{N} = e^{-(1/2)\beta\epsilon}. \qquad \text{(i)}$$

4.25 *Number of interstitial atoms in a solid (exact analysis)*

Consider the physical situation described in Prob. 4.24. To do the problem exactly, try to find the probability $P(n)$ that precisely n interstitial positions are occupied by atoms. There are then, of course, also n normal positions unoccupied by atoms.

(*a*) What is the probability of encountering a given situation where n interstitial atoms are distributed in *one* particular way and the n empty normal positions are also distributed in *one* particular way?

(*b*) In how many ways is it possible to distribute the n atoms over the N possible interstitial positions? In how many ways is it possible to distribute the n missing atoms over the N normal positions?

(*c*) Combining the results of parts (*a*) and (*b*), show that

$$P(n) \propto \left[\frac{N!}{n!(N-n)!}\right]^2 e^{-\beta n\epsilon}. \qquad \text{(i)}$$

(*d*) The probability $P(n)$ has a sharp maximum for some value $n = \tilde{n}$. To find this value $\tilde{n}$, consider $\ln P(n)$ and satisfy the condition $(\partial \ln P/\partial n) = 0$. Since all factorials are large, Stirling's approximation (M.10) is applicable. Show thus that, for $\tilde{n} \ll N$, one obtains the result

$$\frac{\tilde{n}}{N} = e^{-(1/2)\beta\epsilon}. \qquad \text{(ii)}$$

4.26 *Thermal dissociation of an atom

An ideal gas of atoms is confined within a box with edge lengths L_x, L_y, and L_z. The whole system is in equilibrium at some absolute temperature T. The mass of an atom is M. An atom A can dissociate into a positive ion A^+ and an electron e^- according to the scheme

$$A \rightleftarrows A^+ + e^-$$

An *ionization energy* u is required to overcome the electron binding and achieve ionization.

Focusing attention on an individual atom, it can thus be in two possible configurations, call them U and D.

(U): The atom is undissociated. Its energy E is given by

$$E = \epsilon$$

where ϵ is the kinetic energy of its center of mass. The translational state of motion of the atom can, as usual, be specified by the set of quantum numbers $\{n_x, n_y, n_z\}$.

(D): The atom is dissociated into an electron of mass m and a positive ion of mass nearly equal to M (since $m \ll M$). The interaction between the ion and electron after dissociation can be assumed to be negligible. The total energy of the dissociated system consisting of two separate particles is then equal to

$$E = \epsilon^+ + \epsilon^- + u \tag{i}$$

where ϵ^+ is the kinetic energy of the ion, ϵ^- is the kinetic energy of the electron, and u is the ionization energy. The translational state of this dissociated system can then be specified by six quantum numbers, the set of quantum numbers $\{n_x^+, n_y^+, n_z^+\}$ of the ion and the set of quantum numbers $\{n_x^-, n_y^-, n_z^-\}$ of the electron.

(*a*) Using the canonical distribution find, to within a constant of proportionality C, the probability P_U that the atom is found among those of its states where it is in the undissociated configuration (U).

(*b*) Using the canonical distribution find, to within the same constant of proportionality C, the probability P_D that the atom is found among those of its states where it is in the dissociated configuration (D). [Note that the required sum over all the relevant states is the one already performed in calculating Z in Eq. (81) of Sec. 4.7.]

(*c*) Find the ratio P_D/P_U. How does it depend on the temperature T and the volume V?

(*d*) Now consider the whole gas containing N atoms. Suppose that a mean number $\bar{n}$ of these are dissociated. Then there exist in the box $\bar{n}$ ions and $\bar{n}$ electrons, while there remain $(N - \bar{n})$ undissociated atoms. A dissociated configuration of an atom can then be realized in $\bar{n} \times \bar{n} = \bar{n}^2$ possible ways, an undissociated one in $(N - \bar{n})$ possible ways. By approximate reasoning similar to that of Prob. 4.24, one can then write

$$\frac{P_D}{P_U} = \frac{\bar{n}^2}{N - \bar{n}} \approx \frac{\bar{n}^2}{N}$$

if $\bar{n} \ll N$. Find thus an explicit expression for $(\bar{n}/N)$ in terms of the absolute temperature T and the density (N/V) of the gas.

(*e*) Ordinarily $kT \ll u$. Under these conditions would you expect most atoms to be dissociated or not?

(*f*) Suppose that $kT \ll u$, but that the volume of the box is made arbitrarily large while keeping the temperature T constant. Would most atoms then be dissociated or not? Give a simple physical explanation to account for this result.

(*g*) The inner part of the sun consists of very hot dense gases, while its outer corona is cooler and less dense. Studies of spectral lines of light from the sun indicate that an atom may be ionized in the corona, and yet be un-ionized in regions closer to the sun where the absolute temperature is much higher. How do you explain these observations?

4.27 *Thermal generation of a plasma

By heating a gas of atoms to sufficiently high temperature, one may generate a *plasma* consisting of an appreciable number of dissociated positive and negative charges. To study the practical possibility of this procedure, apply the results of Prob. 4.26 to cesium vapor. The cesium atom has a rather low ionization energy $u = 3.89$ electron volts and an atomic weight of 132.9.

(*a*) Express the degree of dissociation $\bar{n}/N$ of Prob. 4.26 in terms of T and the mean pressure $\bar{p}$ of the gas.

(*b*) Suppose that one heats cesium vapor to an absolute temperature 4 times as large as room temperature and maintains it at a pressure of 10^3 dynes/cm^2 (i.e., 10^{-3} of atmospheric pressure). Calculate the percentage of vapor ionized under these conditions.

4.28 *Dependence of energy on temperature for an ideal gas*

The number of states $\Omega(E)$ of an ideal gas of N monatomic atoms depends on the total energy E of the gas in a manner derived in Prob. 3.8. Use this result and the definition $\beta = \partial \ln \Omega/\partial E$ to derive a relation expressing the energy E as a function of the absolute temperature $T = (k\beta)^{-1}$. Compare your result with the expression for $\bar{E}(T)$ derived in Sec. 4.7.

4.29 *Dependence of energy on temperature for a spin system*

The number of states $\Omega(E)$ of a system of N spins $\frac{1}{2}$, each having a magnetic moment μ_0 and located in a magnetic field B, has been calculated in Prob. 3.9.

(*a*) Use this result and the definition $\beta = (\partial \ln \Omega/\partial E)$ to derive a relation expressing the energy E of this system as a function of the absolute temperature $T = (k\beta)^{-1}$.

(*b*) Since the total magnetic moment M of this system is simply related to its total energy E, use the answer to part (*a*) to find an expression for M as a function of T and B. Compare this expression with the result derived for $\bar{M}_0$ in (61) and (59).

4.30 *Negative absolute temperature and heat flow in a spin system

A system consists of N spins $\frac{1}{2}$, each having magnetic moment μ_0 and located in an external magnetic field B. The number of states $\Omega(E)$ of this system has already been calculated as a function of its total energy E in Prob. 3.9.

(*a*) Make an approximate sketch showing the behavior of $\ln \Omega$ as a function of E. Note that the lowest energy of the system is $E_0 = -N\mu_0 B$ and its highest energy is $+N\mu_0 B$, and that the curve is symmetric about the value $E = 0$.

(*b*) Use the curve of part (*a*) to make an approximate sketch showing β as a function of E. Note that $\beta = 0$ for $E = 0$.

(*c*) Use the curve of part (*b*) to make an approximate sketch showing the absolute temperature T as a function of E. What happens to T near $E = 0$? What is the sign of T for $E < 0$ and for $E > 0$?

(*d*) Since T suffers a discontinuity near $E = 0$, it is more convenient to work in terms of β. Show that $\partial\beta/\partial E$ is always negative. Hence prove that, when two systems are placed in thermal contact, heat is always absorbed by the system with the larger value of β. Note that this last statement is generally valid for all systems, irrespective of whether their absolute temperatures are positive or negative.

Chapter 5

Microscopic Theory and Macroscopic Measurements

Chapter 5 *Microscopic Theory and Macroscopic Measurements*

We have now made substantial progress in our understanding of macroscopic systems on the basis of their atomic constituents. In the process we have found it useful to introduce several parameters (such as heat, absolute temperature, and entropy) which serve to describe the *macro*scopic behavior of systems consisting of many particles. Although these parameters have all been carefully defined in terms of *micro*scopic concepts, they ought to be amenable to *macro*scopic measurement. Indeed, any comparison between theory and experiment requires that such measurements be made. This is true regardless of whether the predictions of the theory concern relationships between purely macroscopic quantities, or whether they deal with connections between macroscopic quantities and atomic characteristics. As usual, it is the task of any physical theory to suggest the significant quantities which ought to be measured, and to specify the operations to be used for performing such measurements. We shall devote the present chapter to this aspect of the theory. In short, we shall try to build a firm bridge between fairly abstract atomic and statistical concepts on the one hand, and very direct macroscopic observations on the other.

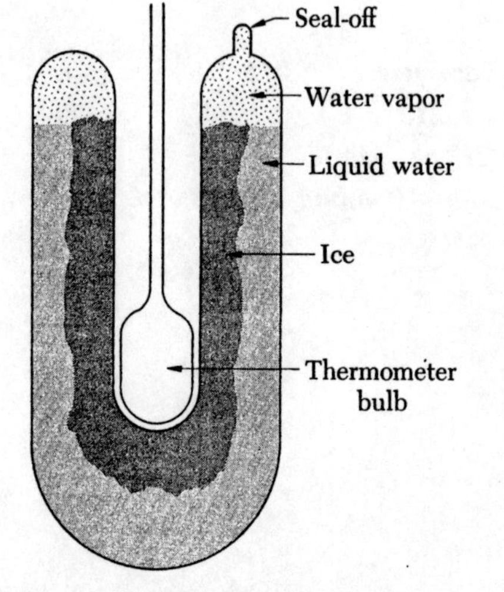

Fig. 5.1 Schematic diagram illustrating a triple-point cell designed to calibrate a thermometer at the triple point of water. A freezing mixture [such as acetone mixed with "dry ice" (i.e., solid CO_2)] is first put into the central well in order to transform some of the water into ice. After the freezing mixture is removed, the thermometer is placed in the well and the system is allowed to come to thermal equilibrium.

5.1 *Determination of the Absolute Temperature*

The absolute temperature is a very important parameter since it appears explicitly in most predictions of the theory. Let us, therefore, examine what procedure might be used for actually measuring the absolute temperature of a system. In principle, such a procedure can be based upon *any* theoretical relation which involves β or T. For example, Eq. (4.65) provides a theoretical relation showing how the susceptibility χ of a paramagnetic substance depends on T. Measurements of the susceptibility of a suitable paramagnetic substance should, therefore, provide a method for measuring the absolute temperature. Another theoretical relation involving T is the equation of state (4.91) of an ideal gas. Any gas, sufficiently dilute to be ideal, can thus be used to measure the absolute temperature. In practice, this is indeed a very convenient method useful in many cases.

To use the equation of state (4.91) as a basis for measuring the absolute temperature T, one can proceed as follows. A small amount of gas is put into a bulb and provisions are made for keeping the volume V of this gas constant irrespective of its pressure.† This constitutes then a constant-volume gas thermometer, of the type shown in Fig. 4.4, whose thermometric parameter is the mean pressure $\bar{p}$ of the gas. Suppose that one has measured the fixed volume V and the number of moles of gas in the bulb (so that one knows the number N of gas molecules). A measurement of $\bar{p}$ then yields, by virtue of (4.91), the pertinent value of kT or β of the gas (and hence also of any other system with which the gas thermometer may be in thermal equilibrium).

The preceding comments complete the discussion of the measurement of β, the absolute temperature parameter of physical importance. The remainder of this section will deal solely with some conventional definitions in common use. If one wants to write β in the form $\beta^{-1} = kT$ and to assign a numerical value to T itself, one must *choose* a particular value for the constant k. The particular choice adopted for this purpose by international convention is motivated by the fact that experimentally it is easier to make a comparison between two absolute temperatures than to measure directly the value of β or kT. It is, therefore, most convenient to specify a procedure for obtaining numerical values of T by means of temperature comparisons, and to let the numerical value of k be determined accordingly.

To achieve the desired temperature comparison, one chooses a standard system in a standard macrostate and assigns to it, by definition, a certain value of the absolute temperature T. By international convention one chooses as this standard system pure water, and as its standard macrostate that where the solid, liquid, and gas forms of water (i.e., ice, liquid water, and water vapor) are in equilibrium with each other. (This macrostate is called the *triple point* of water.) The reason for this choice is that there is only one definite value of pressure and temperature at which all these three forms of water can coexist in equilibrium; as can readily be verified experimentally, the temperature of this system is then unaffected by any changes in the relative amounts of solid, liquid, and gas present under these circumstances. The triple point provides, therefore, a very reproducible standard of temperature. By international convention, adopted in 1954, one then chooses to assign to the absolute temperature T_t of water at its triple point the value

$$\boxed{T_t \equiv 273.16 \text{ exactly.}} \tag{1}$$

† The amount of gas in the bulb of the thermometer must be small enough to guarantee that the gas is sufficiently dilute to be ideal. Experimentally this can be checked by verifying that, if an absolute temperature determination were made with a thermometer containing a smaller amount of gas, the result obtained would be the same.

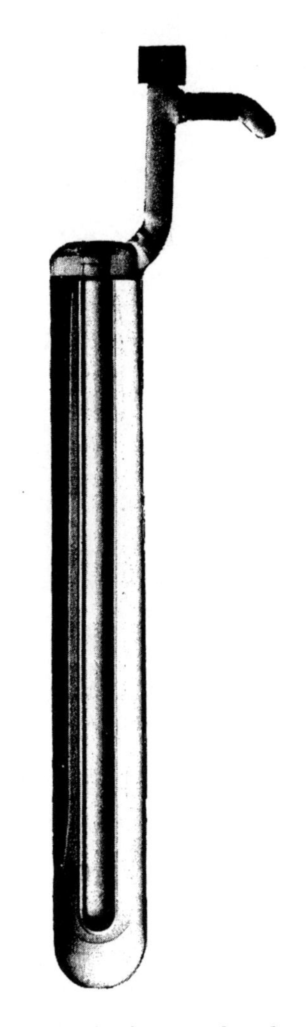

Fig. 5.2 Photograph of a typical triple-point cell used by the National Bureau of Standards to realize the triple point of water. (*Photograph by courtesy of the National Bureau of Standards.*)

This peculiar choice was motivated by the desire that the modern absolute temperature scale thus defined should yield values of T that agree as closely as possible with the less accurate values obtained according to an older and more cumbersome convention.

The numerical value of the absolute temperature of *any* system can then be obtained by comparing it with the temperature T_t of water at its triple point. A numerical value thus determined as a result of the particular choice (1) is said to be expressed in terms of *degrees Kelvin* or simply *degrees K*, commonly abbreviated as "°K." (Indeed, whenever we shall use the term *degree* without further qualifications, we shall always understand it to mean *degree Kelvin.*) The particular choice (1), which serves to define the *Kelvin temperature scale,* then also fixes the value of k. Indeed, if one uses some device (such as a gas thermometer) to measure the value of β or kT at the triple point of water where $T = T_t$, the value of k is immediately determined. Since the quantity $\beta^{-1} = kT$ represents an energy which can be measured in ergs, the constant k can then be expressed in units of ergs/degree.

Let us now illustrate how these conventions can be used to measure absolute temperatures with a constant-volume ideal gas thermometer. By virtue of the equation of state (4.91), the gas pressure $\bar{p}$ measured on this thermometer is directly proportional to the absolute temperature of the gas. The thermometer, therefore, permits ready measurements of absolute temperature *ratios* in terms of pressure ratios. Indeed, suppose that the thermometer is brought into thermal contact with some system A; after it has reached equilibrium, its mean pressure will then have some definite value $\bar{p}_A$. Similarly, suppose that the thermometer is brought into thermal contact with some other system B; after it has reached equilibrium, its mean pressure will then have some definite value $\bar{p}_B$.† The equation of state (3.91) then implies that the absolute temperatures T_A and T_B of A and B are related by

$$\frac{T_A}{T_B} = \frac{\bar{p}_A}{\bar{p}_B}. \tag{2}$$

In particular, suppose that the system B consists of water at its triple point (so that $T_B = T_t$) and that the thermometer in equilibrium with this system indicates a mean pressure $\bar{p}_t$. Using the convention (1), the absolute temperature of A then has the specific value

$$T_A = 273.16 \frac{\bar{p}_A}{\bar{p}_t} \quad \text{degrees Kelvin.} \tag{3}$$

† We assume that the gas thermometer is sufficiently small compared to the systems A and B so that their absolute temperatures are not appreciably affected by being placed in contact with the thermometer.

The absolute temperature of any system thus can be readily determined by measuring the pressure of a constant-volume gas thermometer. This particular method of measuring the absolute temperature is quite convenient provided that the temperature is not so low or so high that the use of gas thermometers becomes impractical.

With the absolute temperature scale fixed by the particular convention (1), one can then use the equation of state of an ideal gas to determine the numerical value of the constant k (or equivalently, of the constant $R \equiv N_a k$, where N_a is Avogadro's number). Taking ν moles of any ideal gas at the triple-point temperature $T_t = 273.16°\text{K}$, one need only measure its volume V (in cm^3) and its corresponding mean pressure $\bar{p}$ (in dynes/cm^2). This information then permits computation of R by (4.93). Careful measurements of this type yield for the *gas constant* R the value†

$$R = (8.31434 \pm 0.00035) \text{ joules mole}^{-1} \text{ deg}^{-1} \tag{4}$$

(where 1 joule $\equiv 10^7$ ergs). But Avogadro's number N_a is known to have the value‡

$$N_a = (6.02252 \pm 0.00009) \times 10^{23} \text{ molecules mole}^{-1}. \tag{5}$$

The definition $R \equiv N_a k$ of the gas constant thus yields for k the value

$$k = (1.38054 \pm 0.00006) \times 10^{-16} \text{ erg deg}^{-1}. \tag{6}$$

As we have pointed out before, k is called *Boltzmann's constant.*§

On the Kelvin temperature scale an energy of 1 electron volt corresponds to an energy kT where $T \approx 11{,}600°\text{K}$. Also, room temperature is approximately 295°K; it corresponds to an energy $kT \approx 1/40$ electron volt, which represents roughly the mean kinetic energy of a gas molecule at room temperature.

Fig. 5.3 Lord Kelvin (born William Thomson) (1824–1907). Born in Scotland and showing very early evidence of intellectual brilliance, he was appointed at the age of 22 to the Chair of Natural Philosophy at Glasgow University where he remained for more than fifty years. He made important contributions to electromagnetism and hydrodynamics. Using purely macroscopic reasoning, he and the German physicist E. Clausius (1822–1888) formulated the "second law of thermodynamics" which served to establish the existence and fundamental properties of the entropy function. His analysis led him also to introduce the concept of absolute temperature. In recognition of his work, he was raised to the peerage and thus assumed the name Lord Kelvin. The absolute temperature scale is named in his honor. *(The photograph, provided by courtesy of the National Portrait Gallery in London, is from a portrait by Elizabeth Thomson King in 1886.)*

† In terms of calories, the value of R is

$$R = (1.98717 \pm 0.00008) \text{ calories mole}^{-1} \text{ deg}^{-1}$$

All indicated errors correspond to one standard deviation.

‡ This value is with respect to the modern *unified* scale of atomic weights where the ^{12}C atom is assigned an atomic weight of *exactly* 12. The best experimental determinations of Avogadro's number are based on electrical measurements of the electrical charge required to decompose by electrolysis a known number of moles of a compound (e.g., water), combined with atomic measurements of the charge of the electron.

§ The numerical values of all these physical constants are those given by E. R. Cohen and J. W. M. DuMond, *Rev. Mod. Phys.* **37**, 590 (1965). See also the table of numerical constants at the end of this book.

Another temperature scale sometimes used is the Celsius (or centigrade) temperature θ_C *defined* in terms of the absolute Kelvin temperature T by the relation

$$\theta_C \equiv (T - 273.15) \qquad \text{degrees Celsius} \tag{7}$$

(abbreviated as "°C"). On this scale, water at atmospheric pressure freezes at approximately 0°C and boils at approximately 100°C.†

5.2 *High and Low Absolute Temperatures*

In order to gain some intuitive appreciation of the absolute temperature scale, a few representative temperatures are listed in Table 5.1. Here the *melting point* of a substance is that temperature where the solid and liquid forms of the substance coexist in equilibrium (at a pressure of 1 atmosphere). Above that temperature the substance is a liquid. The *boiling point* of a substance is that temperature where the liquid and gaseous forms of the substance coexist in equilibrium (at a pressure of 1 atmosphere). Above that temperature the substance at this pressure forms a gas. E.g., at the melting point, water turns from ice to liquid water; at the boiling point, it turns from liquid water into water vapor, i.e., into a gas.

Surface temperature of the sun	5500°K
Boiling point of tungsten (W)	5800°K
Melting point of tungsten	3650°K
Boiling point of gold (Au)	3090°K
Melting point of gold	1340°K
Boiling point of lead (Pb)	2020°K
Melting point of lead	600°K
Boiling point of water (H_2O)	373°K
Melting point of water	273°K
Human body temperature	310°K
Room temperature (approximate)	295°K
Boiling point of nitrogen (N_2)	77°K
Melting point of nitrogen	63°K
Boiling point of hydrogen (H_2)	20.3°K
Melting point of hydrogen	13.8°K
Boiling point of helium (He)	4.2°K

Table 5.1 Some representative temperatures.

Let us consider any ordinary macroscopic system. Its absolute temperature T is then positive‡ and kT has a magnitude of the order of the mean energy (above its ground state E_0) per degree of freedom of the system; i.e., in accordance with (4.30),

$$kT \sim \frac{\bar{E} - E_0}{f}. \tag{8}$$

Since every system has a lowest possible energy, the energy E_0 of its ground state, it follows that the absolute temperature has a minimum possible value $T = 0$ which the system attains when its energy approaches that of its ground state. As the energy of the system increases above E_0, its absolute temperature also increases. There is no upper

† The Fahrenheit temperature θ_F, still used in everyday life in the United States, is defined in terms of θ_C as

$$\theta_F \equiv 32 + 1.8\,\theta_C \text{ degrees Fahrenheit.}$$

‡ The special case of a spin system with energy so high that its absolute temperature is negative was discussed in Prob. 4.29.

limit on how high the absolute temperature can be; this corresponds to the fact that there is no upper limit on the possible magnitude of the kinetic energy of the particles in any ordinary system. For example, temperatures of the order of 10^7 °K can occur in stars or in nuclear fusion explosions on earth.

The preceding comments are consequences of the definition of the absolute temperature,

$$\frac{1}{kT} \equiv \beta \equiv \frac{\partial \ln \Omega}{\partial E} \tag{9}$$

and of the behavior of $\ln \Omega$ as a function of E, as illustrated in Fig. 4.5. Let us look more closely at the limiting case where $E \to E_0$, i.e., where the energy of the system approaches its lowest possible ground state value. The number $\Omega(E)$ of states accessible to the system in any small energy interval between E and $E + \delta E$ then approaches a value Ω_0 which is very small. Indeed, as already pointed out in Sec. 3.1, a system has only one quantum state (or at most a relatively small number of such states) corresponding to its lowest possible energy. Even if the number of states of the system in the energy interval δE near E_0 were as large as f, $\ln \Omega_0$ would only be of the order of $\ln f$; it would thus still be utterly negligible compared to its value at higher energies which is of the order of f in accordance with (3.41). The entropy $S = k \ln \Omega$ of the system near its ground state energy E_0 is thus vanishingly small compared to its value at larger energies. Hence we are led to the following conclusion: As the energy of a system is decreased toward its lowest possible value, the entropy of the system tends to become negligibly small; or in symbols,

$$\text{as } E \to E_0, \qquad S \to 0. \tag{10}$$

The number of states increases very rapidly as the energy of the system is increased above its ground state; by (4.29), one finds approximately

$$\beta \equiv \frac{\partial \ln \Omega}{\partial E} \sim \frac{f}{E - E_0}.$$

As the energy E decreases to its lowest possible value E_0, β becomes exceedingly large and $T \propto \beta^{-1} \to 0$. The limiting relation (10), valid for any system, then can equally well be written in the form

$$\boxed{\text{as } T \to 0, \qquad S \to 0.} \tag{11}$$

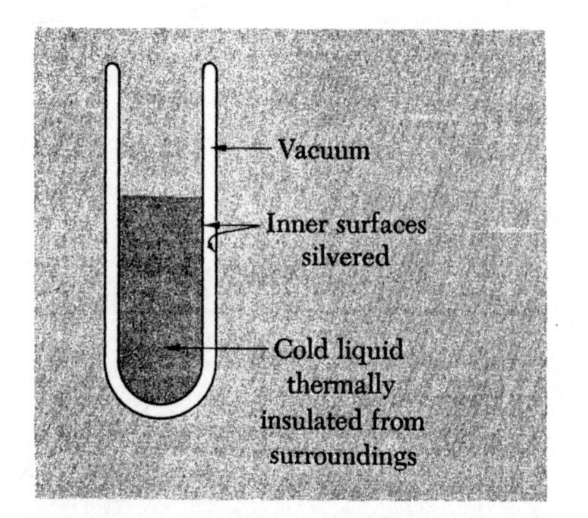

Fig. 5.4 A *dewar flask* of the type used for work at low temperatures. Such a dewar [named after Sir James Dewar (1842–1923) who first liquefied hydrogen in 1898] is similar to the thermos bottle contained in many lunch boxes. It can be made of either glass or metal (e.g., stainless steel) and serves to keep a liquid contained within it thermally insulated from the external surroundings. The thermal insulation is provided by the evacuated space between the two walls. In the case of a dewar made of glass, it is desirable to coat the glass with a reflecting silver layer so as to minimize heat influx in the form of radiation.

The statement (11) is called the *third law of thermodynamics.* In working at temperatures near $T \approx 0$ (near *absolute zero,* in common terminology) one must, however, be careful that the system under consideration is actually in equilibrium; this is particularly true since the rate of attaining equilibrium may become quite slow at such low temperatures. Furthermore, one must have sufficient understanding about the system to interpret the limiting statement (11) properly, i.e., to know how small the temperature must be in practice to justify application of (11). The following remark provides a specific example.

Remark on nuclear spin entropy

Nuclear magnetic moments are so very small that it would be necessary (in the absence of large external magnetic fields) to go to temperatures less than 10^{-6} °K before the mutual interaction between these nuclei could lead to nonrandom orientations of their spins.† Even at a temperature T_0 as low as 10^{-3} °K, the nuclear spins would thus be randomly oriented in the same manner as at any higher temperature. In accordance with (11), the entropy S_0 associated with all degrees of freedom *not* involving the nuclear spins would indeed have become negligibly small at the temperature T_0; the *total* entropy, however, would then still have the large value $S_0 = k \ln \Omega_s$ associated with the total number Ω_s of states corresponding to the possible orientations of the nuclear spins. One would thus obtain, instead of (11), the statement

$$\text{as } T \to 0_+, \qquad S \to S_0. \tag{12}$$

Here $T \to 0_+$ denotes a limiting temperature (such as $T_0 = 10^{-3}$ °K) which is very small, yet large enough so that the spins remain randomly oriented. The statement (12) is still very useful because S_0 is a definite constant which depends only on the kinds of atomic nuclei contained within the system, but which is completely independent of any details involving the energy levels of the system. In short, S_0 is a constant completely independent of the structure of the system, i.e., independent of the spatial arrangement of its atoms, of the nature of their chemical combination, or of the interactions between them. For example, consider a system A consisting of one mole of metallic lead (Pb) and one mole of sulfur (S); consider also another system A' consisting of one mole of the compound lead sulfide (PbS). The properties of these two systems are very different, but they do consist of the same numbers and kinds of atoms. In the limit as $T \to 0_+$, the entropies of both systems should, therefore, be the same.

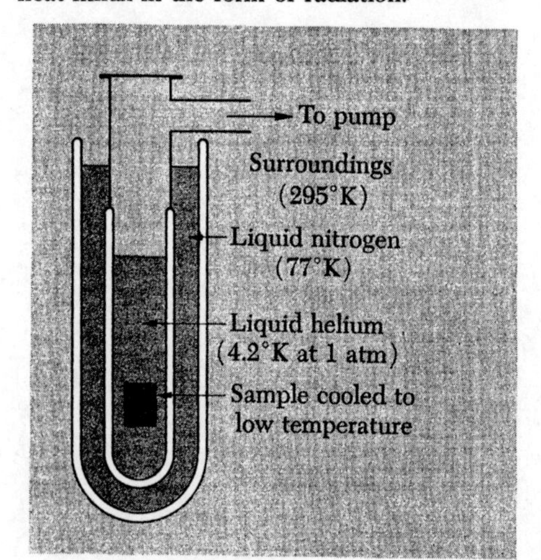

Fig. 5.5 A typical double-dewar apparatus commonly used for work near 1°K. The dewar filled with liquid helium is immersed in another dewar filled with liquid nitrogen in order to minimize the amount of heat influx into the liquid helium.

The study of a system at low absolute temperatures is often of great interest precisely because its entropy is then quite small. Correspondingly, the system is found distributed among a relatively small number of states only. The system manifests, therefore, a much higher degree

† See Prob. 5.2.

of order (or much lower degree of randomness) than would be the case at higher temperatures. Because of this high degree of order, many substances at very low temperatures exhibit quite remarkable properties. The following represent a few striking examples: The electronic spins of some materials can become almost perfectly oriented in the same direction so that these materials become permanent magnets. The conduction electrons in many metals, such as lead or tin, can move without any trace of friction if such a metal is cooled below a sharply defined temperature (7.2°K in the case of lead); such a metal then exhibits current flow without any trace of electrical resistance and is said to be *superconducting*. Similarly, liquid helium (which at atmospheric pressure remains a liquid even down to $T \to 0$) exhibits below a temperature of 2.18°K completely frictionless flow and an uncanny ability to pass with great ease through holes even less than 10^{-6} cm in size; the liquid is then said to be *superfluid*. The region of very low temperatures thus contains a wealth of interesting phenomena. Since any system near $T = 0$ is in states very close to its ground state, quantum mechanics is essential to the understanding of its properties. Indeed, the degree of randomness at these low temperatures is so small that discrete quantum effects can be observed on a *macro*scopic scale. The preceding comments should suffice to indicate why *low temperature physics* is a very active field of contemporary research.

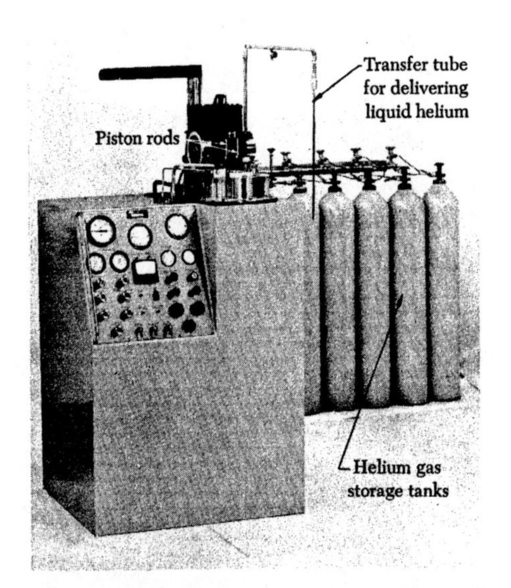

Fig. 5.6 Photograph of a commercially available liquefier for producing liquid helium starting from helium gas at room temperature. With the necessary accessory equipment, consisting of a compressor and gas holder, such a liquefier can produce several liters of liquid helium per hour. The initial cooling of the gas is obtained by letting the thermally insulated gas do mechanical work on some pistons (the piston rods of which can be seen at the top of the machine); in the process the mean energy, and hence also absolute temperature, of the gas is correspondingly reduced. Helium was first liquefied by the Dutch physicist Kamerlingh Onnes in 1908. *(Photograph by courtesy of Arthur D. Little, Inc.)*

The following interesting question arises quite naturally: In practice, how close can a macroscopic system be brought to its ground state, i.e., to how low an absolute temperature can it be cooled? With modern techniques temperatures down to 1°K can be readily achieved by immersing the system under investigation in a bath consisting of liquid helium. The boiling point of this liquid can be reduced to about 1°K by reducing the pressure of the vapor over the liquid with a suitable pump.† By the same method applied to pure liquid ^{3}He (consisting entirely of the rare isotope ^{3}He instead of the abundant isotope ^{4}He), one can attain temperatures down to 0.3°K without excessive difficulty. With appreciably greater efforts it is feasible to work at absolute temperatures as low as 0.01°K, or even 0.001°K, by using methods involving the performance of magnetic work on a thermally insulated system of spins. In this way it has even been possible to attain temperatures as low as 10^{-6} °K.

† The principle of the method should be familiar to hikers ambitious enough to have cooked on mountain expeditions. The boiling point of water on a mountain top is lower than at sea level because of the reduced atmospheric pressure.

5.3 Work, Internal Energy, and Heat

The concepts of heat and work were introduced in Sec. 3.7. The discussion there was summarized by the basic relation (3.53),

$$\Delta \bar{E} = W + Q, \tag{13}$$

which connects the increase in the mean energy $\bar{E}$ of any system to the macroscopic work W done on it and the heat Q absorbed by it. This relation provides the basis for the macroscopic measurement of all the quantities appearing in it. Indeed, it suggests the following method of approach: The macroscopic work is a quantity familiar from mechanics. It is readily measured since essentially it is merely the product of some macroscopic force multiplied by some macroscopic displacement. By insulating the system thermally, one can guarantee that $Q = 0$ in (13); the measurement of the mean energy $\bar{E}$ of the system can thus be reduced to the measurement of work. When the system is not thermally insulated, the heat Q absorbed by it can be determined by (13) if one makes use of the information previously obtained about its mean energy $\bar{E}$ and measures any work W that may be done upon it.

Now that we have outlined the general procedure, let us consider the measurement of the various quantities in greater detail and give specific examples.

Work

In accordance with its definition (3.51), the macroscopic work done on a system is given by the increase in mean energy of the system when the latter is thermally insulated (or *adiabatically isolated*) and some external parameter is changed. This increase in mean energy can then be calculated from primitive notions of mechanics; i.e., it can ultimately be reduced to the product of a force multiplied by the displacement through which it acts. Strictly speaking, the calculation of the mean energy change requires the computation of the mean value of this product for the systems of a statistical ensemble. In dealing with a macroscopic system, however, the force and displacement are macroscopic quantities which are almost always equal to their mean values since they exhibit fluctuations of negligible magnitude. In practice, a measurement on a single system is therefore sufficient for the determination of a mean energy change and the corresponding work.

The following examples illustrate some common procedures used for measuring work.

Example (i) Mechanical work

Figure 5.7 shows a system A consisting of a container filled with water, a thermometer, and a paddle wheel. This system can interact with the rather simple system A' consisting of a weight and the earth which exerts a known gravitational force w upon this weight. The two systems can interact since the falling weight causes the paddle wheel to rotate and thus to churn the water. This interaction is adiabatic since the only connection between the systems is the string which transmits negligible heat. The external parameter describing the system A' is the distance s of the weight below the level of the pulley. If the weight descends a distance Δs without change of velocity, the mean energy of the system A' is reduced by an amount $w\,\Delta s$, the decrease in potential energy of the weight (resulting from the work done upon it by gravity).† Since the combined system consisting of A and A' is isolated, the mean energy of the system A must then *increase* by an amount $w\,\Delta s$ in the process; i.e., an amount of work $w\,\Delta s$ is done upon the adiabatically isolated system A by the falling weight of A'. The measurement of the work done on A is thus reduced to the measurement of the distance Δs.

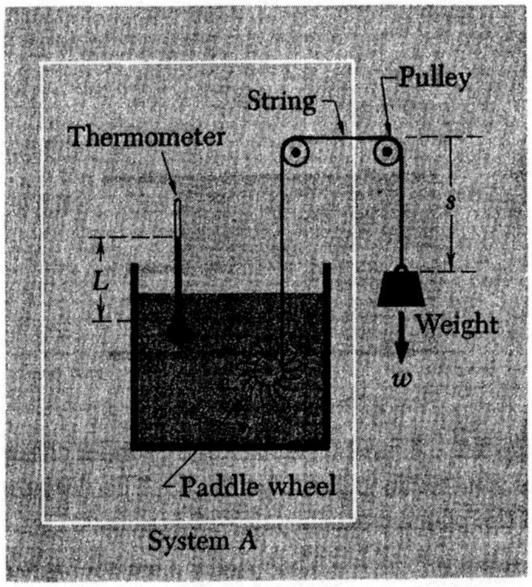

Fig. 5.7 A system A consisting of a container filled with water, a thermometer, and a paddle wheel. Work can be done on this system by the falling weight.

Example (ii) Electrical work

Work can usually be done most conveniently and measured most accurately by electrical means.‡ Figure 5.8 shows such an arrangement, one which is completely analogous to that of Fig. 5.7. Here the system A consists of a container filled with water, a thermometer, and an electrical resistor. A battery of known emf V can be connected to the resistor by wires thin enough to keep the system A thermally insulated from the battery. The charge q which the battery can deliver is its external parameter. When the battery delivers a charge Δq which then passes through the resistor, the work performed by the battery on A in this process is simply $V\,\Delta q$. This charge Δq can readily be measured by determining the time Δt during which a known current i passes through the battery; thus $\Delta q = i\,\Delta t$. The resistor here plays a role completely analogous to that of the paddle wheel in the preceding example, both being merely convenient devices on which work can be done.

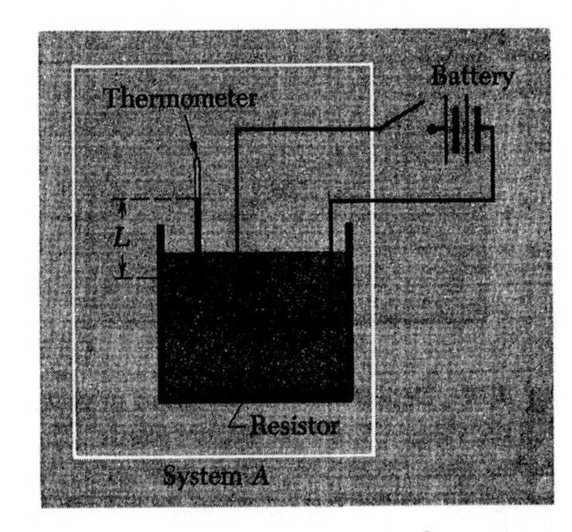

Fig. 5.8 A system A consisting of a container filled with water, a thermometer, and an electrical resistor. Work can be done on this system by the battery.

It is always particularly simple to discuss a process which is *quasi-static,* i.e., which is carried out sufficiently slowly so that the system under consideration is at all times arbitrarily close to equilibrium. Let us thus consider the important case of a fluid (i.e., of a substance which is either a gas or a liquid) and derive an expression for the work done on this fluid in a quasi-static process. Since the fluid then is always essentially in equilibrium, the nonuniformities of density and other complications present in rapidly changing situations are absent; in-

† The weight usually descends with constant velocity since it reaches its terminal velocity very quickly. If the velocity of the weight were changing, the change in the mean energy of A' would be given by the change in the sum of the potential and kinetic energies of the weight.

‡ The work is then, of course, ultimately still mechanical, but involves electrical forces.

stead, the fluid is always characterized by a well-defined mean pressure $\bar{p}$ uniform throughout the fluid. For the sake of simplicity, let us imagine that the fluid is contained in a cylinder closed by a piston of area $\mathcal{A}$, as shown in Fig. 5.9. The external parameter of this system is the distance s of the piston from the left wall or, equivalently, the volume $V = \mathcal{A}s$ of the fluid. Since pressure is defined as force per unit area, the mean force exerted by the fluid on the piston is $\bar{p}\mathcal{A}$ to the right; correspondingly, the mean force exerted by the piston on the fluid is $\bar{p}\mathcal{A}$ to the left. Suppose now that the piston is moved very slowly to the right by an amount ds (so that the volume of the fluid is changed by an amount $dV = \mathcal{A}\, ds$). The work done on the fluid is then simply given by

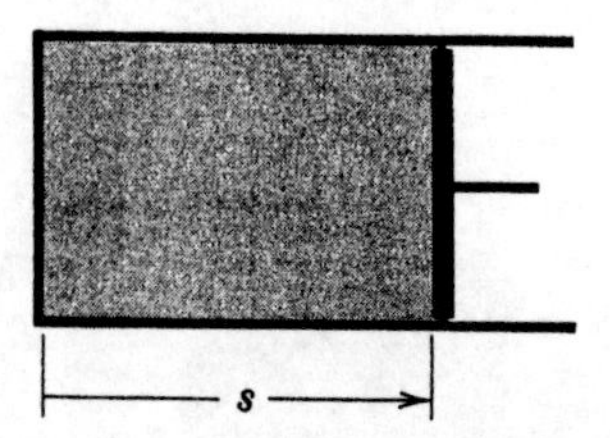

Fig. 5.9 A fluid contained in a cylinder closed by a movable piston of area $\mathcal{A}$. The distance of the piston from the left wall is denoted by s.

$$đW = (-\bar{p}\mathcal{A})\, ds = -\bar{p}(\mathcal{A}\, ds)$$

or

$$\boxed{đW = -\bar{p}\, dV} \tag{14}$$

where the minus sign occurs since the displacement ds and the force $\bar{p}\mathcal{A}$ acting on the gas have opposite directions.†

If the volume of the fluid is changed quasi-statically from some initial volume V_i to some final volume V_f, its pressure $\bar{p}$ at any stage of this process will be some function of its volume and temperature. The total work W done on the fluid in the process is then simply obtained by adding all the infinitesimal works (14); thus

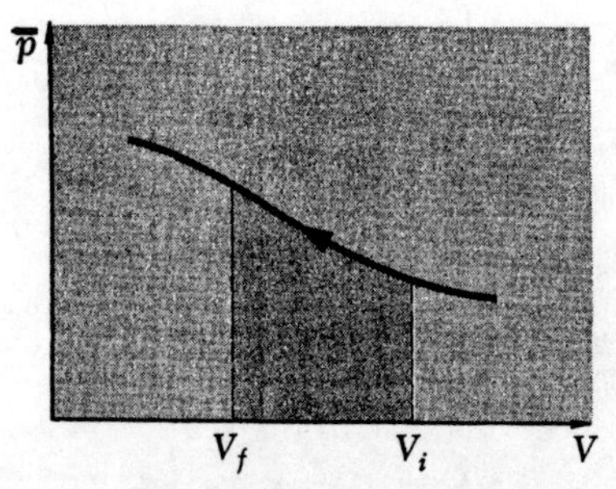

Fig. 5.10 Dependence of the mean pressure $\bar{p}$ on the volume V of a particular system. The shaded area under the curve represents the work done on the system when its volume is changed quasi-statically from V_i to V_f.

$$W = -\int_{V_i}^{V_f} \bar{p}\, dV = \int_{V_f}^{V_i} \bar{p}\, dV. \tag{15}$$

The work done *on* the fluid is positive if $V_f < V_i$ and negative if $V_f > V_i$. By virtue of (15), its magnitude is equal to the shaded area contained below the curve of Fig. 5.10.

Internal energy

Let us now turn our attention to the determination of the *internal energy* E of a macroscopic system (i.e., the total energy of all its particles in the frame of reference where the center of mass of the system

† It can readily be shown that the relation (14) is generally valid for a fluid enclosed in a container of volume V and of *arbitrary* shape. See, for example, F. Reif, *Fundamentals of Statistical and Thermal Physics*, p. 77 (McGraw-Hill Book Company, New York, 1965).

is at rest).† We recall from mechanics that the energy of a system (in particular, its potential energy) is always undefined to within an arbitrary constant. The same remark applies, of course, to the mean internal energy $\bar{E}$ of a macroscopic system. The value of $\bar{E}$ of the system in a given macrostate has significance only when measured with respect to its value in some standard macrostate of this system. Only *differences* in mean energy are thus relevant physically and such energy differences can always be measured by the performance of work if the system is kept adiabatically isolated. The following example will illustrate the procedure.

Example (iii) Electrical measurement of internal energy

Consider the system A of Fig. 5.8. Its macrostate can be specified by a single macroscopic parameter, its temperature, since all its other macroscopic parameters (such as its pressure) are kept constant. This temperature need *not* be the absolute temperature of the system; indeed, we shall suppose it to be merely the length L of the column of liquid of the arbitrary thermometer in thermal contact with the liquid. We shall denote by $\bar{E}$ the mean internal energy of the system when it is in equilibrium in a macrostate characterized by a temperature reading L. We shall denote by $\bar{E}_a$ the mean internal energy of the system when it is in equilibrium in some standard macrostate a characterized by the particular temperature reading L_a. (The value of $\bar{E}_a$ may be chosen to be zero without loss of generality.) The question of interest is then the following: What is the value of the mean internal energy $\bar{E} - \bar{E}_a$ of the system, measured with respect to its standard macrostate a, when this system is in any macrostate characterized by a temperature L?

To answer this question, we keep the system A thermally insulated, as it is in Fig. 5.8. Starting with the system in its macrostate a, we now perform on the system a certain amount of work $W = V\,\Delta q$ by passing a measured amount of total charge Δq through the resistor. We then let the system come to equilibrium and measure its temperature parameter L. By virtue of the relation (13) with $Q = 0$, the mean energy $\bar{E}$ of the system in its new macrostate is then given by

$$\bar{E} - \bar{E}_a = W = V\,\Delta q.$$

Thus we have found the value of $\bar{E}$ corresponding to the particular temperature L.

We can now proceed to repeat this type of experiment many times, doing a different amount of work on the system in each experiment. Similarly we can obtain information about macrostates having a mean energy $\bar{E}$ less than $\bar{E}_a$; we need merely start in such a macrostate characterized by a temperature L and measure the amount of work necessary to bring the system to its standard macrostate of temperature L_a. As a result of this series of experiments, we obtain a set of values of $\bar{E}$ corresponding to various values of the temperature parameter L. This information can be presented in the form of a graph of the type shown in Fig. 5.11. Our task has then been accomplished. Indeed, if the system is in equilibrium in a macrostate specified by a temperature L, its mean internal energy (with respect to the standard macrostate a) can now be ascertained immediately from the graph.

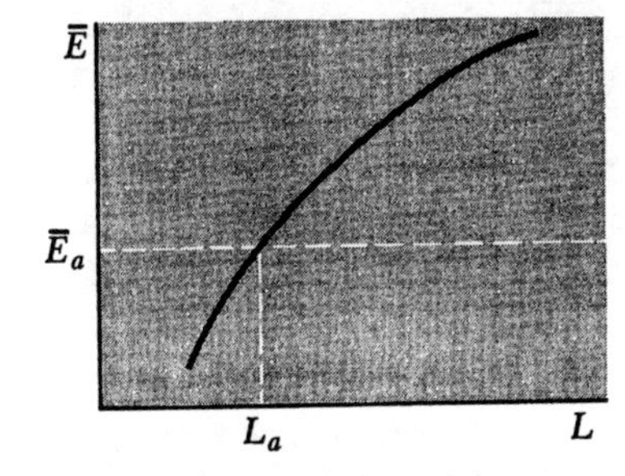

Fig. 5.11 Graph showing schematically how the mean internal energy $\bar{E}$ of the system A of Fig. 5.8 depends on the thermometer reading L.

† The internal energy is, of course, the total energy in the usual case where the system as a whole is at rest in the laboratory. If the whole system were moving, its total energy would differ from its internal energy merely by the kinetic energy associated with its center of mass.

Heat

The measurement of heat (commonly called *calorimetry*) is, by virtue of (13), ultimately reducible to the measurement of work. The heat Q absorbed by a system thus can be measured in two slightly different ways: either by measuring it directly in terms of work, or by comparing it with the known change of internal energy of some other system which gives off the heat Q. The two methods are illustrated by the following examples:

Example (iv) Direct measurement of heat in terms of work

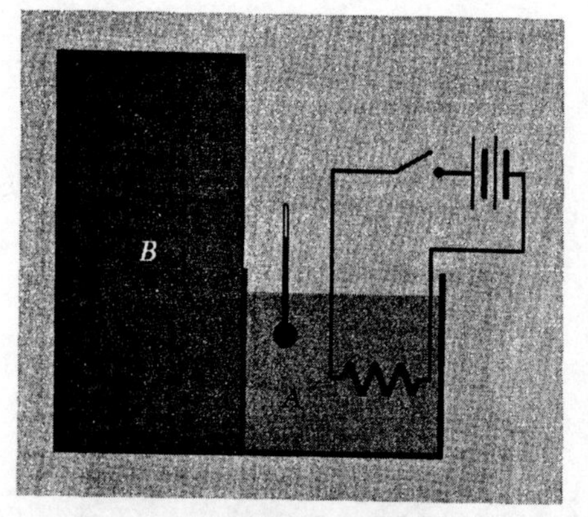

Fig. 5.12 Direct measurement in terms of work of the heat Q_B absorbed by a system B. In practice the auxiliary system A, containing the resistor and thermometer, is usually much smaller than the system B on which the measurement is performed.

Figure 5.12 shows a system B in thermal contact with the system A of Fig. 5.8. Here B can be any macroscopic system, such as a copper block or a container filled with water. The external parameters of B are supposed to be fixed so that it can do no work. Hence it can only interact with A by absorbing from A some amount of heat Q_B. Suppose that one starts from some initial macrostate a where the entire system $A + B$ is in equilibrium and the thermometer reading is L_a. After the battery has done a certain amount of work W, the entire system then attains a final equilibrium situation b where the thermometer reading is L_b. How large is the amount of heat Q_B absorbed by B in this process?

The combined system $A + B$ is thermally insulated. Hence it follows by (13) that the work W done on this system merely serves to increase its mean energy, i.e.,

$$W = \Delta \bar{E}_A + \Delta \bar{E}_B \tag{16}$$

where $\Delta \bar{E}_A$ is the increase in mean energy of A and $\Delta \bar{E}_B$ that of B. But since no work is done on B itself, (13) applied to B implies simply that

$$\Delta \bar{E}_B = Q_B, \tag{17}$$

i.e., that the mean energy of B increases merely by virtue of the heat it absorbs from A. Hence (16) and (17) yield

$$Q_B = W - \Delta \bar{E}_A. \tag{18}$$

Here the work W done by the battery can be directly measured. In practice, the auxiliary system A containing the resistor and thermometer is usually small compared to the system B of interest. In this case the mean energy change of A is negligible (in the sense that $\Delta \bar{E}_A \ll W$ or $\Delta \bar{E}_A \ll Q_B$) and (18) gives simply $Q_B = W$. In the more general case, one can make use of prior measurements on the system A alone to find from the graph in Fig. 5.11 the mean energy change $\Delta \bar{E}_A$ corresponding to the temperature change from L_a to L_b. The relation (18) then yields the heat absorbed by B.

Note that a set of measurements of the type just described allows one to determine the mean internal energy $\bar{E}_B$ of B as a function of its macroscopic parameters.

Example (v) Measurement of heat by comparison

It is also possible to measure the heat Q_C absorbed by any system C by simply comparing the heat Q_C with the heat given off by some other system, such as B, whose internal energy is already known as a function of its temperature. For example, the system C might be a copper block and the system B (discussed in our preceding example) might be a container of water with a thermometer. Now suppose that B and C are simply brought into contact, e.g., by immersing the copper block in the water. The entire system $B + C$ is supposed to be thermally insulated and all external parameters remain unchanged. The conservation of energy applied to the thermal interaction between B and C then requires that

$$Q_C + Q_B = 0 \tag{19}$$

where Q_C is the heat absorbed by C and Q_B is that absorbed by B. But we can read the thermometer in the initial equilibrium situation of B (before B was brought into contact with C), and can read it also in the final situation where B and C are in equilibrium with each other. Hence we know the mean energy change $\Delta \bar{E}_B$ of B in this process and thus the heat $Q_B = \Delta \bar{E}_B$ which it has absorbed. Hence (19) yields immediately the heat Q_C absorbed by C.

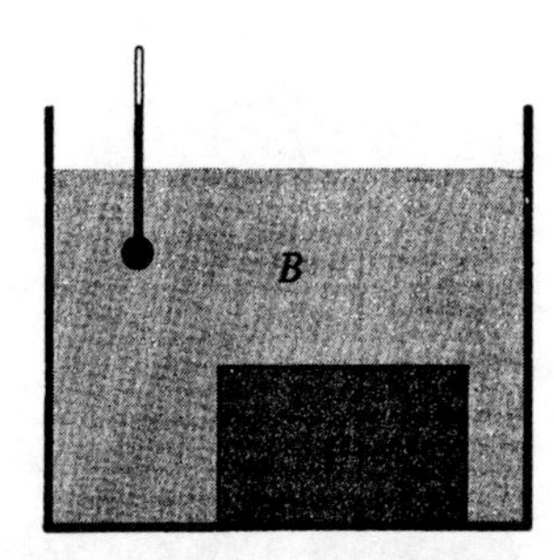

Fig. 5.13 The heat absorbed by a copper block C is measured by comparing it with the heat given off by a known system B consisting of a container of water and a thermometer.

In conclusion, it is worth emphasizing that the entire discussion of this section was based merely upon the conservation of energy and upon the equation (13) which defines the concepts of heat and work. The specific experimental procedures which we have illustrated by various examples can perhaps best be summarized by the following homely analogy due to H. B. Callen†:

> A certain gentleman owns a little pond, fed by one stream and drained by another. The pond also receives water from an occasional rainfall and loses it by evaporation, which we shall consider as "negative rain." In the analogy we wish to pursue the pond is our system, the water within it is the internal energy, water transferred by the streams is work, and water transferred as rain is heat.
>
> The first thing to be noted is that no examination of the pond at any time can indicate how much of the water within it came by way of the stream and how much came by way of rain. The term rain refers only to a method of water transfer.
>
> Let us suppose that the owner of the pond wishes to measure the amount of water in the pond. He can purchase flow meters to be inserted in the streams, and with these flow meters he can measure the amount of stream water entering and leaving the pond. But he cannot purchase a rain meter. However, he can throw a tarpaulin over the pond, enclosing the pond in a wall impermeable to rain (an *adiabatic wall*). The pond owner conse-

† H. B. Callen, *Thermodynamics*, pp. 19–20 (John Wiley & Sons, Inc., New York, 1960). The passage is quoted here with permission of the publisher.

quently puts a vertical pole into the pond, covers the pond with his tarpaulin, and inserts his flow meters into the streams. By damming one stream and then the other, he varies the level in the pond at will, and by consulting his flow meters he is able to calibrate the pond level, as read on his vertical stick, with total water content ($\bar{E}$). Thus, by carrying out processes on the system enclosed by an adiabatic wall, he is able to measure the total water content of any state of his pond.

Our obliging pond owner now removes his tarpaulin to permit rain as well as stream water to enter and leave the pond. He is then asked to ascertain the amount of rain entering his pond during a particular day. He proceeds simply: he reads the difference in water content from his vertical stick, and from this he deducts the total flux of stream water, as registered by his flow meters. The difference is a quantitative measure of the rain.

Fig. 5.14 James Prescott Joule (1818–1889). The son of an English brewer to whose business he succeeded, Joule undertook a systematic investigation to measure heat directly in terms of work. In his experiments he used paddle wheels and electrical resistors to do work in the manner described in our examples (i) and (ii). His careful and precise measurements, first published in 1843 and extending over a period of about 25 years, served to establish conclusively that heat is a form of energy and that the principle of conservation of energy is generally valid. The energy unit of the *joule* is, of course, named in Joule's honor. *(From G. Holton and D. Roller, "Foundations of Modern Physical Science," Addison-Wesley Publishing Co., Inc., Cambridge, Mass., 1958. By permission of the publishers.)*

5.4 *Heat Capacity*

Consider a macroscopic system whose macrostate can be specified by its absolute temperature T and by some other set of macroscopic parameters denoted collectively by y. For example, y might be the volume or the mean pressure of the system. Suppose that, starting with the system at a temperature T, an infinitesimal amount of heat $đQ$ is added to the system while all its other parameters y are kept fixed. As a result, the temperature of the system will change by an infinitesimal amount dT which depends on the nature of the system under consideration and which usually depends also on the parameters T and y specifying the initial macrostate of this system. The ratio

$$C_y \equiv \left(\frac{đQ}{dT}\right)_y \tag{20}$$

is called the *heat capacity* of the system.† Here we have used the subscript y to denote explicitly the parameters kept constant in the process of adding heat. The heat capacity C_y is an easily measured property of a system. Note that it depends ordinarily not only on the nature of the system, but also on the parameters T and y specifying the macrostate of this system; i.e., in general, $C_y = C_y(T,y)$.

The amount of heat $đQ$ which must be added to a homogeneous system to produce a given temperature change dT is expected to be proportional to the total number of particles in this system. Hence it is convenient to define a related quantity, the *specific heat,* which depends only on the nature of the substance under consideration, but

† Note that the right side of (20) is ordinarily *not* a derivative since the heat $đQ$ does not, in general, denote an infinitesimal difference between two quantities.

not on the amount present. This can be achieved by dividing the heat capacity C_y of ν moles (or m grams) of the substance by the corresponding number of moles (or of grams). The heat capacity per mole or *specific heat per mole* is thus defined as

$$c_y \equiv \frac{1}{\nu} C_y = \frac{1}{\nu} \left(\frac{đQ}{dT} \right)_y. \tag{21}$$

Equivalently, the *specific heat per gram* is defined as

$$c_y' \equiv \frac{1}{m} C_y = \frac{1}{m} \left(\frac{đQ}{dT} \right)_y. \tag{22}$$

The cgs units of the molar specific heat are thus, by (21), ergs degree^{-1} mole^{-1}.

The simplest situation is that where all the *external* parameters of the system (such as its volume V) are kept constant in the process of adding heat. In this case no work is done on the system so that $đQ = d\bar{E}$; i.e., the absorbed heat serves then merely to increase the internal energy of the system. Denoting the *external* parameters collectively by the symbol x, we can write

$$C_x \equiv \left(\frac{đQ}{dT} \right)_x = \left(\frac{\partial \bar{E}}{\partial T} \right)_x. \tag{23}$$

The last expression is a derivative since $d\bar{E}$ is a genuine differential quantity; we have written it as a *partial* derivative to indicate that all the external parameters x are supposed to be kept constant. Note that the heat capacity must always be positive by virtue of (4.35), i.e.,

$$\boxed{C_x > 0.} \tag{24}$$

To indicate a typical order of magnitude, the specific heat of water† at room temperature is experimentally found to be 4.18 joules degree^{-1} gram^{-1}.

In Sec. 4.7 we discussed the case of a gas sufficiently dilute to be ideal and nondegenerate. If the gas is *monatomic,* the results (4.83) and (4.85) yield for the mean energy per mole of such a gas

$$\bar{E} = \tfrac{3}{2} N_a kT = \tfrac{3}{2} RT \tag{25}$$

where N_a is Avogadro's number and $R \equiv N_a k$ is the gas constant.

† Historically, this specific heat was, by definition, assigned the value of one calorie degree^{-1} gram^{-1}. This is the reason for the modern definition of the calorie as being a unit of heat such that 1 calorie $\equiv$ 4.184 joules.

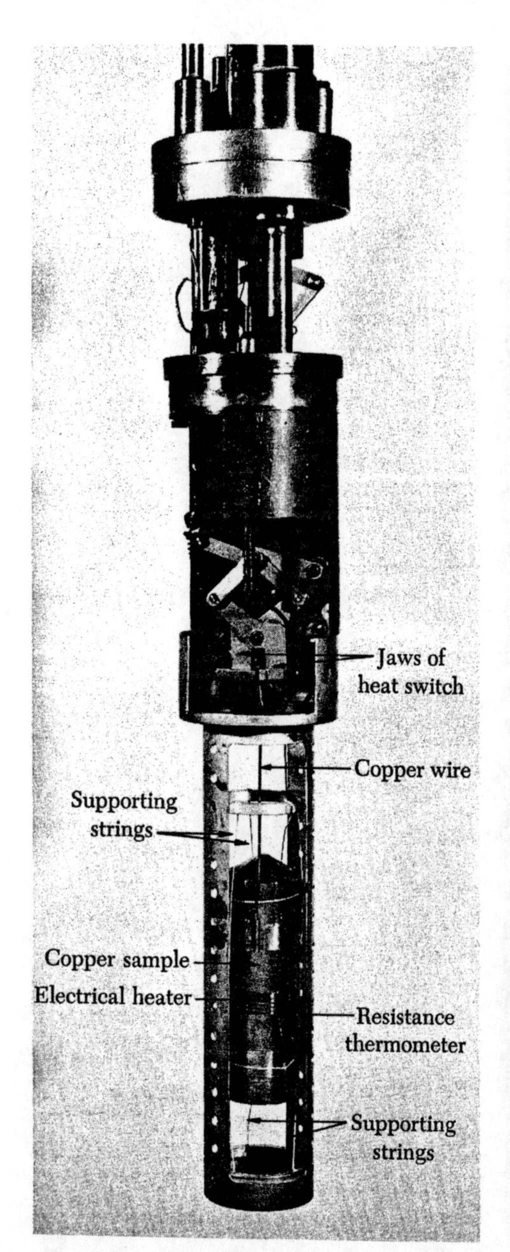

Fig. 5.15 The essential inner part of an apparatus used for measuring specific heats down to temperatures as low as 0.1°K. (In principle, this apparatus is similar to the arrangement of Fig. 5.12.) Here the copper sample is the system B whose heat capacity is to be measured. This system B is in thermal contact with an auxiliary system A comprising an electrical resistance heater (made of a few turns of manganin wire) and an electrical resistance thermometer. The combined system $A + B$ is thermally insulated by being suspended from some thin strings and by being then enclosed in a can (shown in Fig. 5.16) which is evacuated. The sample is initially cooled to the desired low temperatures by clamping the jaws of the heat switch onto the copper wire extending from the sample; this establishes the needed thermal contact with the refrigerant in the top of the apparatus. (*Photograph by courtesy of Professor Norman E. Phillips, University of California, Berkeley.*)

Fig. 5.16 The complete apparatus used for measuring specific heats down to 0.1°K. The inner part of the apparatus (enlarged in Fig. 5.15) is here shown hanging from the bottom of an array of stainless steel tubes through which one can pump on the apparatus and bring down the necessary electrical leads. The evacuated can which ordinarily surrounds the inner part of the apparatus is shown separately. To perform the measurements, the whole apparatus is immersed in the dewar assembly shown on the left. (*Photograph by courtesy of Professor Norman E. Phillips, University of California, Berkeley.*)

Hence it follows by (23) that the predicted molar specific heat c_V at constant volume should be,

$$\boxed{\begin{array}{l}\text{for a monatomic ideal gas,}\\ c_V = \left(\frac{\partial E}{\partial T}\right)_V = \frac{3}{2}R.\end{array}} \tag{26}$$

Note that this result is independent of the volume, the temperature, or the nature of the gas. Using for R its numerical value (4), we then obtain by (26)

$$c_V = 12.47 \text{ joules deg}^{-1} \text{ mole}^{-1}. \tag{27}$$

This result is in excellent agreement with the experimentally measured specific heats of monatomic gases such as helium or argon.

5.5 *Entropy*

The relation (4.42),

$$dS = \frac{đQ}{T}, \tag{28}$$

suggests that it should be possible to determine the entropy S of a system by appropriate measurements of heat and absolute temperature. Indeed, if the heat capacity of the system is known as a function of its temperature, the calculation of the entropy should become straightforward. To verify this surmise, let us assume that all the external parameters x of the system remain fixed. Suppose then that the system is in equilibrium at the absolute temperature T and that an infinitesimal amount of heat $đQ$ is added to this system by bringing it into contact with a heat reservoir at a temperature infinitesimally different from T (so that the equilibrium is disturbed by a negligible amount and the temperature T of the system remains well-defined). Then the resulting entropy change of the system is, by (28), equal to

$$dS = \frac{đQ}{T} = \frac{C_x(T)\, dT}{T} \tag{29}$$

where the last step exploits merely the definition (23) of the heat capacity C_x.

Now suppose that one wishes to compare the entropy of the system in two different macrostates where the values of the external parameters of the system are the same. Assume that the absolute temperature is T_a in one macrostate and T_b in the other. The system then has a well-defined entropy $S_a \equiv S(T_a)$ in the first macrostate and a well-defined entropy $S_b \equiv S(T_b)$ in the second macrostate. It should be possible to calculate the entropy difference $S_b - S_a$ by imagining that the system is brought from the initial temperature T_a to the final temperature T_b in many successive infinitesimal steps. This can be done by placing the system in successive contacts with a series of heat reservoirs of infinitesimally different temperatures. At all these stages the system would then be arbitrarily close to equilibrium and thus would always have a well-defined temperature T. Thus the result (29) can be applied successively to yield

$$\boxed{S_b - S_a = \int_{T_a}^{T_b} \frac{đQ}{T} = \int_{T_a}^{T_b} \frac{C_x(T)}{T}\, dT.} \tag{30}$$

If the heat capacity C_x is independent of temperature in the temperature range between T_a and T_b, (30) becomes simply

$$S_b - S_a = C_x(\ln T_b - \ln T_a) = C_x \ln \frac{T_b}{T_a}. \tag{31}$$

The relation (30) allows the calculation of entropy *differences*. To get the absolute magnitude of the entropy, one need only consider the limiting case where $T_a \to 0$, since the entropy S_a is then known to approach the value $S_a = 0$ by virtue of (11) [or the value $S_a = S_0$, due to nuclear spin orientations, by virtue of (12)].

The relation (30) permits us to deduce an interesting limiting property of the heat capacity. Note that the entropy difference on the left side of (30) must always be some finite number, since the number of accessible states is always finite. The integral on the right side, therefore, cannot become infinite when $T_a = 0$. Hence, to guarantee that the integral remain finite despite the factor T in the denominator, it is necessary that the temperature dependence of the heat capacity be such that,

$$\boxed{\text{as } T \to 0, \qquad C_x(T) \to 0.} \tag{32}$$

This is a general property which must be satisfied by the heat capacity of any substance.†

† The expression (26) for the heat capacity of an ideal gas does not contradict this property since it was derived on the basis of the assumption that the gas is nondegenerate. This assumption breaks down at a sufficiently low temperature, although this temperature is exceedingly low if the gas is dilute.

The relation (30) is extremely interesting since it provides an explicit connection between two very different types of information about the system under consideration. On the one hand, (30) involves the heat capacity $C_x(T)$ which can be obtained from purely *macroscopic* measurements of heat and temperature. On the other hand, it involves the entropy $S = k \ln \Omega$ which is a quantity obtainable from a *microscopic* knowledge of the quantum states of the system; i.e., either it can be calculated from first principles, or it can be obtained by using spectroscopic data to deduce the energy levels of the system.

Example

As a simple illustration, consider a system of N magnetic atoms, each having spin $\frac{1}{2}$. Suppose that this system is known to become *ferromagnetic* at sufficiently low temperatures. This means that the interaction between the spins is such that they tend to align themselves parallel to each other so that they all point in the same direction; the substance then acts like a permanent magnet. As $T \to 0$, it thus follows that the system is in a single state, the one where all spins point in a given direction; thus $\Omega \to 1$ or $\ln \Omega \to 0$. But at sufficiently high temperatures, all spins must be completely randomly oriented. There are then two possible states per spin (either up or down) and $\Omega = 2^N$ possible states for the whole system; thus $S = kN \ln 2$. Hence it follows that this system must have associated with its spins a heat capacity $C(T)$ which satisfies, by (30), the equation

$$\int_0^\infty \frac{C(T)\,dT}{T} = kN \ln 2.$$

This relation must *always* be valid, irrespective of the details of the interactions which bring about ferromagnetic behavior and irrespective of the detailed temperature dependence of $C(T)$.

5.6 *Intensive and Extensive Parameters*

Before concluding this chapter, it is worth pointing out briefly how the various macroscopic parameters which we have discussed depend on the size of the system under consideration. Roughly speaking, these parameters are of two types: (i) those which are independent of the size of the system (these are called *intensive*); and (ii) those which are proportional to the size of the system (these are called *extensive*). More precisely, these two types of parameters may be characterized by considering a homogeneous macroscopic system in equilibrium and imagining the system to be divided into two parts (e.g., by introducing a partition). Suppose that a macroscopic parameter y characterizing

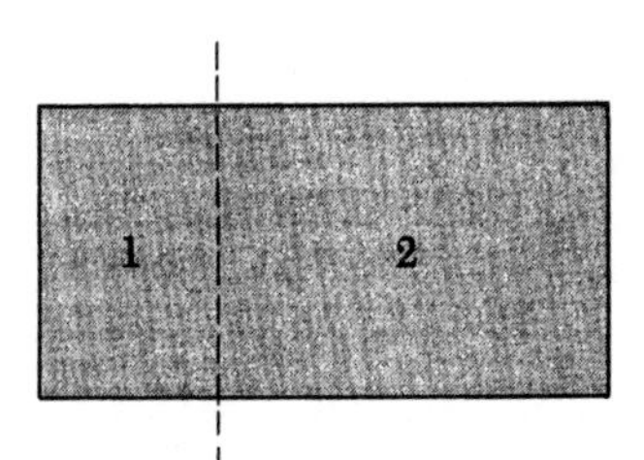

Fig. 5.17 Subdivision of a homogeneous macroscopic system into two parts.

the entire system assumes the values y_1 and y_2 for the two resulting subsystems. Then

(i) the parameter y is said to be *intensive* if

$$y = y_1 = y_2;$$

(ii) the parameter y is said to be *extensive* if

$$y = y_1 + y_2.$$

For example, the mean pressure of a system is an intensive parameter, since both parts of a system, after subdivision, will have the same mean pressure as before. Similarly, the temperature of a system is an intensive parameter.

On the other hand, the volume V of a system is an extensive parameter, as is the total mass M of a system. The density ρ of a system, $\rho = M/V$, is then an intensive parameter. Indeed, it is clear that the ratio of any two extensive parameters is an intensive parameter.

The internal energy $\bar{E}$ of a system is an extensive quantity. Indeed, no work is required to subdivide the system into two parts *if* one neglects the work involved in creating the two new surfaces. (This work is negligible for large systems for which the ratio of the number of molecules near the boundary to the number of molecules in the bulk of the system is very small.) Thus the total energy of the system is the same after subdivision as it was before, i.e., $\bar{E} = \bar{E}_1 + \bar{E}_2$.

The heat capacity C, being the ratio of an energy increase divided by a fixed small temperature increment, is similarly an extensive quantity. On the other hand, the specific heat per mole, by its definition C/ν (where ν is the number of moles in the system), is obviously an intensive quantity.

The entropy S is also an extensive quantity. This follows from the relation $\Delta S = \int đQ/T$, since the heat absorbed $đQ = C\,dT$ is an extensive quantity. It also follows from the statistical definition $S = k \ln \Omega$ since the number Ω of states accessible to the entire system is essentially equal to the product $\Omega_1\Omega_2$ of the number of states accessible to its two parts.

When dealing with an extensive quantity, it is often convenient to introduce the quantity per mole, which is an intensive parameter independent of the size of the system. This was, for example, the motivation for introducing the concept of the specific heat.

Summary of Definitions

triple point That macrostate of a pure substance where its solid, liquid, and gaseous forms can coexist in equilibrium.

Kelvin temperature The absolute temperature T expressed on a scale where the absolute temperature of the triple point of water is assigned the value of 273.16 degrees.

absolute zero Zero absolute temperature.

Celsius temperature The Celsius temperature θ_C is defined in terms of the absolute Kelvin temperature T by the relation

$$\theta_C \equiv T - 273.15.$$

quasi-static process A process carried out sufficiently slowly so that the system under consideration remains at all times arbitrarily close to equilibrium.

heat capacity If the addition of an infinitesimal amount of heat $đQ$ to a system results in an increase dT of its temperature while all its other macroscopic parameters y remain fixed, the heat capacity C_y of the system (for fixed values of y) is defined as

$$C_y \equiv \left(\frac{đQ}{dT}\right)_y.$$

molar specific heat The heat capacity per mole of the substance under consideration.

intensive parameter A macroscopic parameter describing a system in equilibrium and having the same value for any part of the system.

extensive parameter A macroscopic parameter describing a system in equilibrium and having a value equal to the sum of its values for each part of the system.

Important Relations

Limiting property of the entropy:

$$\text{as } T \to 0_+, \qquad S \to S_0 \tag{i}$$

where S_0 is a constant independent of the structure of the system.

Limiting property of the heat capacity:

$$\text{as } T \to 0, \qquad C \to 0. \tag{ii}$$

Suggestions for Supplementary Reading

M. W. Zemansky, *Temperatures Very Low and Very High* (Momentum Books, D. Van Nostrand Company, Inc., Princeton, N.J., 1964).

D. K. C. MacDonald, *Near Zero* (Anchor Books, Doubleday & Company, Inc., New York, 1961). An elementary account of low-temperature phenomena.

K. Mendelssohn, *The Quest for Absolute Zero* (World University Library, McGraw-Hill Book Company, New York, 1966). A historical and well-illustrated account of low-temperature physics up to the present day.

N. Kurti, *Physics Today,* 13, 26–29 (October 1960). A simple account describing the attainment of temperatures close to 10^{-6} °K.

Scientific American, vol. 191 (September 1954). This issue of the magazine is devoted entirely to the subject of heat and contains several articles about high temperatures.

M. W. Zemansky, *Heat and Thermodynamics*, 4th ed., chaps. 3 and 4 (McGraw-Hill Book Company, New York, 1957). Macroscopic discussion of work, heat, and internal energy.

Historical and biographical accounts:

D. K. C. MacDonald, *Faraday, Maxwell, and Kelvin* (Anchor Books, Doubleday & Company, Inc., New York, 1964). The last part of this book contains a short account of Lord Kelvin's life and work.

A. P. Young, *Lord Kelvin* (Longmans, Green & Co., Ltd., London, 1948).

M. H. Shamos, *Great Experiments in Physics*, chap. 12 (Holt, Rinehart and Winston, Inc., New York, 1962). A description of Joule's experiments in his own words.

Problems

5.1 *Temperatures necessary for producing spin polarization*

Consider the numerical implications of the polarization experiments previously examined in problems 4.4 and 4.5. Suppose that a magnetic field as high as 50,000 gauss is available in the laboratory. It is desired to use this field to polarize a sample containing particles of spin $\frac{1}{2}$ so that the number of spins pointing in one direction is at least 3 times as large as the number pointing in the opposite direction.

(*a*) To how low an absolute temperature must the sample be cooled if the spins are electronic with a magnetic moment $\mu_0 \approx 10^{-20}$ erg/gauss?

(*b*) To how low an absolute temperature must the sample be cooled if the particles are protons having a nuclear magnetic moment $\mu_0 \approx 1.4 \times 10^{-23}$ erg/gauss?

(*c*) Comment on the ease and feasibility of these two experiments.

5.2 *Temperature necessary for removal of nuclear spin entropy*

Consider any solid, such as silver, whose nuclei have spin. The magnetic moment μ_0 of each nucleus is of the order of 5×10^{-24} erg/gauss and the spatial separation r between adjacent nuclei is of the order of 2×10^{-8} cm. No externally applied magnetic field is present. Adjacent nuclei can interact, however, by virtue of the internal magnetic field B_i produced by the magnetic moment of one nucleus at the position of a neighboring one.

(*a*) Estimate the magnitude of B_i by using your elementary knowledge of the magnetic field produced by a bar magnet.

(*b*) How low must the temperature T of the solid be so that a nucleus, subject to the magnetic field B_i due to its neighbors, has significantly different probabilities of pointing in opposite directions?

(*c*) Estimate numerically the magnitude of the absolute temperature below which an appreciably nonrandom orientation of the nuclear spins might be expected.

5.3 *Work done in compressing a gas at constant temperature*

Consider ν moles of an ideal gas contained in a cylinder closed by a piston. Find the work that must be done on the gas to compress it very slowly from some initial volume V_1 to some final volume V_2 while it is maintained at a constant temperature T (by being kept in contact with a heat reservoir at this temperature).

5.4 *Work done in an adiabatic process*

A gas has a well-defined mean energy $\bar{E}$ when its volume is V and mean pressure is $\bar{p}$. If the volume of the gas is changed quasi-statically, the mean pressure $\bar{p}$ (and energy $\bar{E}$) of the gas will then change accordingly. Suppose that the gas is taken very slowly from a to b (see Fig. 5.18) while the gas is kept thermally insulated. In this case $\bar{p}$ is found to depend on the volume V in accordance with the relation

$$\bar{p} \propto V^{-5/3}.$$

What is the work done on the gas in this process?

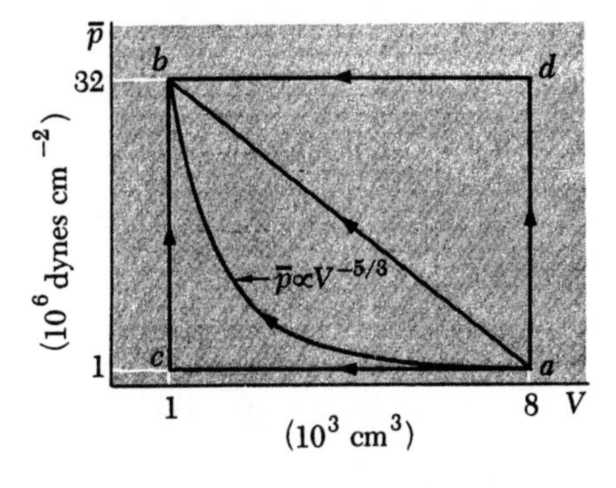

Fig. 5.18 Various processes illustrated on a diagram of mean pressure $\bar{p}$ versus volume V.

5.5 *Work done in alternative processes connecting the same macrostates*

The gas of Prob. 5.4 can also be brought quasi-statically from a to b in various other ways. In particular, consider the following processes and calculate for each the total work W done on the system and the total heat Q absorbed by the system when it is brought quasi-statically from a to b. (See Fig. 5.18.)

Process $a \to c \to b$. The system is compressed from its original to its final volume, heat being removed to maintain the pressure constant. The volume is then kept constant and heat is added to increase the mean pressure to 32×10^6 dynes cm^{-2}.

Process $a \to d \to b$. The two steps of the preceding process are performed in the opposite order.

Process $a \to b$. The volume is decreased and heat is supplied so that the mean pressure varies linearly with the volume.

5.6 *Work done in a cyclic process*

A system consisting of a fluid is subjected to a quasi-static process which can be described by a curve showing successive values of the volume V and corresponding mean pressure $\bar{p}$ of the fluid. The process is such that at its end the system is in the same macrostate as it was at the beginning. (A process of this kind is called *cyclic.*) The curve describing this process is closed, as shown in Fig. 5.19. Show that the work done on the system in this process is given by the area contained within this closed curve.

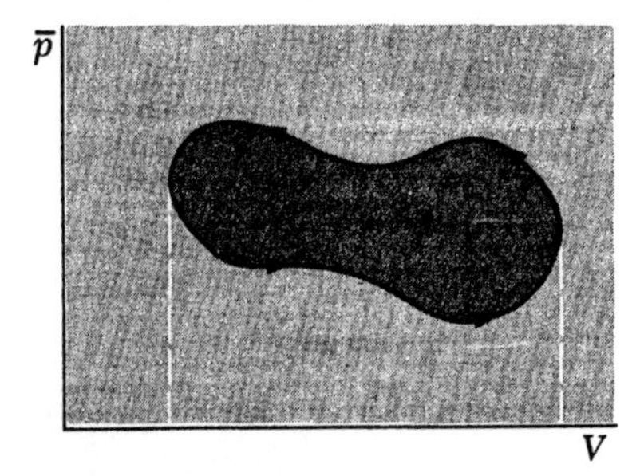

Fig. 5.19 A cyclic process.

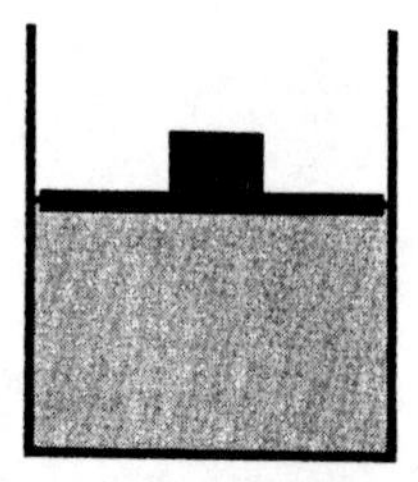

Fig. 5.20 A system contained within a cylinder closed by a movable piston.

5.7 *Heat absorbed by a system at constant pressure*

Consider a system, such as a gas or liquid, whose only external parameter is its volume V. If the volume is kept fixed and an amount of heat Q is added to the system, then no work gets done and

$$Q = \Delta \bar{E} \tag{i}$$

where $\Delta \bar{E}$ denotes the increase in mean energy of the system. Suppose, however, that the system is always maintained at a constant pressure p_0 by being enclosed in a cylinder of the type shown in Fig. 5.20. Here the pressure p_0 is always determined by the weight of the piston, but the volume V of the gas is free to adjust itself. If an amount of heat Q is now added to the system, the relation (i) is no longer valid. Show that it must be replaced by the relation

$$Q = \Delta H \tag{ii}$$

where ΔH denotes the change in the quantity $H \equiv \bar{E} + p_0 V$ of the system. (The quantity H is called the *enthalpy* of the system.)

5.8 *A mechanical process involving an ideal gas*

A vertical cylinder contains ν moles of a monatomic ideal gas and is closed off by a piston of mass M and area A. The whole system is thermally insulated. The downward acceleration due to gravity is g. Initially the piston is clamped in position so that the gas has a volume V_0 and an absolute temperature T_0. The piston is now released and, after some oscillations, comes to rest in a final equilibrium position corresponding to some smaller volume V of the gas where it has a temperature T. Neglect any frictional forces which might prevent the piston from sliding freely within the cylinder. Neglect also the heat capacities of the piston and of the cylinder.

(*a*) What must be the final mean pressure of the gas?

(*b*) By considering the work done on the gas and using your knowledge of the properties of an ideal monatomic gas, calculate the final temperature T and volume V of the gas in terms of T_0, V_0, the gas constant R, and the quantities ν, M, A, g.

5.9 *An experiment in calorimetry*

A container is partially filled with water, in which there are immersed an electrical resistor and a thermometer consisting of mercury in a glass tube. The whole system is thermally insulated. When the system is initially in equilibrium at room temperature, the length L of the mercury column in the thermometer is 5.00 cm. If a 12-volt storage battery is connected by a switch to the resistor, a current of 5 amperes flows through it.

In the first set of experiments, the switch is closed for 3 min and then opened again. After equilibrium has been attained, the thermometer reading is $L = 9.00$ cm. The switch is then again closed for 3 min before being opened; the final equilibrium reading of the thermometer is then $L = 13.00$ cm.

In the second set of experiments, an additional 100 gm of water is added to the container. The initial thermometer reading is again 5.00 cm. The switch is closed for 3 min, and then opened. After equilibrium has been attained, the thermometer reading is $L = 7.52$ cm. The switch is then closed again for 3 min before being opened. After equilibrium has been attained, $L = 10.04$ cm.

(*a*) Make a plot of the internal energy of 100 gm of water as a function of the thermometer reading L.

(*b*) In the temperature range investigated, what is the change of internal energy of 1 gm of water when the thermometer reading L changes by 1 cm?

5.10 ***An experiment in calorimetry by comparison***

A vessel contains 150 gm of water and a thermometer of the type described in Prob. 5.9. The whole system is thermally insulated. Initially the thermometer reading of this system in equilibrium corresponds to a length $L = 6.00$ cm of the mercury column. To this system is added 200 gm of water at an initial temperature corresponding to a thermometer reading of 13.00 cm. After equilibrium has been reached, the thermometer reading is $L = 9.66$ cm.

After this preliminary experiment, a second experiment is performed. A copper block of 500 gm is immersed in the original vessel containing 150 gm of water and the thermometer. The initial thermometer reading is again $L = 6.00$ cm. To this system is added 200 gm of water at an initial temperature corresponding to a thermometer reading of $L = 13.00$ cm. After equilibrium has been reached, the final thermometer reading is $L = 8.92$ cm.

Use the information obtained in Prob. 5.9 to answer the following questions:

(*a*) In the preliminary experiment, calculate the heat absorbed by the system consisting of the vessel, the water, and the thermometer.

(*b*) In the temperature range of interest, what is the change of internal energy of 1 gm of copper if its temperature changes by an amount corresponding to a 1-cm change of thermometer reading?

5.11 ***Schottky specific heat anomaly***

Consider a system which consists of N weakly interacting particles and suppose that each of these can be in either one of two states with respective energies ϵ_1 and ϵ_2, where $\epsilon_1 < \epsilon_2$.

(*a*) Without explicit calculation, make a qualitative plot of the mean energy $\bar{E}$ of the system as a function of its absolute temperature T. Use this plot (previously constructed in Prob. 4.8) to make a qualitative plot of the heat capacity C of this system as a function of T (assuming that all external parameters remain fixed). Show that this plot exhibits a maximum and estimate roughly the value of the temperature at which this maximum occurs.

(*b*) Calculate explicitly the mean energy $\bar{E}(T)$ and heat capacity $C(T)$ of this system. Verify that your expressions exhibit the qualitative features discussed in part (*a*).

Cases where two discrete energy levels of a system become important in a given temperature range do occur in practice; the accompanying heat capacity behavior is called a *Schottky anomaly*.

5.12 *Heat capacity of a spin system*

A system of N atoms, each having spin $\frac{1}{2}$ and magnetic moment μ_0, is located in an external magnetic field $\mathbf{B}$ and is in equilibrium at the absolute temperature T. Focus attention on the spins only and answer the following questions:

(*a*) Without doing any computations, find the limiting values of the mean energy $\bar{E}(T)$ of this system as $T \to 0$ and as $T \to \infty$.

(*b*) Without doing any computations, find the limiting values of the heat capacity $C(T)$, at constant magnetic field, as $T \to 0$ and as $T \to \infty$.

(*c*) Calculate the mean energy $\bar{E}(T)$ of this system as a function of the temperature T. Make an approximate sketch of $\bar{E}$ versus T.

(*d*) Calculate the heat capacity $C(T)$ of this system. Make an approximate sketch of C versus T.

5.13 *Thermal effects due to nonspherical nuclei*

The nuclei of atoms in a certain crystalline solid have spin 1. According to quantum theory, each nucleus therefore can be in any one of three quantum states labeled by the quantum number m, where $m = 1, 0$, or -1. This quantum number measures the projection of the nuclear spin along a crystal axis of the solid. Since the electric charge distribution in the nucleus is not spherically symmetrical, but ellipsoidal, the energy of a nucleus depends on its spin orientation with respect to the nonuniform internal electric field existing at its location. Thus a nucleus has the same energy $E = \epsilon$ in the state $m = 1$ and the state $m = -1$, compared with an energy $E = 0$ in the state $m = 0$.

(*a*) Find an expression, as a function of the absolute temperature T, of the nuclear contribution to the mean internal energy per mole of the solid.

(*b*) Make a qualitative graph showing the temperature dependence of the nuclear contribution to the molar specific heat of the solid. Calculate explicitly its temperature dependence. What is its temperature dependence for large values of T?

Although the thermal effects just discussed are small, they can become important when making heat capacity measurements at very low temperatures in some substances (e.g., in indium metal, since the ^{115}In nucleus departs appreciably from spherical symmetry).

5.14 *Thermal interaction between two systems*

Consider a system A (e.g., a copper block) and a system B (e.g., a container filled with water) which initially are in equilibrium at the temperatures T_A and T_B, respectively. In the temperature range of interest, the volumes of the systems remain essentially unchanged and their respective heat capacities C_A and C_B are essentially temperature-independent. The systems are now placed in thermal contact with each other and one waits until the systems attain their final equilibrium situation at some temperature T.

(*a*) Use the condition of conservation of energy to find the final temperature T. Express your answer in terms of T_A, T_B, C_A, and C_B.

(*b*) Use Eq. (31) to calculate the entropy change ΔS_A of A and the entropy change ΔS_B of B. Use these results to calculate the total entropy change $\Delta S = \Delta S_A + \Delta S_B$ of the combined system in going from the initial situation where the systems are separately in equilibrium to the final situation where they are in thermal equilibrium with each other.

(*c*) Show explicitly that ΔS can never be negative, and that it will be zero only if $T_A = T_B$. (*Suggestion:* You may find it useful to exploit the inequality $\ln x \leq x - 1$ derived in (M.15), or equivalently, the inequality $\ln (x^{-1}) \geq -x + 1$.)

5.15 ***Entropy changes for various methods of adding heat***

The specific heat of water is 4.18 joules gram^{-1} degree^{-1}.

(*a*) One kilogram of water at 0°C is brought into contact with a large heat reservoir at 100°C. When the water has reached 100°C, what has been the change in entropy of the water? Of the heat reservoir? Of the entire system consisting of both water and heat reservoir?

(*b*) If the water had been heated from 0°C to 100°C by first bringing it in contact with a reservoir at 50°C and then with a reservoir at 100°C, what would have been the change in entropy of the entire system?

(*c*) Show how the water might be heated from 0°C to 100°C with no change in the entropy of the entire system.

5.16 ***Entropy of melting***

Ice and water coexist in equilibrium at a temperature of 0°C (273°K). It requires 6000 joules of heat to melt 1 mole of ice at this temperature.

(*a*) Calculate the difference in entropy between 1 mole of water and 1 mole of ice at this temperature.

(*b*) Find the ratio of the number of states accessible to water to that accessible to ice at this temperature.

5.17 ***A practical problem in calorimetry***

Consider a *calorimeter* (apparatus designed to measure heat) consisting essentially of a 750-gm copper can. This can contains 200 gm of water and is in equilibrium at a temperature of 20°C. An experimenter now places 30 gm of ice at 0°C in the calorimeter and encloses the latter with a heat-insulating shield. It is known that the specific heat of water is 4.18 joules gm^{-1} deg^{-1} and that the specific heat of copper is 0.418 joules gm^{-1} deg^{-1}. It is also known that the heat of melting of ice (i.e., the heat required to convert one gram of ice to water at 0°C) is 333 joules gm^{-1}.

(*a*) What will be the temperature of the water after all the ice has melted and equilibrium has been reached?

(*b*) Compute the total entropy change resulting from the process of part (*a*).

(*c*) After all the ice has melted and equilibrium has been reached, how many joules of work must be done on the system (e.g., by means of a stirring rod) to restore all the water to 20°C?

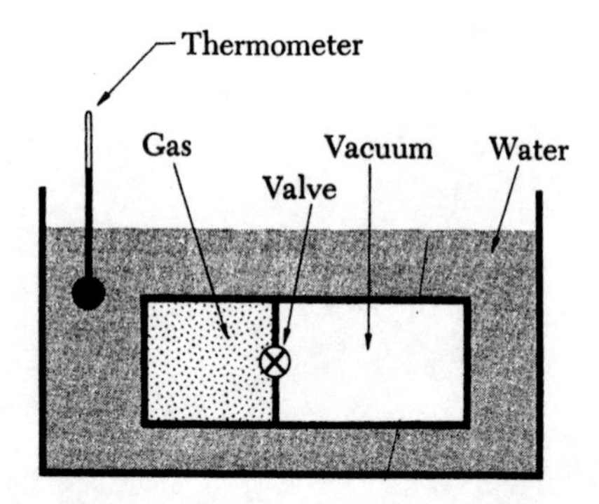

Fig. 5.21 Apparatus for studying the free expansion of a gas.

5.18 *Free expansion of a gas*

Figure 5.21 shows schematically an experimental arrangement used by Joule to study how the internal energy $\bar{E}$ of a gas depends on its volume. Consider the system A consisting of the closed container subdivided by a partition and containing a gas in one side only. The experiment consists simply in opening the valve and allowing the gas to come to equilibrium throughout the entire container. Suppose that the thermometer indicates that the temperature of the water remains unchanged in this process.

(*a*) What is the work done on the system A in this process? (The container walls are rigid and do not move.)

(*b*) What is the heat absorbed by A in the process?

(*c*) What is the change in the internal energy of A in the process?

(*d*) Since the temperature of the gas is unchanged, what conclusion does the experiment allow one to draw about the dependence of the internal energy of the gas on its volume at a fixed temperature?

5.19 *Entropy arguments applied to the heat capacity of a superconducting metal*

5.19 The heat capacity C_n of a normal metal at a very low absolute temperature is of the form $C_n = \gamma T$ where γ is a constant characteristic of the metal. If such a metal is superconducting below a critical temperature T_c, then its heat capacity C_s in the superconducting state in the temperature range $0 \leq T \leq T_c$ is approximately given by the relation $C_s = \alpha T^3$ where α is some constant. No heat is absorbed or given off when a metal is transformed from its normal to its superconducting state at the critical temperature T_c. Hence it follows that at this temperature $S_n = S_s$, where S_n and S_s denote the entropies of the metal in its normal and superconducting states, respectively.

(*a*) What statements can you make about the entropies S_n and S_s in the limit as $T \to 0$?

(*b*) Use the answer to part (*a*), and the connection between heat capacity and entropy, to find a relation between C_s and C_n at the critical temperature T_c.

5.20 *Heat capacity of an assembly of harmonic oscillators*

Consider an assembly of N weakly interacting simple harmonic oscillators at an absolute temperature T. (Such an assembly of oscillators provides an approximate model for the atoms in a solid.) Suppose that the classical angular frequency of oscillation of each oscillator is ω.

(*a*) Use the result for the mean energy calculated in Prob. 4.22 to find the heat capacity C (with all external parameters fixed) of this assembly of oscillators.

(*b*) Make a sketch showing how the heat capacity C depends on the absolute temperature T.

(*c*) What is the heat capacity at temperatures high enough so that $kT \gg \hbar\omega$?

*** 5.21 *Specific heat of a diatomic gas***

Consider an ideal diatomic gas (such as N_2) at an absolute temperature T close to room temperature. This temperature is sufficiently low so that a molecule is always in its lowest vibrational state, but is sufficiently high so that the molecule is distributed over many of its possible rotational states.

(*a*) Use the result of Prob. 4.23 to write down an expression for the mean energy of a diatomic molecule in the gas. This energy should include the kinetic energy of its center-of-mass motion and the energy of rotation of the molecule about its center mass.

(*b*) Use the answer to part (*a*) to find the molar specific heat c_V at constant volume of an ideal diatomic gas. What is the numerical value of c_V?

*** 5.22 *Energy fluctuations of a system in contact with a heat reservoir***

Consider an arbitrary system in contact with a heat reservoir at the absolute temperature $T = (k\beta)^{-1}$. Using the canonical distribution, it has already been shown in Prob. 4.18 that $\bar{E} = -(\partial \ln Z/\partial\beta)$ where

$$Z \equiv \sum_r e^{-\beta E_r} \tag{i}$$

is the sum over all states of the system.

(*a*) Obtain an expression for $\overline{E^2}$ in terms of Z, or preferably, $\ln Z$.

(*b*) The dispersion of the energy $\overline{(\Delta E)^2} \equiv \overline{(E - \bar{E})^2}$ can be written as $\overline{E^2} - \bar{E}^2$. (See Prob. 2.8.) Use this relation and your answer to part (*a*) to show that

$$\overline{(\Delta E)^2} = \frac{\partial^2 \ln Z}{\partial \beta^2} = -\frac{\partial \bar{E}}{\partial \beta}. \tag{ii}$$

(*c*) Show thus that the standard deviation $\underset{\sim}{\Delta} E$ of the energy can be expressed quite generally in terms of the heat capacity C of the system (with external parameters kept fixed) by

$$\underset{\sim}{\Delta} E = T(kC)^{1/2}. \tag{iii}$$

(*d*) Suppose that the system under consideration is an ideal monatomic gas consisting of N molecules. Use the general result (iii) to find an explicit expression for $(\underset{\sim}{\Delta} E/\bar{E})$ in terms of N.

Chapter 6

Canonical Distribution in the Classical Approximation

Chapter 6 Canonical Distribution in the Classical Approximation

The canonical distribution (4.49) represents a simple result of fundamental importance and exceedingly great practical utility. As we showed in Chap. 4, it can be applied directly to calculate the equilibrium properties of the most diverse systems. Thus we illustrated specifically how it can be used to derive the magnetic properties of a system of spins or to calculate the pressure and specific heat of an ideal gas. We also examined several interesting applications in the problems at the end of Chap. 4. It would carry us too far afield to discuss the wide range of other important applications, a task that might easily fill several books. In the present chapter we do, however, want to show how some particularly simple and useful results follow immediately from the canonical distribution when the approximations of classical mechanics are applicable.

6.1 The Classical Approximation

We know that the quantum-mechanical description of a system of particles, under appropriate circumstances, can be approximated by a description in terms of classical mechanics. In this section we shall want to examine the following two questions: (i) Under what conditions can a statistical theory in terms of classical concepts be expected to be a valid approximation? (ii) If the approximation is permissible, how can the statistical theory be formulated in classical terms?

Validity of the classical approximation

The classical approximation can certainly *not* be valid if the absolute temperature is sufficiently low. Indeed, suppose that the typical thermal energy kT is less than (or comparable to) the average spacing ΔE between the energy levels of the system. It is then very significant that the possible energies of the system are quantized so as to be separated by discrete amounts. For example, the canonical distribution (4.49) implies that the probabilities of finding the system in a state of energy E or in a state of next higher possible energy $E + \Delta E$ are then very different. On the other hand, if $kT \gg \Delta E$, the probabilities vary very little from state to state. The fact that the possible energies are discrete rather than continuous then becomes relatively unimportant and a classical description might become possible. The definite conclusion emerging from these comments is that

> a classical description *cannot* be valid if
>
> $$kT \lesssim \Delta E. \tag{1}$$

The classical approximation should certainly be valid, however, if quantum-mechanical effects can be shown to be of negligible importance. The fundamental quantum-mechanical limitation on the meaningful use of classical concepts is expressed by the Heisenberg uncertainty principle. This asserts that a simultaneous determination of a position coordinate q and its corresponding momentum p cannot be accomplished to infinite precision, but that these quantities are subject to minimum uncertainties of respective magnitudes Δq and Δp such that

$$\Delta q\, \Delta p \gtrsim \hbar \tag{2}$$

where $\hbar \equiv h/2\pi$ is Planck's constant divided by 2π. Let us then examine the classical description of a system at a particular temperature. To be meaningful, this classical description must be able to consider a particle of the system localized within some typical minimum distance which we shall denote by s_0. Furthermore, we shall denote by p_0 the typical momentum of the particle. If s_0 and p_0 are large enough so that

$$s_0 p_0 \gg \hbar,$$

the limitations imposed by the Heisenberg uncertainty principle should become of negligible importance and the classical description should, accordingly, be valid. Thus we are led to the conclusion that

> a classical description should be valid
>
> if $$s_0 p_0 \gg \hbar, \tag{3a}$$
>
> i.e., if $$s_0 \gg \lambda\!\!\!^{-}_0. \tag{3b}$$

Here we have introduced the typical length $\lambda\!\!\!^{-}_0$ defined by

$$\lambda\!\!\!^{-}_0 \equiv \frac{\hbar}{p_0} = \frac{1}{2\pi}\frac{h}{p_0}. \tag{4}$$

This is just the de Broglie wavelength h/p_0 divided by 2π. The statement ($3b$) asserts thus merely that quantum effects ought to be negligible if the minimum significant classical dimension s_0 is large compared to the de Broglie wavelength of the particle. The wave properties of the particle then clearly become unimportant.

The classical description

Suppose that a classical discussion of a system of particles is justified. Then the basic questions that must be examined are precisely the same as those which formed the starting point of our quantum-theoretical considerations at the beginning of Chap. 3. In particular, the first question that arises is the following: How does one specify the microscopic state of a system described in terms of classical mechanics?

Let us begin by considering the very simple case of a system consisting of a single particle moving in one dimension. The position of this particle can be described by a single coordinate, call it q. The complete specification of the system in classical mechanics then requires a knowledge of the coordinate q and of its corresponding momentum p.† (A simultaneous knowledge of q and p at any one time is classically possible. It is also required for a complete description so that values of q and p at any other time can be predicted uniquely in accordance with the laws of classical mechanics.) It is possible to represent the situation geometrically by drawing cartesian axes labeled by q and p as shown in Fig. 6.1. Specification of q and p is then equivalent to specifying a point in this two-dimensional space (commonly called *phase space*).

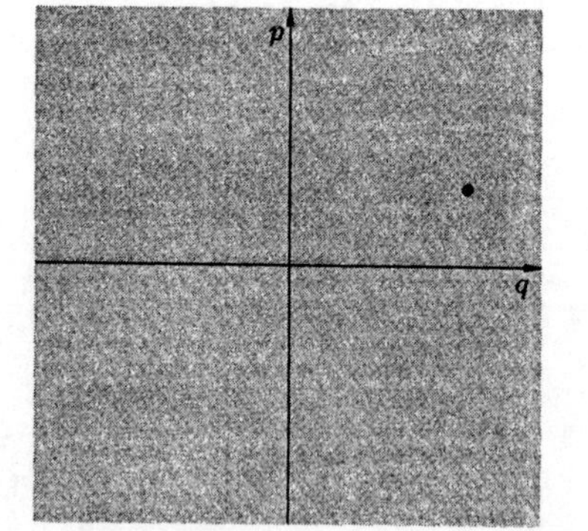

Fig. 6.1 Classical phase space for a single particle in one dimension.

In order to describe the situation involving the continuous variables q and p so that the possible states of the particle are countable, it is convenient to follow the procedure of Sec. 2.6 by subdividing the ranges of the variables q and p into arbitrarily small discrete intervals. For example, one can choose fixed small intervals of size δq for the subdivision of q, and fixed small intervals of size δp for the subdivision of p. Phase space is thus subdivided into small cells of equal size and of two-dimensional "volume" (i.e., area)

$$\delta q \, \delta p = h_0$$

where h_0 is some small constant (having the dimensions of angular momentum). A complete description of the state of the particle can then be given by specifying that its coordinate lies in some particular

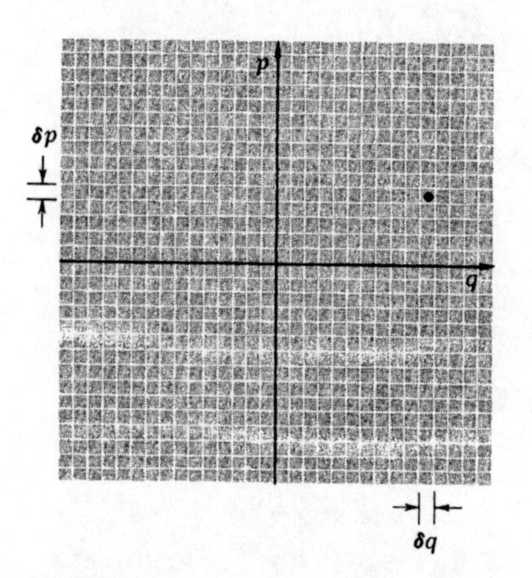

Fig. 6.2 The two-dimensional phase space of the preceding figure is here shown subdivided into equal cells of "volume" $\delta q \, \delta p = h_0$.

† If q denotes an ordinary cartesian coordinate and if no magnetic field is present, the momentum p is simply related to the velocity v of the particle of mass m by the proportionality $p = mv$. The description in terms of the momentum p rather than the velocity v is, however, valid in more general cases and is thus the one commonly used.

interval between q and $q + \delta q$ and that its momentum lies in some particular interval between p and $p + \delta p$, i.e., by specifying that the pair of numbers $\{q,p\}$ lies in some particular range. Geometrically this specifies that the point represented by $\{q,p\}$ lies in a particular cell of phase space.

Remark concerning the size of h_0

The specification of the state of the system is clearly more precise the smaller one chooses the size of the cells into which phase space has been divided, i.e., the smaller one chooses h_0. This constant h_0 can be chosen arbitrarily small in a classical description. The correct quantum-mechanical description imposes, however, a limitation on the maximum accuracy that can be achieved in a simultaneous specification of a coordinate q and its corresponding momentum p. Indeed, q and p can only be determined within uncertainties Δq and Δp whose order of magnitude satisfies the Heisenberg uncertainty principle $\Delta q \, \Delta p > \hbar$. A subdivision of phase space into cells of volume less than $\hbar$ is thus physically meaningless; i.e., a choice of $h_0 < \hbar$ would lead to a specification of the system more precise than is allowed by quantum theory.

The generalization of the previous discussion to an arbitrarily complex system is immediate. Such a system can be described by some set of f coordinates $q_1, \ldots, q_f$ and f corresponding momenta $p_1, \ldots, p_f$, i.e., by a total of $2f$ numbers. (As usual, the number f of independent coordinates needed for the description of the system is called the *number of degrees of freedom* of the system.) In order to deal with these continuous variables in a manner where the possible states of the system are countable, it is again convenient to subdivide the possible values of the ith coordinate q_i into fixed small intervals of magnitude δq_i, and the possible values of the ith momentum p_i into fixed small intervals of magnitude δp_i. For each i, the size of the subdivision interval can be chosen so that the product

$$\delta q_i \, \delta p_i = h_0 \tag{5}$$

where h_0 is some arbitrarily small constant of fixed magnitude independent of i. The state of the system can then be specified by stating that its coordinates and momenta are such that the set of values

$$\{q_1,q_2, \ldots ,q_f;p_1,p_2, \ldots ,p_f\}$$

lies in a particular set of intervals. In a convenient geometrical interpretation this set of values can again be regarded as a "point" in a *phase space* of $2f$ dimensions where each cartesian axis is labeled by one of the coordinates or momenta.† The subdivision into intervals thus divides this space into equal small cells of volume $(\delta q_1\, \delta q_2 \cdots \delta q_f\, \delta p_1\, \delta p_2 \cdots \delta p_f) = h_0{}^f$. The state of the system can then be described by specifying in which particular set of intervals (i.e., in which cell in phase space) the coordinates $q_1, q_2, \ldots, q_f$ and momenta $p_1, p_2, \ldots, p_f$ of the system actually lie. For simplicity each such set of intervals (or cell in phase space) can be labeled by some index r so that all these possible cells can be listed and enumerated in some convenient order $r = 1, 2, 3, \ldots$. Our entire discussion can then be summarized by the observation that

> the state of a system in classical mechanics can be described by specifying the particular cell r in phase space in which the coordinates and momenta of the system are found. (6)

The specification of the state of a system in classical mechanics is thus very similar to that in quantum mechanics, a cell in phase space in the classical description being analogous to a quantum state in the quantum-mechanical description. One distinction, however, is worth noting. In the classical case there exists an element of arbitrariness since the size of a cell in phase space (i.e., the magnitude of the constant h_0) can be chosen at will. On the other hand, in the quantum description a quantum state is an unambiguously defined entity (essentially because quantum theory involves Planck's constant $\hbar$ which has a unique value).

Classical statistical mechanics

The statistical description of a system in terms of classical mechanics now becomes completely analogous to that in quantum mechanics. The difference is one of interpretation: whereas the microstate of a system refers to a particular quantum state of a system in the quantum theory, it refers to a particular cell of phase space in the classical theory. When considering a statistical ensemble of systems, the basic postulates introduced in the classical theory are the same as the corresponding postulates (3.17) and (3.18) of quantum theory. In particular, the statement (3.19) of these postulates asserts in classical terms the fol-

† Except for the fact that it is not so readily visualized by our three-dimensional minds, this phase space is completely analogous to the two-dimensional phase space of Fig. 6.2.

lowing: *If an isolated system is in equilibrium, it is found with equal probability in each one of its accessible states, i.e., in each one of its accessible cells in phase space.*†

Example

In order to illustrate the classical notions in a very simple case, let us consider a single particle moving in one dimension under the influence of no forces, but confined within a box of length L. If we denote the position coordinate of this particle by x, the possible positions of the particle are then restricted by a condition of the form $0 < x < L$. The energy E of the particle of mass m is merely its kinetic energy so that

$$E = \frac{1}{2}mv^2 = \frac{1}{2}\frac{p^2}{m}$$

where v is the velocity and $p = mv$ is the momentum of the particle. Suppose that the particle is isolated and is thus known to have a constant energy in some small range between E and $E + dE$. Then its momentum must lie in some small range dp about the possible values $p = \pm\sqrt{2mE}$. The region of phase space accessible to this particle is then the one indicated by the dark areas shown in Fig. 6.3. If phase space has been subdivided into small cells of equal size $\delta x\ \delta p = h_0$, this region contains a large number of such cells. These represent the accessible states in which the system can be found.

Suppose that the particle is known to be in equilibrium. Then the statistical postulate asserts that the particle is equally likely to be found with its coordinate x and momentum p in any one of the equal-size cells contained within the dark areas. This implies that the particle is as likely to have a momentum in the range dp near $+\sqrt{2mE}$ as in the range dp near $-\sqrt{2mE}$. It also implies that the position coordinate x of the particle is equally likely to lie anywhere within the length L of the box. For example, the probability that the particle is located in the left third of the box is $\frac{1}{3}$ since the number of accessible cells for which x lies in the range $0 < x < \frac{1}{3}L$ is one-third of the total number of accessible cells.

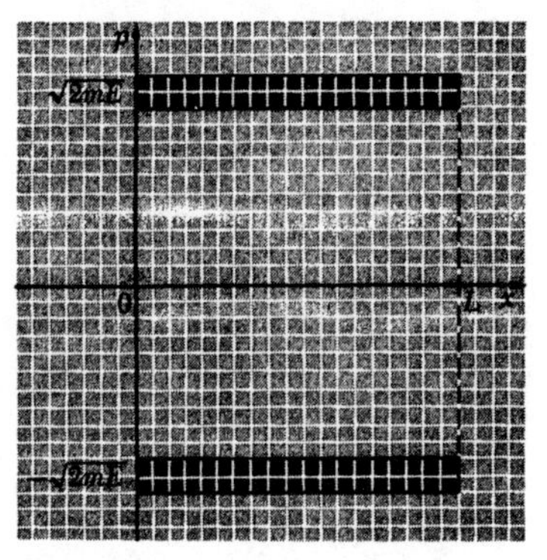

Fig. 6.3 Classical phase space for a single particle free to move in one dimension and confined within a box of length L. The particle, specified by a coordinate x and a momentum p, has an energy in the range between E and $E + \delta E$. The states accessible to the particle are indicated by the cells contained within the dark areas.

The preceding comments make it apparent that any general argument based upon the statistical postulates and the counting of states must remain equally valid in the classical description. In particular, it follows that the derivation of the canonical distribution in Sec. 4.5 remains applicable. If a system A, described classically, is in thermal equilibrium with a heat reservoir at the absolute temperature

† The statistical postulates can, with certain assumptions, be derived from the laws of classical mechanics in the same way that (3.17) and (3.18) could be based upon the laws of quantum mechanics. Such a classical derivation shows, incidentally, that a description in terms of coordinates and momenta (rather than coordinates and velocities) is the appropriate one in the most general case. In all the simple cases discussed in this book the distinction is trivial, however, since the momentum **p** and velocity **v** of a particle of mass m in these cases will always be related by the simple proportionality $\mathbf{p} = m\mathbf{v}$.

$T = (k\beta)^{-1}$, the probability P_r of finding this system in a particular state r of energy E_r is thus given by (4.49) so that

$$P_r \propto e^{-\beta E_r}. \tag{7}$$

Here the state r refers to a particular cell of phase space where the coordinates and momenta of A have particular values $\{q_1, \ldots, q_f; p_1, \ldots, p_f\}$. Correspondingly, the energy E_r of A denotes the energy E of this system when its coordinates and momenta have these particular values, i.e.,

$$E_r = E(q_1, \ldots, q_f; p_1, \ldots, p_f) \tag{8}$$

since the energy of A is a function of its coordinates and momenta.

It is convenient to express the canonical distribution (7) in terms of a probability *density* by proceeding in the general manner familiar from Sec. 2.6. Let us thus try to find the following probability

$\mathcal{P}(q_1, \ldots, q_f; p_1, \ldots, p_f)\, dq_1 \cdots dq_f\, dp_1 \cdots dp_f$

$\equiv$ the probability that the system A in contact with the heat reservoir is found to have its first coordinate in the range between q_1 and $dq_1, \ldots,$ its fth coordinate in the range between q_f and $q_f + dq_f$, its first momentum in the range between p_1 and $p_1 + dp_1, \ldots,$ and its fth momentum in the range between p_f and $p_f + dp_f$. (9)

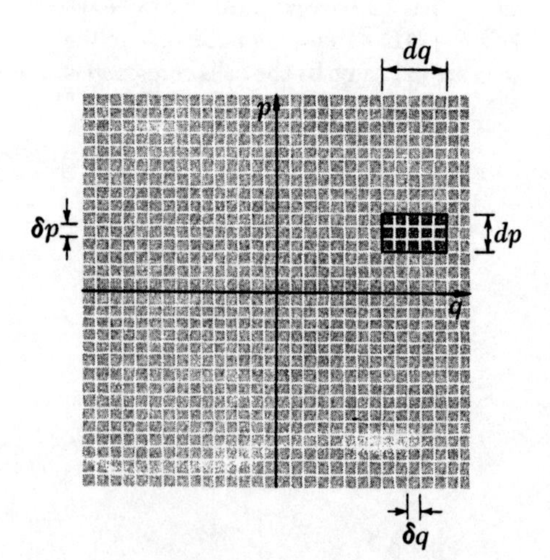

Fig. 6.4 A two-dimensional example of phase space subdivided into small cells of equal "volume" $\delta q\, \delta p = h_0$. The dark region indicates an element of volume having a size $dq\, dp$ and containing many cells.

Here the ranges dq_i and dp_i are supposed to be small in the sense that the energy E of A does not vary appreciably when q_i changes by an amount dq_i or p_i changes by an amount dp_i. They are, however, supposed to be large compared to the intervals used in the subdivision of phase space, i.e., $dq_i \gg \delta q_i$ and $dp_i \gg \delta p_i$. The element of volume $(dq_1 \cdots dq_f\, dp_1 \cdots dp_f)$ of phase space thus contains many cells, each of volume $(\delta q_1 \cdots \delta q_f\, \delta p_1 \cdots \delta p_f) = h_0{}^f$. (See Fig. 6.4.) In each of these cells the energy of the system A, and hence also its probability (7), is then nearly the same. Hence the desired probability (9) is found simply by multiplying the probability (7) of finding A in a given cell of phase space by the total number $(dq_1 \cdots dp_f)/h_0{}^f$ of such cells, i.e.,

$$\mathcal{P}(q_1, \ldots, p_f)\, dq_1 \cdots dp_f \propto e^{-\beta E_r} \frac{dq_1 \cdots dp_f}{h_0{}^f}$$

or
$$\boxed{\mathcal{P}(q_1, \ldots, p_f)\, dq_1 \cdots dp_f = C\, e^{-\beta E(q_1, \ldots, p_f)}\, dq_1 \cdots dp_f} \tag{10}$$

where C is merely some constant of proportionality (which includes the constant $h_0{}^f$). The value of this constant is, of course, determined by the normalization requirement that the sum of the probability (10) over all accessible coordinates and momenta of the system A be unity, i.e., that

$$\int \mathcal{P}(q_1 \cdots p_f)\, dq_1 \cdots dp_f = 1$$

where the integral extends over the entire region of phase space accessible to the system A. Thus it follows immediately that

$$C^{-1} = \int e^{-\beta E(q_1, \ldots, p_f)}\, dq_1 \cdots dp_f. \tag{11}$$

These general considerations will be illustrated in the next section by applying them to a simple case of great importance, that of a single molecule moving in three dimensions.

6.2 *Maxwell Velocity Distribution*

Consider an ideal gas confined within a container of volume V and in equilibrium at the absolute temperature T. This gas may consist of several different types of molecules. We shall assume that conditions are such that a classical treatment of the molecules in the gas is permissible. At the end of our discussion we shall examine the range of conditions within which such classical considerations can be expected to be valid. Let us think thus in classical terms and focus attention on any one of the molecules of the gas. This molecule then constitutes a distinct small system in thermal contact with a heat reservoir which consists of all the other molecules and has a temperature T. The canonical distribution is thus immediately applicable. Suppose, for the time being, that the molecule is monatomic. If one neglects any external force fields (such as gravity), the energy ϵ of this molecule then is simply its kinetic energy

$$\epsilon = \frac{1}{2} m\mathbf{v}^2 = \frac{1}{2} \frac{\mathbf{p}^2}{m} \tag{12}$$

where $\mathbf{v}$ is the velocity and $\mathbf{p} = m\mathbf{v}$ is the momentum of the molecule of mass m. Here we have assumed that the gas is sufficiently dilute to be ideal; hence any potential energy of interaction with other molecules is supposed to be negligible. The energy of the molecule anywhere within the container is thus independent of the position vector $\mathbf{r}$ of the molecule.

The state of the molecule is described classically in terms of the three position coordinates x,y,z of the molecule and its corresponding three momentum components p_x,p_y,p_z. We can then ask for the probability that the position of the molecule lies in the range between $\mathbf{r}$ and $\mathbf{r} + d\mathbf{r}$ (i.e., that its x coordinate lies between x and $x + dx$, its y coordinate lies between y and $y + dy$, and its z coordinate lies between z and $z + dz$) and that simultaneously its momentum lies in the range between $\mathbf{p}$ and $\mathbf{p} + d\mathbf{p}$ (i.e., that its x component of momentum lies between p_x and $p_x + dp_x$, its y component of momentum lies between p_y and $p_y + dp_y$, and its z component of momentum lies between p_z and $p_z + dp_z$). This range of position and momentum variables corresponds to a "volume" of phase space of size $(dx\,dy\,dz\,dp_x\,dp_y\,dp_z) \equiv d^3\mathbf{r}\,d^3\mathbf{p}$. Here we have introduced the convenient abbreviations

$$d^3\mathbf{r} \equiv dx\,dy\,dz$$

and
$$d^3\mathbf{p} \equiv dp_x\,dp_y\,dp_z \tag{13}$$

for an element of volume of real space and an element of volume of momentum space, respectively. Applying the canonical distribution (10), we then immediately obtain for the desired probability $\mathcal{P}(\mathbf{r},\mathbf{p})\,d^3\mathbf{r}\,d^3\mathbf{p}$ that the molecule has a position between $\mathbf{r}$ and $\mathbf{r} + d\mathbf{r}$ and a momentum between $\mathbf{p}$ and $\mathbf{p} + d\mathbf{p}$ the result

$$\mathcal{P}(\mathbf{r},\mathbf{p})\,d^3\mathbf{r}\,d^3\mathbf{p} \propto e^{-\beta(p^2/2m)}\,d^3\mathbf{r}\,d^3\mathbf{p} \tag{14}$$

where $\beta = (kT)^{-1}$. Here we have used the expression (12) for the energy of the molecule and have written $p^2 = \mathbf{p}^2$. Equivalently we can express this result in terms of the velocity $\mathbf{v} = \mathbf{p}/m$ of the molecule to find the probability $\mathcal{P}'(\mathbf{r},\mathbf{v})\,d^3\mathbf{r}\,d^3\mathbf{v}$ that the molecule has a position between $\mathbf{r}$ and $\mathbf{r} + d\mathbf{r}$ and a velocity between $\mathbf{v}$ and $\mathbf{v} + d\mathbf{v}$. Thus

$$\boxed{\mathcal{P}'(\mathbf{r},\mathbf{v})\,d^3\mathbf{r}\,d^3\mathbf{v} \propto e^{-(1/2)\beta m v^2}\,d^3\mathbf{r}\,d^3\mathbf{v}} \tag{15}$$

where $d^3\mathbf{v} \equiv dv_x\,dv_y\,dv_z$ and where $v^2 = \mathbf{v}^2$.

The probability (15) is a very general result which provides detailed information about the position and velocity of any molecule in the gas. It allows us to deduce readily a variety of more special results. For example, we may ask how many molecules have a velocity in a specified range. Or more generally, if the gas consists of a mixture of different kinds of molecules having different masses (e.g., helium and argon molecules), we may ask how many molecules of a given kind have a velocity in any specified range. Focusing attention on the molecules of a particular kind, we can thus try to find

$f(\mathbf{v})\,d^3\mathbf{v} \equiv$ mean number of molecules (of the specified kind), *per unit volume*, which have a velocity between $\mathbf{v}$ and $\mathbf{v} + d\mathbf{v}$. (16)

Since the N molecules of the ideal gas move independently without appreciable mutual interaction, the gas constitutes a statistical ensemble of molecules of which a fraction given by the probability (15) have a position between $\mathbf{r}$ and $\mathbf{r} + d\mathbf{r}$ and a velocity between $\mathbf{v}$ and $\mathbf{v} + d\mathbf{v}$. The mean number $f(\mathbf{v})\,d^3\mathbf{v}$ of (16) is thus simply obtained by multiplying the probability (15) by N, the total number of molecules of this kind, and dividing by the volume element $d^3\mathbf{r}$. Thus

$$f(\mathbf{v})\,d^3\mathbf{v} = \frac{N\,\mathcal{P}'(\mathbf{r},\mathbf{v})\,d^3\mathbf{r}\,d^3\mathbf{v}}{d^3\mathbf{r}}$$

or

$$\boxed{f(\mathbf{v})\,d^3\mathbf{v} = C\,e^{-(1/2)\beta m v^2}\,d^3\mathbf{v}} \tag{17}$$

where C is a constant of proportionality and $\beta \equiv (kT)^{-1}$. The result (17) is called the *Maxwell velocity distribution* since it was first derived by Maxwell in 1859 (by the use of less general arguments).

Note that the probability $\mathcal{P}'$ of (15) [or the mean number f of (17)] does not depend on the position $\mathbf{r}$ of the molecule. This result must, of course, be true by virtue of symmetry considerations since a molecule can have no preferred position in space in the absence of external force fields. Note also that $\mathcal{P}'$ (or f) depends only on the magnitude of $\mathbf{v}$ and not on its direction; i.e.,

$$f(\mathbf{v}) = f(v) \tag{18}$$

where $v = |\mathbf{v}|$. Again this is obvious by symmetry, since there is no preferred direction in a situation where the container (and thus also the center of mass of the whole gas) is considered to be at rest.

Determination of the constant C

The constant C can be determined by the requirement that the sum of (17) over all possible velocities must yield the total mean number n of molecules (of the kind considered) per unit volume. Thus

$$C\int e^{-(1/2)\beta mv^2}\, d^3\mathbf{v} = n \tag{19}$$

or $C\iiint e^{-(1/2)\beta m(v_x{}^2+v_y{}^2+v_z{}^2)}\, dv_x\, dv_y\, dv_z = n.$

Thus $C\iiint e^{-(1/2)\beta mv_x{}^2} e^{-(1/2)\beta mv_y{}^2} e^{-(1/2)\beta mv_z{}^2}\, dv_x\, dv_y\, dv_z = n,$

or $C\int_{-\infty}^{\infty} e^{-(1/2)\beta mv_x{}^2}\, dv_x \int_{-\infty}^{\infty} e^{-(1/2)\beta mv_y{}^2}\, dv_y \int_{-\infty}^{\infty} e^{-(1/2)\beta mv_z{}^2}\, dv_z = n.$

Each of these integrals has, by (M.23), the same value

$$\int_{-\infty}^{\infty} e^{-(1/2)\beta mv_x{}^2}\, dv_x = \left(\frac{\pi}{\frac{1}{2}\beta m}\right)^{1/2} = \left(\frac{2\pi}{\beta m}\right)^{1/2}.$$

Hence $$C = n\left(\frac{\beta m}{2\pi}\right)^{3/2} \tag{20}$$

and $$\boxed{f(\mathbf{v})\, d^3\mathbf{v} = n\left(\frac{\beta m}{2\pi}\right)^{3/2} e^{-(1/2)\beta mv^2}\, d^3\mathbf{v}.} \tag{21}$$

Validity of the results for polyatomic molecules

Suppose that the gas under consideration contains molecules which are *not* monatomic. Under the conditions contemplated in the preceding paragraphs, the motion of the center of mass of such a molecule can still be treated by the classical approximation, although the intramolecular motion of rotation and vibration about its center of mass must ordinarily be discussed in terms of quantum mechanics. The state of the molecule can then be described by the position $\mathbf{r}$ and momentum $\mathbf{p}$ of its center of mass, and by specifying the particular quantum state s describing its intramolecular motion. The energy of the molecule is then

$$\epsilon = \frac{\mathbf{p}^2}{2m} + \epsilon_s{}^{(i)} \tag{22}$$

where the first term on the right is the kinetic energy of its center-of-mass motion and the second term is its intramolecular energy of rotation and vibration in the state s. The canonical distribution allows us to write down immediately an expression for the probability $\mathcal{P}_s(\mathbf{r},\mathbf{p})\, d^3\mathbf{r}\, d^3\mathbf{p}$ that the molecule is found in the state where its center-of-mass position is between $\mathbf{r}$ and $\mathbf{r} + d\mathbf{r}$, its center-of-mass momentum is between $\mathbf{p}$ and $\mathbf{p} + d\mathbf{p}$, and its intramolecular motion is specified by s. Thus

$$\begin{aligned}\mathcal{P}_s(\mathbf{r},\mathbf{p})\, d^3\mathbf{r}\, d^3\mathbf{p} &\propto e^{-\beta[p^2/2m+\epsilon_s{}^{(i)}]}\, d^3\mathbf{r}\, d^3\mathbf{p}\\ &\propto e^{-\beta p^2/2m}\, d^3\mathbf{r}\, d^3\mathbf{p}\, e^{-\beta\epsilon_s{}^{(i)}}.\end{aligned} \tag{23}$$

To find the probability $\mathcal{P}(\mathbf{r},\mathbf{p})\, d^3\mathbf{r}\, d^3\mathbf{p}$ that the center of mass has a position between $\mathbf{r}$ and $\mathbf{r} + d\mathbf{r}$ and a momentum between $\mathbf{p}$ and $\mathbf{p} + d\mathbf{p}$, *irrespective* of the molecule's state of intramolecular motion, it is only necessary to sum (23) over all possible intramolecular states s. But since the expression (23) is simply a product of two factors, the sum over all possible values of the second factor merely yields some constant multiplying the first factor. The result (23) reduces thus to an expression of the form (14) describing the center of mass of the molecule. Hence (15) and the Maxwell velocity distribution (17) are very general results which remain valid also in describing the center-of-mass motion of a polyatomic molecule in a gas.

6.3 *Discussion of the Maxwell Distribution*

The Maxwell velocity distribution (17) allows us to deduce immediately some related velocity distributions, in particular the distribution of molecular speeds in a gas. As we shall see later, some of these results can be tested very directly by experiment. Let us explore some of these consequences of the Maxwell distribution and then examine the conditions under which the Maxwell distribution may be expected to be valid.

Distribution of a velocity component

Suppose that we focus attention on the component of a molecule's velocity along a particular direction, say the x direction. Then we are concerned with the following quantity describing a given kind of molecule,

$g(v_x)\,dv_x \equiv$ the mean number of molecules, per unit volume, which have an x component of velocity in the range between v_x and $v_x + dv_x$ (irrespective of the values of the other velocity components).

We obtain this number by simply adding all the molecules which have an x component of velocity in this range. Thus

$$g(v_x)\,dv_x = \int_{(v_y)}\int_{(v_z)} f(\mathbf{v})\,d^3\mathbf{v}$$

where the sums (i.e., integrations) extend over all the possible y and z velocity components of the molecules. Hence (17) yields

$$g(v_x)\,dv_x = C\int_{(v_y)}\int_{(v_z)} e^{-(1/2)\beta m(v_x{}^2+v_y{}^2+v_z{}^2)}\,dv_x\,dv_y\,dv_z$$

$$= C\,e^{-(1/2)\beta m v_x{}^2}\,dv_x\int_{-\infty}^{\infty}\int_{-\infty}^{\infty} e^{-(1/2)\beta m(v_y{}^2+v_z{}^2)}\,dv_y\,dv_z,$$

or

$$\boxed{g(v_x)\,dv_x = C'\,e^{-(1/2)\beta m v_x{}^2}\,dv_x} \tag{24}$$

since the integration over all values of v_y and v_z gives merely some constant which can be absorbed into C', a new constant of proportionality.† The constant C' can again be determined by the requirement that the total mean number of molecules per unit volume is properly equal to n, i.e., by the condition

$$\int_{-\infty}^{\infty} g(v_x)\,dv_x = C'\int_{-\infty}^{\infty} e^{-(1/2)\beta m v_x{}^2}\,dv_x = n.$$

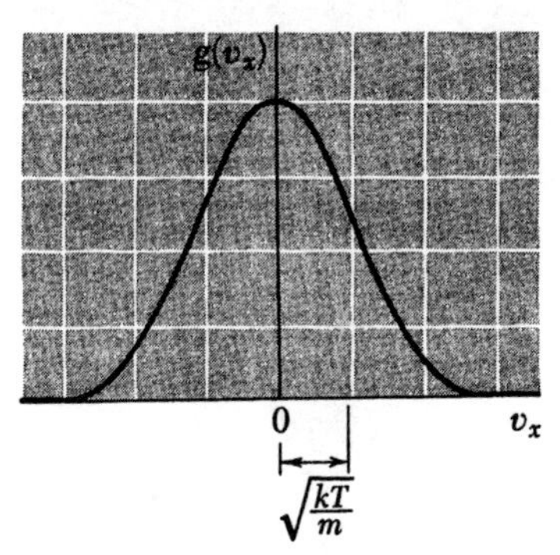

Fig. 6.5 Maxwell distribution showing the mean number $g(v_x)\,dv_x$ of molecules, per unit volume, having an x component of velocity between v_x and $v_x + dv_x$.

† Note that (24) is a simple Gaussian distribution of the kind discussed in Appendix A.1.

This gives
$$C' = n\left(\frac{\beta m}{2\pi}\right)^{1/2}. \tag{25}$$

The result (24) shows that the velocity component v_x is distributed symmetrically about the value $v_x = 0$. The *mean* value of any velocity component of a molecule must therefore always vanish, i.e.,

$$\bar{v}_x = 0. \tag{26}$$

This is physically clear by symmetry since the x component of velocity of a molecule is as likely to be positive as negative. Mathematically this result follows from the definition of the average†

$$\bar{v}_x \equiv \frac{1}{n}\int_{-\infty}^{\infty} g(v_x)\, v_x\, dv_x.$$

Here the integrand is an odd function of v_x (i.e., reverses its sign when v_x reverses sign) because $g(v_x)$ is an even function of v_x (i.e., remains unchanged under this operation since it depends only on v_x^2). Thus contributions to the integrand from $+v_x$ and $-v_x$ cancel each other.

Note that $g(v_x)$ has its maximum value when $v_x = 0$ and decreases rapidly as $|v_x|$ increases. It becomes negligibly small when $|\beta m v_x^2| \gg 1$; that is,

$$\text{if } |v_x| \gg (kT/m)^{1/2}, \qquad g(v_x) \to 0. \tag{27}$$

The distribution $g(v_x)$ thus becomes increasingly sharply peaked near $v_x = 0$ if the absolute temperature T is reduced. This merely reflects the fact that the mean kinetic energy of a molecule becomes increasingly small as $T \to 0$.

Needless to say, exactly similar results hold for the velocity components v_y and v_z since all velocity components are, by the symmetry of the situation, completely equivalent.

Distribution of molecular speeds

Considering a given kind of molecule, let us now investigate the quantity

$F(v)\,dv \equiv$ the mean number of molecules, per unit volume, which have a speed $v \equiv |\mathbf{v}|$ in the range between v and $v + dv$.

We can obtain this number by adding all molecules having speeds in this range, irrespective of the *directions* of their velocity. Thus

$$F(v)\,dv = \int' f(\mathbf{v})\, d^3\mathbf{v} \tag{28}$$

† The average can here be written as an integral in the manner of Eq. (2.78).

where the prime on the integral indicates that the integration is over all velocities satisfying the condition that

$$v < |\mathbf{v}| < v + dv,$$

i.e., over all velocity vectors which terminate in velocity space within a spherical shell of inner radius v and outer radius $v + dv$. Since dv is infinitesimal and $f(\mathbf{v})$ depends only on the magnitude of $\mathbf{v}$, the function $f(\mathbf{v})$ has essentially the constant value $f(v)$ over the entire domain of integration of (28) and can thus be taken outside the integral. The remaining integral represents merely the volume in velocity space of a spherical shell of radius v and thickness dv, a volume equal to the area $4\pi v^2$ of this shell multiplied by its thickness dv. Hence (28) becomes simply

$$\boxed{F(v)\,dv = 4\pi f(v)\,v^2\,dv.} \tag{29}$$

Using (17) this becomes explicitly

$$\boxed{F(v)\,dv = 4\pi C\,e^{-(1/2)\beta m v^2} v^2\,dv} \tag{30}$$

where C is given by (20). The relation (30) is the Maxwell distribution of speeds. Note that it has a maximum for the same reason that is responsible for the maxima encountered in our general discussion of statistical mechanics. As v increases, the exponential factor *decreases*, but the volume of phase space available to the molecule is proportional to v^2 and *increases;* the net result is a gentle maximum.

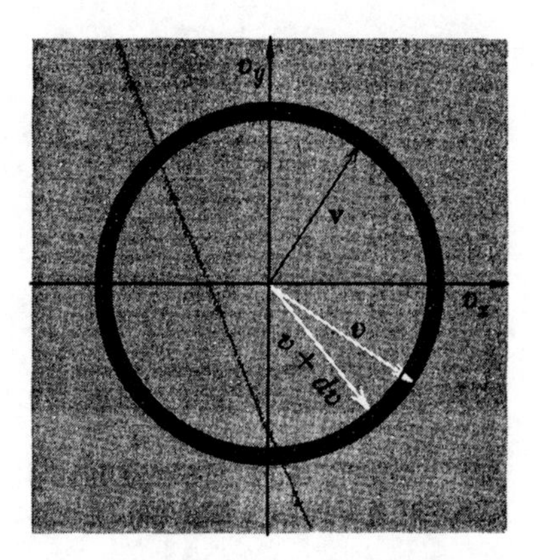

Fig. 6.6 Velocity space indicated in two dimensions, the v_z axis pointing out of the paper. The spherical shell contains all molecules having a velocity $\mathbf{v}$ such that $v < |\mathbf{v}| < v + dv$.

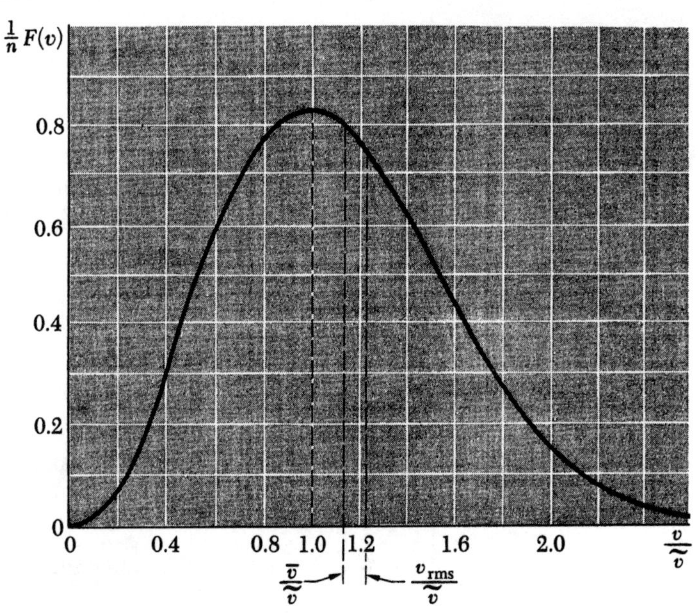

Fig. 6.7 Maxwell distribution showing the mean number $F(v)\,dv$ of molecules, per unit volume, having a speed between v and $v + dv$. The speed v is here expressed in terms of the most probable speed $\tilde{v} = (2kT/m)^{1/2}$. Also shown are the mean speed $\bar{v}$ and the root-mean-square speed $v_{rms} \equiv (\overline{v^2})^{1/2}$.

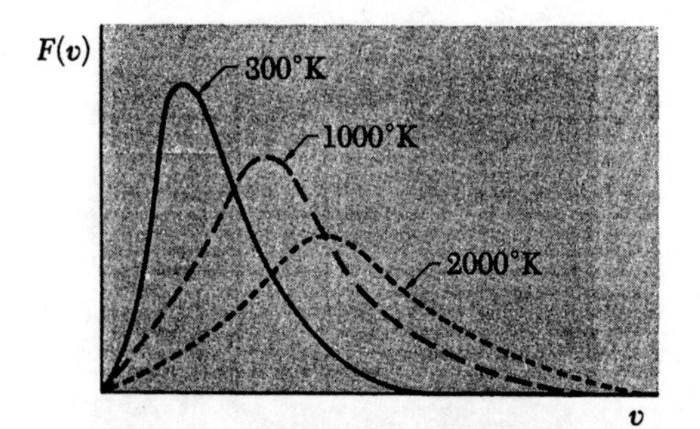

Fig. 6.8 Behavior of the Maxwell distribution of molecular speeds as a function of temperature.

Of course, if $F(v)\,dv$ is summed over all possible speeds $v = |\mathbf{v}|$, the result must again be the total mean number n of molecules per unit volume, i.e.,

$$\int_0^\infty F(v)\,dv = n. \tag{31}$$

The lower limit of the integral reflects the fact that, by its definition, the speed $v \equiv |v|$ of a molecule cannot be negative.

A plot of $F(v)$ as a function of the speed v is shown in Fig. 6.7. The particular speed $v = \tilde{v}$ where $F(v)$ has its maximum value is called the *most probable speed.* It can be found by setting $dF/dv = 0$. By using (30), this condition becomes

$$(-\beta m v\, e^{-(1/2)\beta m v^2})v^2 + e^{-(1/2)\beta m v^2}(2v) = 0$$

so that

$$\tilde{v} = \sqrt{\frac{2}{\beta m}} = \sqrt{\frac{2kT}{m}}. \tag{32}$$

Let us, for example, consider nitrogen (N_2) gas at room temperature so that $T \approx 300°\text{K}$. Since the molecular weight of N_2 is 28 and Avogadro's number is 6×10^{23} molecules/mole, the mass of a N_2 molecule is $m \approx 28/(6 \times 10^{23}) \approx 4.6 \times 10^{-23}$ gm. Hence (32) yields for the most probable speed of such a N_2 molecule

$$\tilde{v} \approx 4.2 \times 10^4 \text{ cm/sec} = 420 \text{ m/sec}, \tag{33}$$

a number of the order of the speed of sound in the gas.

Validity of the classical discussion of a gas

Now let us examine under what conditions our classical discussion of the ideal gas, and hence also the Maxwell velocity distribution, can be expected to be valid. Our criterion of validity is the condition (3) following from the Heisenberg uncertainty principle. If the condition (3) is satisfied, the classical description ought to be adequate since it makes no statements that might violate limitations imposed by quantum ideas.

Since we are merely interested in typical orders of magnitude, we can be content with approximate estimates of the pertinent quantities in (3). The typical magnitude p_0 of the momentum of a molecule of mass m in a gas at temperature T can be found from the most probable speed $\tilde{v}$ of such a molecule. Thus, by (32),

$$p_0 \approx m\tilde{v} = \sqrt{2mkT}.$$

The corresponding typical de Broglie wavelength $\lambda\!\!\!^{-}_0$ of the molecule

is then
$$\lambda\!\!\!^{-}_0 \equiv \frac{\hbar}{p_0} \approx \frac{\hbar}{\sqrt{2mkT}}. \tag{34}$$

The classical description considers the molecules as distinguishable particles traveling along well-defined trajectories. This point of view ought certainly to be valid if there exist no quantum mechanical limitations preventing a molecule from being localized within a distance no greater than the typical separation s_0 between nearest molecules. This requires, in accordance with (3), that

$$\boxed{s_0 \gg \lambda\!\!\!^{-}_0.} \tag{35}$$

(The quantum-mechanical discussion shows that quantum effects *do* indeed become significant when the condition (35) is violated, precisely because the essential indistinguishability of the molecules is then of paramount importance). To estimate the typical separation s_0 between nearest molecules, imagine each molecule to be at the center of a little cube of side s_0, these cubes filling the volume V available to the gas consisting of N molecules. Then

$$s_0{}^3 N = V$$

or
$$s_0 = \left(\frac{V}{N}\right)^{1/3} = n^{-1/3} \tag{36}$$

where $n \equiv N/V$ is the number of molecules per unit volume. The condition (35) for the validity of the classical approximation thus becomes

$$\frac{\lambda\!\!\!^{-}_0}{s_0} \approx \hbar \frac{n^{1/3}}{\sqrt{2mkT}} \ll 1. \tag{37}$$

This shows that the classical approximation ought to be applicable if the gas is sufficiently dilute so that n is small, if the temperature T is sufficiently high, and if the mass m of a molecule is not too small.

Fig. 6.9 James Clerk Maxwell (1831–1879). Although best known for his fundamental work in electromagnetic theory, Maxwell also contributed significantly to macroscopic thermodynamics and to the atomic theory of gases. He derived the molecular velocity distribution in 1859. After spending the earlier part of his career at the University of Aberdeen in Scotland, Maxwell became professor at Cambridge University in 1871. *(From G. Holton and D. Roller, "Foundations of Modern Physical Science," Addison-Wesley Publishing Co., Inc., Cambridge, Mass., 1958. By permission of the publishers.)*

Numerical estimates

To estimate the typical magnitudes numerically, consider helium (He) gas at room temperature and atmospheric pressure (760 mm Hg). The relevant parameters are then:

mean pressure $\bar{p}$
$= 760$ mm Hg $\approx 10^6$ dynes/cm^2;

temperature T
$\approx 300°$K; hence $kT \approx 4.1 \times 10^{-14}$ erg;

molecular mass m

$$= \frac{4}{6 \times 10^{23}} \approx 6.6 \times 10^{-24} \text{ gm}.$$

The equation of state of an ideal gas gives

$$n = \frac{\bar{p}}{kT} = 2.5 \times 10^{19} \text{ molecules/cm}^3.$$

Hence (34) and (36) yield the estimates

$$\lambda\!\!\!^{-}_0 \approx 0.14 \text{ Å}$$

and

$$s_0 \approx 33 \text{ Å}$$

where $1 \text{ Å} = 10^{-8}$ cm. Here the condition (35) is well satisfied and the classical approximation ought to be very good. Most gases have larger molecular weights and thus smaller de Broglie wavelengths; the condition (35) is then even better satisfied.

On the other hand, consider the conduction electrons in a typical metal such as copper. In first approximation, interactions between these electrons can be neglected so that they can be treated as an ideal gas. But the numerical values of the significant parameters are then quite different. The mass of an electron is very small, about 10^{-27} gm or 7300 times less than that of a He atom. This makes the de Broglie wavelength of the electron much longer,

$$\lambda\!\!\!^{-}_0 \approx (0.14) \times \sqrt{7300} \approx 12 \text{ Å}.$$

Furthermore, since there is about one conduction electron per atom in the metal and since a typical interatomic spacing is about 2 Å,

$$s_0 \approx 2 \text{ Å}.$$

The distance between particles is thus much smaller than in the case of He gas, i.e., the electrons in a metal form a very dense gas. These estimates show that the condition (35) is not satisfied by the electrons in a metal. Hence there exists no justification for discussing such electrons by classical statistical mechanics. In fact, it is then essential to give a completely quantum mechanical treatment which takes into account the Pauli exclusion principle obeyed by the electrons.

6.4 *Effusion and Molecular Beams*

Consider a gas in equilibrium inside some container. Suppose that a small hole of diameter D (or a narrow slit of width D) is now made in one of the walls of this container. If the hole is sufficiently small, the equilibrium of the gas inside the container should be disturbed to a negligible extent. The few molecules escaping through the hole into a vacuum surrounding the container should then constitute a representative sample of the molecules of the gas in its equilibrium state. Indeed, the molecules that have thus escaped can be collimated by slits to form a well-defined beam and, since they are few in number, their mutual interaction in this beam is insignificant. The molecules in such a beam can then be studied very effectively with two possible aims: (i) It may be of interest to study the properties of the molecules in the gas in equilibrium inside the container. For example, one may want to check whether the molecules in the container have velocities distributed in accordance with the predictions of the Maxwell distribution. (ii) It may be of interest to study the properties of essentially isolated molecules or atoms in order to investigate fundamental atomic or nuclear properties. Several Nobel prizes attest to the fruitfulness of this technique. We need only mention the fundamental experiments of Stern and Gerlach which led to the discovery of the spin and associated magnetic moment of the electron, those of Rabi and coworkers which led to precision measurements of nuclear magnetic moments, and those of Kusch and Lamb which helped to lead to our modern understanding of the quantum theory of electromagnetic interactions.†

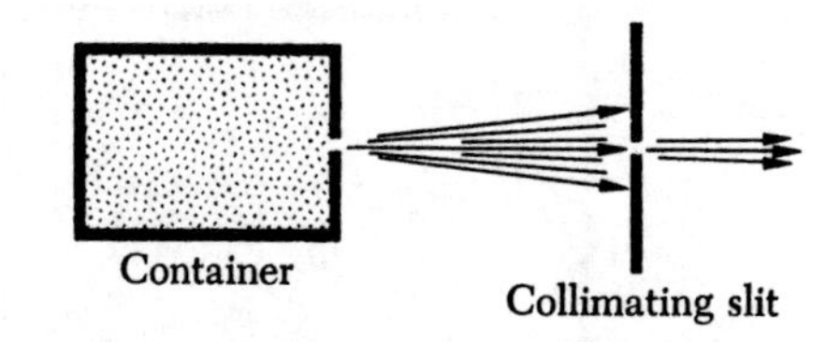

Fig. 6.10 Formation of a molecular beam by molecules escaping through a small hole in a container.

How small must be the size D of the hole so that the equilibrium of the gas inside the container is disturbed to a negligible extent? The hole must be sufficiently small so that the few molecules which are in the vicinity of the hole (and thus manage to escape through it) do not affect appreciably the large number of molecules in the remainder of the gas. This condition is satisfied if, during the time that a molecule spends in the vicinity of the hole, it suffers practically no collisions with other molecules. But if the mean speed of a molecule is $\bar{v}$, the time that a molecule spends near the hole is of the order of $D/\bar{v}$. On the other hand, the time between successive collisions of a molecule with other molecules is of the order of $l/\bar{v}$ if l denotes the mean free path of a molecule in the gas.‡ The preceding condition is thus equivalent to the statement that

$$\frac{D}{\bar{v}} \ll \frac{l}{\bar{v}}$$

or

$$D \ll l. \tag{38}$$

If this condition is satisfied, the molecules inside the container remain essentially in equilibrium (although their total number decreases slowly) and the escape of the molecules through the hole is said to constitute *effusion*.

Remark

The situation is appreciably different if $D \gg l$ so that molecules suffer frequent collisions with each other near the hole. When some molecules emerge through the hole (as in Fig. 6.10), the molecules behind them are then appreciably affected. They no longer continue colliding with the molecules on the right which have just escaped through the hole, but they still suffer frequent collisions with the molecules on the left. These collisions cause the molecules near the hole to experience an unbalanced force to the right and thus to acquire a net velocity in the direction toward the hole. The resultant motion of all these molecules moving together as a group is then analogous to the flow of water through the hole of a tank. In this case one has not effusion, but *hydrodynamic flow*.

† A good and readable review of molecular beam experiments can be found in an article by O. R. Frisch in *Sci. American*, vol. 212, p. 58 (May 1965).

‡ As discussed in Sec. 1.6, the mean free path is defined as the mean distance traveled by a molecule in the gas before it collides with some other molecule.

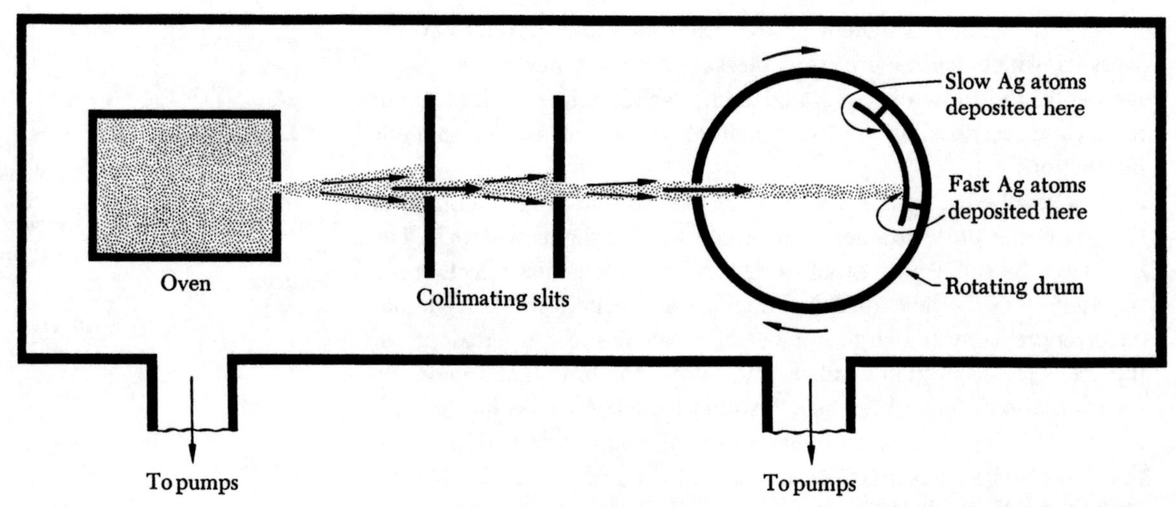

Fig. 6.11 A molecular-beam apparatus for studying the velocity distribution of silver (Ag) atoms. The Ag atoms stick to the drum surface upon impact.

If the hole is small enough so that the condition (38) is satisfied, the equilibrium of the gas is not affected appreciably by the presence of the hole. Accordingly the mean number $\mathcal{J}_0$ of molecules escaping per unit time through the hole is the same as the mean total number of molecules that would, per unit time, strike the area occupied by the hole if this hole had never been made. Hence $\mathcal{J}_0$ is simply given by the approximate expression (1.18) obtained in Sec. 1.6, i.e.,

$$\mathcal{J}_0 \approx \tfrac{1}{6} n\bar{v} \tag{39}$$

where n is the mean number of molecules per unit volume and $\bar{v}$ is their mean speed.† If one wants to focus attention only on those molecules having a speed between v and $v + dv$, the mean number $\mathcal{J}(v)\,dv$ of such molecules escaping through the hole per unit time is, analogously to (39), approximately given by

$$\mathcal{J}(v)\,dv \approx \tfrac{1}{6}[F(v)\,dv]v \tag{40}$$

where $F(v)\,dv$ denotes the mean number of molecules having a speed between v and $v + dv$. Using the Maxwell distribution of speeds (30), one then obtains the proportionality

$$\mathcal{J}(v)\,dv \propto v^3 e^{-(1/2)\beta m v^2}. \tag{41}$$

The last factor v in (40) expresses merely the fact that a fast molecule is more likely to escape through the hole than a slow one.

By measuring the relative number of molecules having various speeds in the molecular beam emerging from the hole in the container,

† An exact calculation yields, instead of (39), the result $\mathcal{J}_0 = \frac{1}{4} n\bar{v}$. See Appendix A.4 where the exact calculation is discussed.

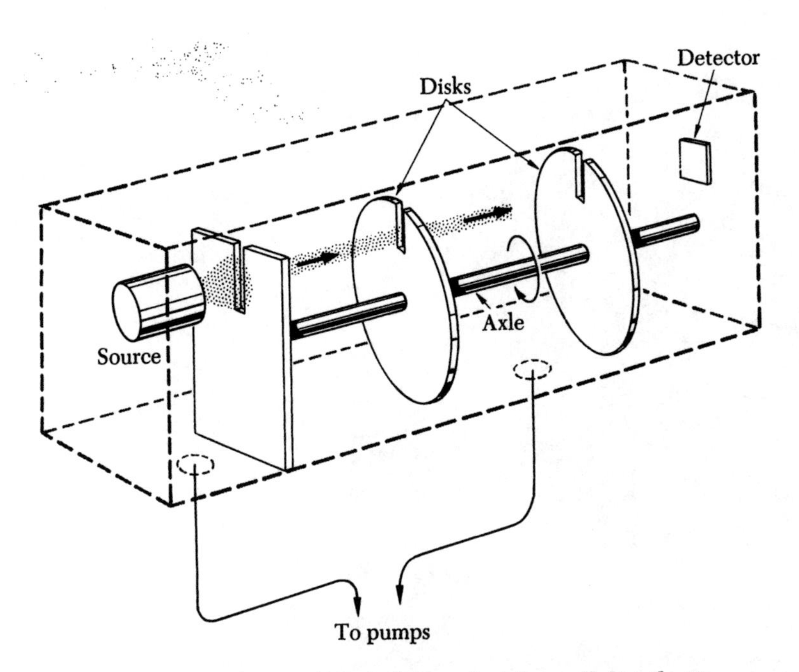

Fig. 6.12 A molecular-beam apparatus for studying the molecular velocity distribution by a velocity selector. By the time the beam of molecules reaches the second disk, the slot in this disk has ordinarily moved to a position such that the beam cannot pass through this disk. An exception occurs only if this disk has rotated by one revolution (or an integral number of revolutions) in the time required by the molecules to traverse the distance between the two disks. (A more effective velocity selector results if more than two similar disks are mounted on the same axle.)

one can test the prediction (41) and thus the Maxwell distribution upon which it is based. One experimental arrangement for achieving this purpose is shown in Fig. 6.11. Here silver is heated in an oven to generate a gas of silver (Ag) atoms, some of which emerge through a narrow slit to form an atomic beam. A hollow cylindrical drum, which has a slit in it and which rotates rapidly about its axis, is located in front of the beam. When the Ag atoms enter the slit in the drum, they require different times to reach the opposite side of the drum, a fast atom requiring less time than a slow one. Thus, since the drum is rotating, Ag atoms with different speeds strike the inside surface of the drum at different places and then stick to it. A subsequent measurement of the thickness of the deposited silver layer as a function of distance along the inside drum surface thus provides a measurement of the atomic velocity distribution.

A more accurate method for determining the velocity distribution uses a device which selects molecules having a particular velocity. (See Fig. 6.12. The method is analogous to the toothed-wheel method used by Fizeau to measure the speed of light.) In this method the molecular beam emerges from a hole and is detected at the other end of the apparatus. The velocity selector, which is placed between the source and the detector, consists in the simplest case of a pair of disks mounted on a common axle that can be rotated with known angular velocity. Both disks are identical and each has a slot cut into its periphery. The rotating disks act thus as two shutters which are alternately opened

Fig. 6.13 Photograph of a modern molecular-beam apparatus used to study the properties of hydrogen molecules and atoms. (*Photograph by courtesy of Professor Norman F. Ramsey, Harvard University.*)

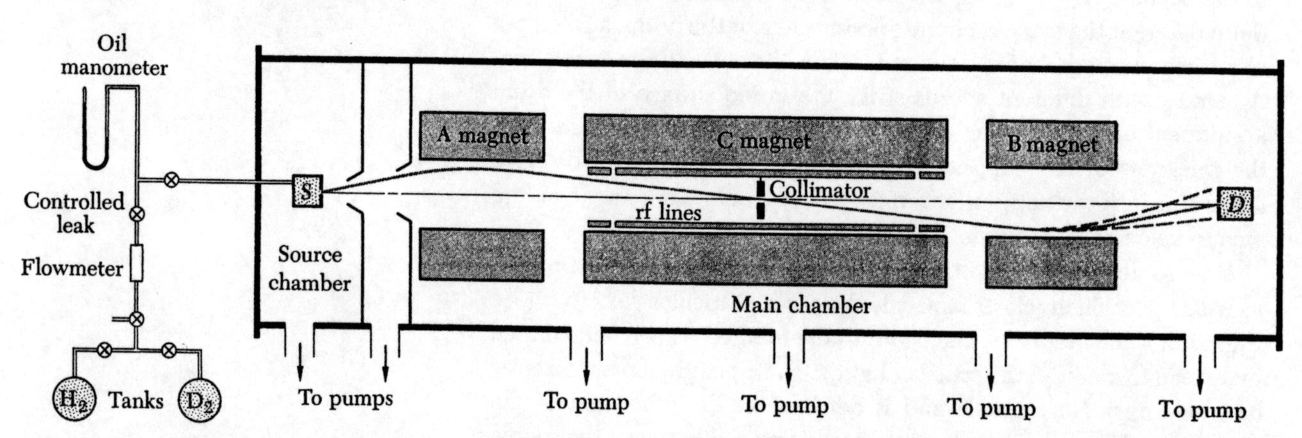

Fig. 6.14 Schematic diagram showing the essential ingredients of the molecular-beam apparatus of the preceding photograph. The source of molecules is the container S, while *D* denotes some kind of device detecting the molecules arriving at the other end of the apparatus. The inhomogeneous magnetic fields produced by the A and B magnets exert forces on the very small magnetic moments of the molecules and thus deflect them along the indicated paths. The experiments detect the effect of radio-frequency radiation acting on the molecules in the experimental region of the C magnet.

and closed. When the disks are properly aligned and not rotating, all molecules can reach the detector by passing through the slots in both disks. But when the disks are rotating, molecules passing through the slot in the first disk can only reach the detector if their velocity is such that the time of flight required for them to travel to the second disk is equal to the time required for one revolution of the disks (or an integral number of such revolutions).† Otherwise they strike the solid part of the second disk and are stopped. Hence different angular velocities of rotation of the disks allow molecules of different speeds to reach the detector. Measurements of the relative number of molecules arriving at the detector per second thus allow a direct check of the molecular velocity distribution. The validity of the Maxwellian distribution has been well confirmed by such experiments.

The phenomenon of effusion also has several practical applications outside the realm of molecular beams. Returning to the relation (39), note that a knowledge of the absolute temperature T and mean pressure $\bar{p}$ of the gas allows us to calculate n and $\bar{v}$. Thus the equation of state of an ideal gas gives $n = \bar{p}/kT$. Furthermore, the mean speed $\bar{v}$ of a molecule is approximately equal to its most probable speed (32); thus $\bar{v} \propto (kT/m)^{1/2}$. Hence (39) yields

$$\mathcal{J}_0 \propto \frac{\bar{p}}{\sqrt{mT}}. \tag{42}$$

The rate of effusion $\mathcal{J}_0$ is seen to depend on the mass of the molecule because a lighter molecule has a higher mean speed than a heavier one and thus effuses more rapidly. This property can be exploited to achieve a practical method for the separation of isotopes. Suppose that a container is closed off by a membrane which has very many small holes through which molecules can effuse. If this container is surrounded by a vacuum on the outside and is filled with a gas mixture of two isotopes at some initial time, then the relative concentration of the isotope of larger molecular weight will increase in the container as time goes on. Similarly, the gas pumped off from the surrounding vacuum will be more concentrated in the lighter isotope. This method of isotope separation has been of practical importance in obtaining uranium rich in ^{235}U, the isotope which undergoes ready nuclear fission and is thus of great importance for the operation of nuclear power reactors (or the manufacture of nuclear weapons). Ordinary uranium consists mostly of the isotope ^{238}U, but by using the chemical compound uranium hexafluoride (UF_6), which is a gas at room tempera-

† Experimentally one can readily distinguish between observations corresponding to such different integral numbers.

ture, one can separate by effusion the slightly lighter $^{235}UF_6$ molecules from the much more abundant and slightly heavier $^{238}UF_6$ molecules. Since the difference in mass between these molecules is so small, the process of effusion must be repeated many times in succession to achieve an appreciably increased concentration of the ^{235}U isotope.

6.5 *The Equipartition Theorem*

The canonical distribution in its classical form (10) is a function of coordinates and momenta which are continuous variables. Any calculation of mean values can thus be reduced to a calculation of integrals rather than of discrete sums. Under certain conditions the mean energy of a system can then be computed in a particularly simple way.

To be specific, consider any system described classically in terms of f coordinates $q_1, \ldots, q_f$ and f corresponding momenta $p_1, \ldots, p_f$. Its energy E is then a function of these variables, i.e., $E = E(q_1, \ldots, p_f)$. It is frequently the case that this energy is of the form

$$E = \epsilon_i(p_i) + E'(q_1, \ldots, p_f) \tag{43}$$

where ϵ_i is a function of the particular momentum p_i only and where E' may depend on all coordinates and momenta *except* p_i. (For example, the functional form (43) may arise because the kinetic energy of a particle depends only on its momentum components while its potential energy depends only on its position.) Suppose that the system under consideration is in thermal equilibrium with a heat reservoir at an absolute temperature T. What then is the mean value of the energy contribution ϵ_i in (43)?

The probability of finding the system with its coordinates and momenta in a range near $\{q_1, \ldots, q_f, p_1, \ldots, p_f\}$ is given by the canonical distribution (10) with the constant C determined by (11). By its definition, the mean value of ϵ_i is found by performing the appropriate sum (or integral) over all possible states of the system, i.e.,

$$\bar{\epsilon}_i = \frac{\int e^{-\beta E(q_1, \ldots, p_f)} \epsilon_i \, dq_1 \cdots dp_f}{\int e^{-\beta E(q_1, \ldots, p_f)} \, dq_1 \cdots dp_f} \tag{44}$$

where the integrals extend over all possible values of all the coordinates $q_1, \ldots, q_f$ and all the momenta $p_1, \ldots, p_f$. By virtue of (43), the expression (44) becomes

$$\bar{\epsilon}_i = \frac{\int e^{-\beta(\epsilon_i + E')} \epsilon_i \, dq_1 \cdots dp_f}{\int e^{-\beta(\epsilon_i + E')} \, dq_1 \cdots dp_f}$$

$$= \frac{\int e^{-\beta\epsilon_i} \epsilon_i \, dp_i \int' e^{-\beta E'} \, dq_1 \cdots dp_f}{\int e^{-\beta\epsilon_i} \, dp_i \int' e^{-\beta E'} \, dq_1 \cdots dp_f}$$

where we have used the multiplicative property of the exponential function. The primes on the last integral signs indicate that these integrals extend over all the coordinate q and momenta p *except* p_i. But since the primed integral in the numerator is the same as that in the denominator, the resulting cancellation leads immediately to the simple result

$$\bar{\epsilon}_i = \frac{\int e^{-\beta\epsilon_i}\epsilon_i \, dp_i}{\int e^{-\beta\epsilon_i} \, dp_i}. \tag{45}$$

In short, since ϵ_i involves only the variable p_i, all the other variables are irrelevant in calculating the mean value of ϵ_i.

The expression (45) can be simplified further by relating the integral in the numerator to that in the denominator. Thus

$$\bar{\epsilon}_i = \frac{-\dfrac{\partial}{\partial\beta}\left(\int e^{-\beta\epsilon_i} \, dp_i\right)}{\int e^{-\beta\epsilon_i} \, dp_i}$$

or

$$\bar{\epsilon}_i = -\frac{\partial}{\partial\beta} \ln\left(\int_{-\infty}^{\infty} e^{-\beta\epsilon_i} \, dp_i\right) \tag{46}$$

where the explicit limits in the integral reflect the fact that the momentum p_i can assume all possible values from $-\infty$ to $+\infty$.

Consider now the specific case where ϵ_i is a quadratic function of p_i, as it would be if it represents a kinetic energy. In short, suppose that ϵ_i is of the form

$$\epsilon_i = bp_i^2 \tag{47}$$

where b is some constant. Then the integral in (46) becomes

$$\int_{-\infty}^{\infty} e^{-\beta\epsilon_i} \, dp_i = \int_{-\infty}^{\infty} e^{-\beta b p_i^2} \, dp_i = \beta^{-(1/2)} \int_{-\infty}^{\infty} e^{-by^2} \, dy$$

where we have introduced the variable $y \equiv \beta^{1/2}p_i$. Hence

$$\ln\left(\int_{-\infty}^{\infty} e^{-\beta\epsilon_i} \, dp_i\right) = -\tfrac{1}{2}\ln\beta + \ln\left(\int_{-\infty}^{\infty} e^{-by^2} \, dy\right).$$

But the integral on the right does *not* involve β at all. The differentiation of (46) yields thus simply

$$\bar{\epsilon}_i = -\frac{\partial}{\partial\beta}\left(-\frac{1}{2}\ln\beta\right) = \frac{1}{2\beta}$$

or

$$\boxed{\bar{\epsilon}_i = \tfrac{1}{2}kT.} \tag{48}$$

Note that, although the starting point (44) of our calculation involved a formidable array of integrals, we were able to derive the final result (48) without evaluating a single one.

If the functional forms (43) and (47) had been the same except for involving a coordinate q_i instead of a momentum p_i, all our previous arguments would be identical and would thus lead again to (48). Hence we have established the following general statement, known as the *equipartition theorem:*

> If a system described by classical statistical mechanics is in equilibrium at the absolute temperature T, every independent quadratic term in its energy has a mean value equal to $\frac{1}{2}kT$. (48*a*)

6.6 *Applications of the Equipartition Theorem*

Specific heat of a monatomic ideal gas

The energy of a molecule in such a gas is simply its kinetic energy (12), i.e.,

$$\epsilon = \frac{1}{2m}(p_x^2 + p_y^2 + p_z^2). \tag{49}$$

By virtue of the equipartition theorem, the mean value of each of the three terms in this expression is equal to $\frac{1}{2}kT$. Hence it follows immediately that

$$\bar{\epsilon} = \tfrac{3}{2}kT. \tag{50}$$

Since one mole of gas consists of Avogadro's number N_a of molecules, the mean energy of the gas is thus,

per mole, $$\bar{E} = N_a(\tfrac{3}{2}kT) = \tfrac{3}{2}RT \tag{51}$$

where $R \equiv N_a k$ is the gas constant. By virtue of (5.23), the molar specific heat c_V at constant volume is then equal to

$$c_V = \left(\frac{\partial \bar{E}}{\partial T}\right)_V = \frac{3}{2}R. \tag{52}$$

This result agrees with that previously obtained in (5.26) on the basis of quantum-mechanical reasoning applied to a gas sufficiently dilute to be ideal and nondegenerate.†

---

† In accordance with (37), quantum effects should indeed be unimportant for a sufficiently dilute gas. The agreement between the classical and quantum results is thus to be expected.

Kinetic energy of a molecule in any gas
Consider any gas, not necessarily ideal. The energy of any molecule of mass m can then be written in the form

$$\epsilon = \epsilon^{(k)} + \epsilon', \qquad \text{where } \epsilon^{(k)} = \frac{1}{2m}(p_x^2 + p_y^2 + p_z^2).$$

The first term $\epsilon^{(k)}$ is the kinetic energy of the molecule and depends on the momentum components p_x, p_y, p_z, of its center of mass. The term ϵ' may involve the position of the center of mass of the molecule (if the molecule is located in an external force field or if it interacts appreciably with other molecules); it may involve coordinates and momenta describing the rotation or vibration of the atoms of the molecule with respect to its center of mass (if the molecule is not monatomic); but it does not involve the center-of-mass momentum $\mathbf{p}$. Hence the equipartition theorem allows us again to conclude immediately that

$$\overline{\frac{1}{2m}p_x^2} = \overline{\frac{1}{2}mv_x^2} = \frac{1}{2}kT \tag{53}$$

or

$$\overline{v_x^2} = \frac{kT}{m}. \tag{54}$$

Since $\bar{v}_x = 0$ by symmetry, as already mentioned in (26), the result (54) represents also the dispersion $\overline{(\Delta v_x)^2}$ of the velocity component v_x. The three quadratic terms of the kinetic energy $\epsilon^{(k)}$ then yield for its mean value, as in (50), the result

$$\overline{\epsilon^{(k)}} = \tfrac{3}{2}kT. \tag{55}$$

Brownian motion
Consider a macroscopic particle of mass m (about a micron in size) suspended in a fluid at an absolute temperature T. The energy of this particle can again be written in the form

$$\epsilon = \frac{1}{2m}(p_x^2 + p_y^2 + p_z^2) + \epsilon'.$$

Here the first term is the kinetic energy involving the velocity $\mathbf{v}$ or momentum $\mathbf{p} = m\mathbf{v}$ of the center-of-mass motion of the particle, while ϵ' is the energy associated with the motion of all the atoms of the particle with respect to its center of mass. The equipartition theorem leads then again to the results (53) and (54) so that

$$\overline{v_x^2} = \frac{kT}{m}. \tag{56}$$

Since the mean value $\overline{v_x} = 0$ vanishes by symmetry, (56) gives directly the dispersion of the velocity component v_x. The result (56) thus shows immediately that the particle does not simply remain at rest, but must always exhibit a fluctuating velocity. The existence of the phenomenon of Brownian motion, discussed in Sec. 1.4, is thus an immediate consequence of our theory. The quantitative statement (56) also shows explicitly that, if the mass m of the particle is sufficiently large, the fluctuations become so small as to be unobservable.

Harmonic oscillator

Consider a particle of mass m performing simple harmonic oscillations in one dimension. Its energy is then given by

$$\epsilon = \frac{1}{2m} p_x^2 + \frac{1}{2} \alpha x^2. \tag{57}$$

Here the first term is the kinetic energy of the particle whose momentum is denoted by p_x. The second term is its potential energy if a displacement x of the particle produces a restoring force $-\alpha x$, where α is a constant (called the *spring constant*). Suppose that the oscillator is in equilibrium with a heat reservoir at a temperature T high enough so that the oscillator can be described in terms of classical mechanics. Then the equipartition theorem (48) can be immediately applied to each of the quadratic terms in (57) to give for the mean energy of the oscillator

$$\bar{\epsilon} = \tfrac{1}{2}kT + \tfrac{1}{2}kT = kT. \tag{58}$$

6.7 *The Specific Heat of Solids*

As a last application of the equipartition theorem, we shall discuss the specific heat of solids at temperatures sufficiently high so that a classical description is valid. Consider thus any simple solid consisting of N atoms; e.g., the solid might be copper, gold, aluminum, or diamond. By virtue of the mutual forces between neighboring atoms, the situation of stable mechanical equilibrium of the solid is that in which its atoms are located at regular positions in a crystal lattice. Each atom is, however, free to move by small amounts about its equilibrium position. The force (due to neighboring atoms) tending to restore the atom to its equilibrium position vanishes, of course, when the atom is actually located at its equilibrium position. Since the displacement of the atom from its equilibrium position is always quite small, the restoring force must, in first approximation, be simply proportional to the atomic displacement. This approximation, which is usually excellent, thus implies that the atom performs about its equilibrium position simple harmonic motion in three dimensions.

With proper choice of orientation of the x, y, z coordinate axes, the motion of an atom along any one of these axes, say along the x direction, is then a simple harmonic motion with an associated energy ϵ_x of the form (57), i.e.,

$$\epsilon_x = \frac{1}{2m} p_x^2 + \frac{1}{2} \alpha x^2. \tag{59}$$

Here p_x denotes the x component of momentum of the atom while x denotes the x component of displacement of the atom from its equilibrium position. In writing (59), we have supposed that the atom has a mass m and is subject to a restoring force of spring constant α. Correspondingly, the (angular) frequency of oscillation of the atom along the x direction is then given by

$$\omega = \sqrt{\frac{\alpha}{m}}. \tag{60}$$

Similar expressions hold for the energies ϵ_y and ϵ_z associated with the motion of the atom in the y and z directions. The total energy of the atom is thus of the form

$$\epsilon = \epsilon_x + \epsilon_y + \epsilon_z. \tag{61}$$

If the solid is in equilibrium at an absolute temperature T which is high enough so that the approximation of classical statistical mechanics is valid, the equipartition theorem is immediately applicable to each of the quadratic terms of (59). The mean value of ϵ_x is thus simply

$$\bar{\epsilon}_x = \tfrac{1}{2}kT + \tfrac{1}{2}kT = kT. \tag{62}$$

Similarly $\bar{\epsilon}_y = \bar{\epsilon}_z = kT$. By (61), the mean energy of an atom is then

$$\bar{\epsilon} = 3kT.$$

The mean energy of one mole of the solid, containing Avogadro's number N_a of atoms, is thus simply

$$\bar{E} = 3N_a kT = 3RT \tag{63}$$

where $R = N_a k$ is the gas constant. In accordance with (5.23), the molar specific heat c_V at constant volume of the solid is then given by

$$c_V = \left(\frac{\partial \bar{E}}{\partial T}\right)_V$$

or

$$\boxed{c_V = 3R.} \tag{64}$$

Using the numerical value (5.4) of R, this becomes†

$$c_V \approx 25 \text{ joules mole}^{-1} \text{ deg}^{-1}. \tag{65}$$

Note the extreme generality of the result (64). It is completely independent of the atomic mass or of the magnitude of the spring constant α. If the solid contained atoms having different masses or having different spring constants, the result (64) would thus still remain valid. If the solid were not isotropic, the restoring force on an atom would have a different magnitude along different directions and the spring constants for the x, y, and z directions would also differ correspondingly; but the value of the mean energy per atom would still be $3kT$ and (64) would still be valid. Indeed, the rigorous analysis of the simultaneous vibration of all the atoms in the solid shows that a description in terms of individual motions of single atoms is inadequate and that simple harmonic motion is really performed by atoms moving together in groups of different sizes.‡ But since (64) is independent of the masses and spring constants, it still remains valid. The only limitation on (64) is thus that the temperature be sufficiently high so that the classical approximation is valid. Hence (64) asserts that:

> At sufficiently high temperatures, all solids have the same temperature-independent molar specific heat c_V equal to $3R$. (66)

We shall see presently that in the case of most solids (diamond being a conspicuous exception) room temperature is sufficiently high for the approximate validity of the classical discussion.

Historically, the validity of statement (66) was first discovered empirically and is known as the law of Dulong and Petit. Table 6.1 lists directly measured values of the molar specific heat c_p (at constant *pressure*) for some solids at room temperature. The values of the molar specific heat c_V (at constant *volume*) can be obtained from the corre-

Solid	c_p	c_V
Aluminum	24.4	23.4
Bismuth	25.6	25.3
Cadmium	26.0	24.6
Carbon (diamond)	6.1	6.1
Copper	24.5	23.8
Germanium	23.4	23.3
Gold	25.4	24.5
Lead	26.8	24.8
Platinum	25.9	25.4
Silicon	19.8	19.8
Silver	25.5	24.4
Sodium	28.2	25.6
Tin (metallic)	26.4	25.4
Tungsten	24.4	24.4

Table 6.1 Values of the molar specific heat c_p at constant pressure and the molar specific heat c_V at constant volume for some simple solids at a temperature $T = 298°$K. The values of c_V were deduced from the directly measured values of c_p by applying some small corrections. All values are expressed in units of joules deg^{-1} mole^{-1}. [Data taken from Dwight E. Gray (ed.), *American Institute of Physics Handbook*, 2d ed., p. 4–48 (McGraw-Hill Book Company, New York, 1963).]

† In terms of calories, $c_V \approx 6$ cal mole^{-1} deg^{-1}.

‡ That is, by the *normal modes* of the solid.

sponding values of c_p after applying some small corrections.† It is seen that the values of c_V listed in the table are, on the whole, in rather good agreement with the value (65) predicted by the classical theory. Bad discrepancies occur, however, in the cases of silicon and, particularly, diamond. The reason is that quantum effects are still important for these solids even at a temperature as high as 300°K.

Validity of the classical approximation

Let us then ask under what circumstances the preceding classical discussion is expected to be valid. The criterion is again provided by the condition (3). Consider thus a vibrating atom which has an energy ϵ_x associated with its motion in the x direction. By virtue of the equipartition theorem applied to (59), the momentum p_x of the atom is then such that

$$\frac{1}{2m}\overline{p_x^2} = \frac{1}{2}kT.$$

The momentum of an atom thus has a typical magnitude p_0 of the order of

$$p_0 \approx \sqrt{\overline{p_x^2}} \approx \sqrt{mkT}. \tag{67}$$

To assure the validity of the classical description, quantum effects should not prevent us from localizing an atom within a typical distance s_0 of the order of the mean amplitude of vibration of this atom. But the equipartition theorem applied to (59) yields

$$\tfrac{1}{2}\alpha\overline{x^2} = \tfrac{1}{2}kT.$$

The typical magnitude s_0 of the atomic displacement is thus of the order of

$$s_0 \approx \sqrt{\overline{x^2}} \approx \sqrt{\frac{kT}{\alpha}}. \tag{68}$$

Hence the condition (3) that the Heisenberg uncertainty principle be of negligible significance becomes simply

† It is quite easy to make specific heat measurements on solids at constant atmospheric pressure, allowing the volume of the solid to change slightly, but very difficult to devise an experimental arrangement guaranteed to prevent the solid from expanding its volume when its temperature is raised. Since the volume of a solid changes only slightly as a result of a temperature change, the difference between c_p and c_V is, however, quite small. It can readily be calculated from a knowledge of the measured macroscopic properties of the particular solid under consideration.

$$s_0 p_0 \approx kT \sqrt{\frac{m}{\alpha}} \gg \hbar$$

or

$$\boxed{kT \gg \hbar\omega} \tag{69}$$

where ω is, by (60), the typical (angular) frequency of oscillation of an atom in the solid. Equivalently the criterion (69) for the validity of the classical approximation can be written in the form

$$T \gg \Theta, \qquad \text{where } \Theta \equiv \frac{\hbar\omega}{k} \tag{70}$$

is a temperature parameter characteristic of the solid under consideration.

Numerical estimates

The magnitude of the atomic frequency of oscillation ω can be estimated from the elastic properties of the solid under consideration. Suppose, for example, that a small pressure Δp is applied to a solid†; as a result, the volume V of the solid will *de*crease by some small amount ΔV. The quantity κ defined by

$$\kappa \equiv -\frac{1}{V}\frac{\Delta V}{\Delta p} \tag{71}$$

is called the *compressibility* of the solid (the minus sign being introduced to make κ positive). It is readily measured and provides some information about the forces between the atoms in the solid.

Let us then try to deduce from the compressibility κ a rough estimate of the net force F acting on an atom when it is displaced from its equilibrium position in the solid. Imagine, for simplicity, that the atoms of the solid are located at the centers of cubes of edge length a so that the interatomic spacing is also equal to a. An excess pressure Δp applied to a surface of the solid corresponds then to a force $F = a^2 \Delta p$ applied to the area a^2 occupied by a single atom (see Fig. 6.15). Furthermore, the fractional volume change of the solid under the influence of the excess pressure Δp is equal to the fractional volume change occupied per atom so that

$$\frac{\Delta V}{V} = \frac{\Delta(a^3)}{a^3} = \frac{3a^2\,\Delta a}{a^3} = \frac{3\Delta a}{a}.$$

Using the definition (71) of the compressibility, the force F on an atom is then related to Δa by

$$F = a^2\,\Delta p = a^2\left(-\frac{1}{\kappa}\frac{\Delta V}{V}\right) = -\frac{a^2}{\kappa}\frac{3\Delta a}{a}$$

or

$$F = -\alpha\,\Delta a$$

where the constant α, relating the force F to the displacement Δa of the atom from its equilibrium position, is given by

$$\alpha = \frac{3a}{\kappa}. \tag{72}$$

With our simple approximation of an assumed cubic lattice of atoms in the solid, the estimated frequency of vibration (60) of an atom in the solid should then be approximately

$$\omega = \sqrt{\frac{\alpha}{m}} \approx \sqrt{\frac{3a}{\kappa m}}. \tag{73}$$

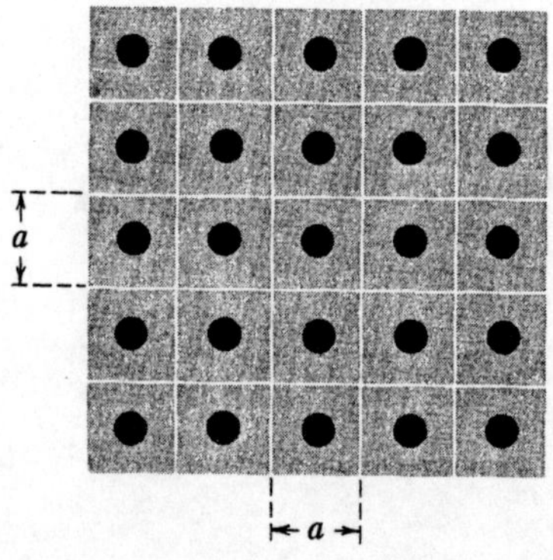

Fig. 6.15 Surface-view of a solid whose atoms are arranged in a simple cubic lattice.

† No confusion should arise between the symbol p used for pressure and the same symbol used elsewhere to denote momentum.

To get an appreciation for the relevant magnitudes, let us estimate ω in the case of copper. The measured parameters of this metal are†:

atomic weight

$$\mu = 63.5,$$

density

$$\rho = 8.95 \text{ gm cm}^{-3},$$

compressibility

$$\kappa = 7.3 \times 10^{-13} \text{ cm}^2 \text{ dyne}^{-1}.$$

From these numbers we find for the atomic mass

$$m = \frac{\mu}{N_a} = \frac{63.5}{6.02 \times 10^{23}}$$
$$= 1.05 \times 10^{-22} \text{ gm}.$$

Since $\rho = m/a^3$, the interatomic distance is equal to

$$a = \left(\frac{m}{\rho}\right)^{1/3} = \left(\frac{1.05 \times 10^{-22}}{8.95}\right)^{1/3}$$
$$= 2.34 \times 10^{-8} \text{ cm}.$$

Hence (73) gives for the angular frequency of vibration

$$\omega \approx \left[\frac{3\,(2.34 \times 10^{-8})}{(7.3 \times 10^{-13})(1.05 \times 10^{-22})}\right]^{1/2}$$
$$= 3.02 \times 10^{13} \text{ radians/sec},$$

or for the corresponding frequency

$$\nu = \frac{\omega}{2\pi} \approx 4.8 \times 10^{12} \text{ cycles/sec.} \tag{74}$$

This is a frequency lying in the infrared region of the electromagnetic spectrum.

The characteristic temperature Θ defined in (70) is then equal to

$$\Theta \equiv \frac{\hbar\omega}{k} = \frac{(1.054 \times 10^{-27})(3.02 \times 10^{13})}{(1.38 \times 10^{-16})}$$
$$\approx 230°\text{K}. \tag{75}$$

Hence the classical result $c_V = 3R$ should be valid for copper if $T \gg 230°$K, i.e., it should begin to be reasonably accurate for temperatures of the order of room temperature or above.

On the other hand, let us consider a solid such as diamond. The atomic mass of one of its carbon atoms is 12, i.e., about 5 times less than that of a copper atom. In addition, diamond is a very hard solid so that its compressibility is very low, about 3 times less than that of copper ($\kappa = 2.26 \times 10^{-13}$ cm^2 dyne^{-1}). Thus the frequency of vibration ω of a carbon atom in diamond is, by (73), much higher than that of a copper atom in copper metal. More precisely, for diamond (density $\rho = 3.52$ gm cm^{-3}) the estimated temperature parameter $\Theta \approx 830°$K. The classical approximation is, therefore, not expected to be applicable to diamond at room temperature and the low value of c_V for diamond in Table 6.1 is by no means surprising.

It is clear that the classical result $c_V = 3R$ must break down at low temperatures where the condition (69) is not satisfied. Indeed, the very general result (5.32) implies that, as the temperature is reduced below the range of validity of (69), the specific heat c_V must ultimately decrease so as to approach zero as $T \to 0$. Any correct quantum-mechanical calculation must yield this limiting result. If every atom in the solid is assumed to vibrate with the same frequency ω, the quantum-mechanical calculation of the specific heat c_V can be carried out quite easily to give an approximate expression for c_V valid for *all* temperatures. Details will be left as an exercise in Prob. 6.21.

---

† The data are taken from Dwight E. Gray (ed.), *American Institute of Physics Handbook*, 2d ed. (McGraw-Hill Book Company, New York, 1963).

Summary of Definitions

phase space A cartesian multidimensional space whose axes are labeled by all the coordinates and momenta describing a system in classical mechanics. A point in this space specifies all the coordinates and momenta of the system.

Maxwell velocity distribution The expression

$$f(\mathbf{v})\, d^3\mathbf{v} \propto e^{-(1/2)\beta m v^2}\, d^3\mathbf{v}$$

giving the mean number of molecules having a velocity between $\mathbf{v}$ and $\mathbf{v} + d\mathbf{v}$ in a gas at the absolute temperature T. It is merely a special case of the canonical distribution.

effusion The outflow of molecules from a container through a small hole whose size is much smaller than the molecular mean free path.

Important Relations

If a system described classically is in equilibrium at the absolute temperature T, every independent quadratic term ϵ_i of its energy has a mean value

$$\bar{\epsilon}_i = \tfrac{1}{2}kT. \qquad \text{(i)}$$

Suggestions for Supplementary Reading

O. R. Frisch, "Molecular Beams," *Sci. American* **212**, 58 (May 1965). A good discussion of the wide range of fundamental physical experiments made possible by molecular beam investigations.

F. Reif, *Fundamentals of Statistical and Thermal Physics*, chap. 7 (McGraw-Hill Book Company, New York, 1965). A somewhat more detailed discussion of the topics of the present chapter.

F. W. Sears, *An Introduction to Thermodynamics, the Kinetic Theory of Gases, and Statistical Mechanics*, 2d ed., chaps. 11 and 12 (Addison-Wesley Publishing Company, Inc., Reading, Mass., 1953).

D. K. C. MacDonald, *Faraday, Maxwell, and Kelvin* (Anchor Books, Doubleday & Company, Inc., Garden City, N.Y., 1964). This book contains a short account of Maxwell's life and scientific work.

Problems

6.1 *Phase space of a classical harmonic oscillator*

The energy of a one-dimensional harmonic oscillator, whose position coordinate is x and whose momentum is p, is given by

$$E = \frac{1}{2m}p^2 + \frac{1}{2}\alpha x^2$$

where the first term on the right is its kinetic and the second term its potential energy. Here m denotes the mass of the oscillating particle and α the spring constant of the restoring force acting on the particle.

Consider an ensemble of such oscillators, the energy of each oscillator being known to lie between E and $E + \delta E$. Treating the situation classically, indicate in the two-dimensional xp phase space the region of states accessible to the oscillator.

6.2 *Ideal gas in a gravitational field*

An ideal gas at the absolute temperature T is in equilibrium in the presence of a gravitational field described by an acceleration g in the downward (or $-z$) direction. The mass of each molecule is m.

(*a*) Use the canonical distribution in its classical form to find the probability $\mathcal{P}(\mathbf{r},\mathbf{p})\, d^3\mathbf{r}\, d^3\mathbf{p}$ that a molecule has a position between $\mathbf{r}$ and $\mathbf{r} + d\mathbf{r}$ and a momentum between $\mathbf{p}$ and $\mathbf{p} + d\mathbf{p}$.

(*b*) Find (to within a trivial constant of proportionality) the probability $\mathcal{P}'(\mathbf{v})\, d^3\mathbf{v}$ that a molecule has a velocity between $\mathbf{v}$ and $\mathbf{v} + d\mathbf{v}$, irrespective of its position in space. Compare this result with the corresponding probability in the absence of a gravitational field.

(*c*) Find (to within a trivial constant of proportionality) the probability $\mathcal{P}''(z)\, dz$ that a molecule is located at a height between z and $z + dz$, irrespective of its velocity or its location in any horizontal plane.

6.3 *Macroscopic discussion of an ideal gas in a gravitational field*

Consider the ideal gas of the last problem from a completely macroscopic point of view. By writing down the condition of mechanical equilibrium for a slice of the gas located between the heights z and $z + dz$, and by using the equation of state (4.92), derive an expression for $n(z)$, the number of molecules per unit volume at a height z. Compare this with the result for $\mathcal{P}''(z)\, dz$ derived in the last problem on the basis of statistical mechanics.

6.4 ***Spatial distribution of electrons in a cylindrical electric field***

A wire of radius r_0 is coincident with the axis of a metal cylinder of radius R and length L. The wire is maintained at a positive potential of V statvolts with respect to the cylinder. The whole system is at some high absolute temperature T. As a result, electrons emitted from the hot metals form a dilute gas filling the cylindrical container and in equilibrium with it. The density of these electrons is so low that their mutual electrostatic interaction can be neglected.

(*a*) Use Gauss's theorem to obtain an expression for the electrostatic field which exists at points at a radial distance r from the wire ($r_0 < r < R$). The cylinder of length L may be assumed to be very long so that end effects are negligible.

(*b*) In thermal equilibrium, the electrons form a gas of variable density which fills the entire space between the wire and cylinder. Using the result of part (*a*), find how the number n of electrons per unit volume depends on the radial distance r.

(*c*) Give an approximate criterion determining how low the temperature T must be so that the electron density is small enough to justify the approximation of neglecting the mutual electrostatic interaction between the electrons.

6.5 ***Determination of large molecular weights by the ultracentrifuge***

Consider a macromolecule (i.e., a very large molecule with a molecular weight of several millions) immersed in an incompressible fluid of density ρ at the absolute temperature T. The volume v occupied by one such molecule can be considered known since the volume occupied by a mole of macromolecules can be determined by volume measurements on a solution of macromolecules. A dilute solution of this type is now placed in an ultracentrifuge rotating with a high angular velocity ω. In the frame of reference rotating with the centrifuge, any particle of mass m at rest with respect to this frame is then acted upon by an outward centrifugal force $m\omega^2 r$, where r denotes the distance of the particle from the axis of rotation.

(*a*) What is the net force acting in this frame of reference on a macromolecule of mass m, if the buoyancy effect of the surrounding fluid is taken into account?

(*b*) Suppose that equilibrium has been attained in this frame of reference so that the mean number $n(r)\,dr$ (per unit volume) of macromolecules located at a distance from the axis of rotation between r and $r + dr$ is independent of time. Apply the canonical distribution to find (to within a constant of proportionality) the number $n(r)\,dr$ as a function of r.

(*c*) Measurements of the relative number $n(r)$ of molecules as a function of r can be made by measuring the absorption of light by the solution. Show how such measurements can be used to deduce the mass m of a macromolecule.

6.6 *Spatial separation of magnetic atoms in an inhomogeneous magnetic field

An aqueous solution at room temperature T contains a small concentration of magnetic atoms, each of which has a spin $\frac{1}{2}$ and magnetic moment μ_0. The solution is placed in an external magnetic field. The magnitude of this field is inhomogeneous over the volume of the solution. To be specific, the z component B of this magnetic field is a uniformly increasing function of z, assuming a value B_1 at the bottom of the solution where $z = z_1$ and a larger value B_2 at the top of the solution where $z = z_2$.

(*a*) Let $n_+(z)\,dz$ denote the mean number of magnetic atoms whose magnetic moment points along the z direction and which are located between z and $z + dz$. What is the ratio $n_+(z_2)/n_+(z_1)$?

(*b*) Let $n(z)\,dz$ denote the *total* mean number of magnetic atoms (of both directions of spin orientation) located between z and $z + dz$. What is the ratio $n(z_2)/n(z_1)$? Is it less than, equal to, or greater than unity?

(*c*) Make use of the fact that $\mu_0 B \ll kT$ to simplify the answers to the preceding questions.

(*d*) Estimate numerically the magnitude of the ratio $n(z_2)/n(z_1)$ at room temperature if $\mu_0 \approx 10^{-20}$ erg/gauss is of the order of a Bohr magneton, $B_1 = 0$, and $B_2 = 5 \times 10^4$ gauss.

6.7 *Most probable energy of a molecule in a gas*

What is the most probable kinetic energy $\tilde{\epsilon}$ of a molecule described by a Maxwellian velocity distribution? Is it equal to $\frac{1}{2}m\tilde{v}^2$, where $\tilde{v}$ is the most probable speed of the molecule?

6.8 *Temperature dependence of effusion*

Molecules of a gas enclosed within a container effuse through a small hole into a surrounding vacuum. Suppose that the absolute temperature of the gas in the container is doubled while its pressure is kept constant.

(*a*) By what factor does the number of molecules escaping per second through the hole change?

(*b*) By what factor does the force change which is exerted on a vane suspended at some distance in front of the hole?

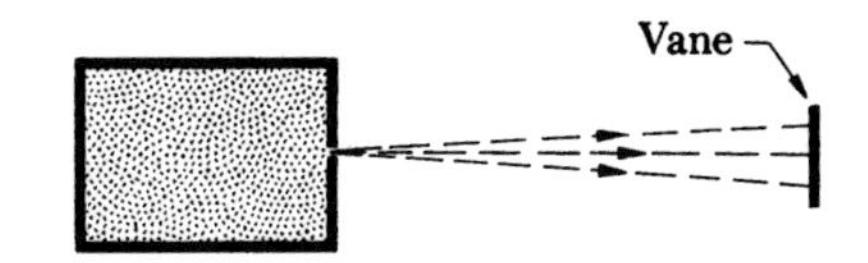

Fig. 6.16 An effusing beam impinging upon a vane.

6.9 *Mean kinetic energy of an effusing molecule*

The molecules of a monatomic ideal gas are escaping by effusion through a small hole in a wall of an enclosure maintained at an absolute temperature T. By physical reasoning (without actual calculation) do you expect the *mean* kinetic energy $\bar{\epsilon}_0$ of a molecule in the effusing beam to be equal to, greater than, or less than the mean kinetic energy $\bar{\epsilon}_i$ of a molecule inside the enclosure?

6.10 *Pressure drop in gas container having a small leak*

A thin-walled vessel of volume V, kept at constant temperature T, contains a gas which slowly leaks out through a small hole of area A. The outside pressure is low enough so that leakage back into the vessel is negligible. Estimate the time required for the pressure in the vessel to decrease to half of its original value. Express your answer in terms of A, V, and the mean molecular speed $\bar{v}$.

6.11 *Cryogenic (i.e., low-temperature) pumping*

Gases can be removed from a container by lowering the temperature of some of its walls. This is a method commonly used to achieve the good vacuums necessary for many physical experiments. To illustrate the principle of the method, consider a spherical bulb 10 cm in radius which is maintained at room temperature (300°K) except for a 1-cm^2 spot which is kept at liquid nitrogen temperature (77°K). The bulb contains water vapor originally at a pressure of 0.1 mm of mercury. Assuming that every water molecule striking the cold area condenses and sticks to the surface, estimate the time required for the pressure to decrease to 10^{-6} mm of mercury.

6.12 *Separation of isotopes by effusion*

A vessel has porous walls containing very many tiny holes. Gas molecules can pass through these holes by effusion and then be pumped off to some collecting chamber. The vessel is filled with a dilute gas consisting of two kinds of molecules which have different masses m_1 and m_2 by virtue of the fact that they contain two different isotopes of the same atom. Let us denote by c_1 the concentration of the first type of molecules in the vessel, and by c_2 the concentration of the second type. (The concentration c_i is the ratio of the number of molecules of type i to the total number of molecules.) These concentrations can be kept constant in the vessel by providing a steady flow of fresh gas through it so as to replenish any gas that has effused.

(*a*) Let c_1' and c_2' denote the concentrations of the two types of molecules in the collecting chamber. What is the ratio c_2'/c_1'?

(*b*) By using the gas UF_6, one can attempt to separate ^{235}U from ^{238}U, the first of these isotopes being the one useful in the initiation of nuclear-fission reactions. The molecules in the vessel are then $^{238}U^{19}F_6$ and $^{235}U^{19}F_6$. (The concentrations of these molecules, corresponding to the natural abundance of the two uranium isotopes, are $c_{238} = 99.3$ per cent and $c_{235} = 0.7$ per cent.) Calculate the corresponding ratio c'_{235}/c'_{238} of the molecules collected after effusion. Express the result in terms of the original concentration ratio c_{235}/c_{238}.

6.13 *Change of concentration as a result of effusion*

A container has as one of its walls a membrane containing many small holes. If the container is filled with gas at some moderate pressure $\bar{p}$, this gas will escape by effusion into the vacuum surrounding the container. It is found that when the container is filled with helium gas at room temperature and at a pressure $\bar{p}$, the pressure will have fallen to $\frac{1}{2}\bar{p}$ after one hour.

Suppose that the container is filled at room temperature and at a total pressure $\bar{p}$ with a mixture of helium (He) and neon (Ne), the atomic concentrations of both gases being 50 per cent (i.e., 50 per cent of the atoms are He and 50 per cent of them are Ne). What will be the ratio n_{Ne}/n_{He} of the atomic concentrations of Ne to He after one hour? Express your answer in terms of the atomic weights μ_{Ne} of neon and μ_{He} of helium.

6.14 *Calculation of mean values for a molecule in a gas*

A gas of molecules, each having mass m, is at rest in thermal equilibrium at the absolute temperature T. Denote the velocity of a molecule by $\mathbf{v}$, its three cartesian components of velocity by v_x, v_y, and v_z, and its speed by v. Find the following mean values:

(*a*) $\overline{v_x}$,
(*b*) $\overline{v_x^2}$,
(*c*) $\overline{v^2 v_x}$,
(*d*) $\overline{v_x^2 v_y}$,
(*e*) $\overline{(v_x + bv_y)^2}$, where b is a constant.

(*Suggestion:* Symmetry arguments and the equipartition theorem should suffice to answer all these questions without any significant calculation.)

6.15 *Doppler broadening of spectral lines*

A gas of atoms, each of mass m, is maintained at the absolute temperature T inside an enclosure. The atoms emit light which passes (in the x direction) through a window of the enclosure and can then be observed as a spectral line in a spectroscope. A stationary atom would emit light at the sharply defined frequency ν_0. But, because of the Doppler effect, the frequency of the light observed from an atom having an x component of velocity v_x is not simply equal to the frequency ν_0, but is given approximately by

$$\nu = \nu_0\left(1 + \frac{v_x}{c}\right)$$

where c is the velocity of light. As a result, not all of the light arriving at the spectroscope is at the frequency ν_0; instead it is characterized by some intensity distribution $I(\nu)\,d\nu$ giving the fraction of light intensity lying in the frequency range between ν and $\nu + d\nu$.

(*a*) Calculate the mean frequency $\bar{\nu}$ of the light observed in the spectroscope.

(*b*) Calculate the dispersion $\overline{(\Delta\nu)^2} = \overline{(\nu - \bar{\nu})^2}$ in the frequency of the light observed in the spectroscope.

(*c*) Show how measurements of the width $\Delta\nu \equiv [\overline{(\Delta\nu)^2}]^{1/2}$ of a spectral line observed in the light coming from a star allow one to determine the temperature of that star.

6.16 *Specific heat of an adsorbed mobile monolayer*

If the surface of some solid is maintained in a reasonably good vacuum, a single layer of molecules, one molecular diameter thick, can form on this surface. (The molecules are then said to be adsorbed on the surface.) The molecules are held to this surface by forces exerted on them by the atoms of the solid, but they may be quite free to move in two dimensions on this surface. They form then, to good approximation, a classical two-dimensional gas. If the molecules are monatomic and the absolute temperature is T, what is the specific heat per mole of molecules thus adsorbed on a surface of fixed size?

6.17 *Temperature dependence of the electrical resistivity of a metal*

The electrical resistivity ρ of a metal is proportional to the probability that an electron is scattered by the vibrating atoms in the lattice, and this probability is in turn proportional to the mean square amplitude of vibration of these atoms. How does the electrical resistivity ρ of the metal depend on its absolute temperature in the range near room temperature, or above, where classical statistical mechanics can validly be applied to discuss the vibrations of the atoms in the metal?

6.18 *Theoretical limiting accuracy of a weight measurement*

A very sensitive spring balance consists of a quartz spring suspended from a fixed support. The spring constant is α, i.e., the restoring force of the spring is $-\alpha x$ if the spring is stretched by an amount x. The balance is at an absolute temperature T in a location where the acceleration due to gravity is g.

(*a*) If a very small object of mass M is suspended from the spring, what is the mean resultant elongation $\bar{x}$ of the spring?

(*b*) What is the magnitude $\overline{(\Delta x)^2} \equiv \overline{(x - \bar{x})^2}$ of the thermal fluctuations of the object about its equilibrium position?

(*c*) It becomes impracticable to measure the mass of an object when the fluctuations are so large that $[\overline{(\Delta x)^2}]^{1/2} \gtrsim \bar{x}$. What is the minimum mass M which can be measured with this balance?

6.19 *Specific heat of anharmonic oscillators*

Consider a one-dimensional oscillator (*not* simple harmonic) which is described by a position coordinate x and by a momentum p and whose energy is given by

$$\epsilon = \frac{p^2}{2m} + bx^4 \tag{i}$$

where the first term on the right is its kinetic energy and the second term is its potential energy. Here m denotes the mass of the oscillator and b is some constant. Suppose that this oscillator is in thermal equilibrium with a heat reservoir at a temperature T high enough so that the approximation of classical mechanics is a good one.

(*a*) What is the mean kinetic energy of this oscillator?

(*b*) What is its mean potential energy?

(*c*) What is its mean total energy?

(*d*) Consider an assembly of weakly interacting particles, each vibrating in one dimension so that its energy is given by (i). What is the specific heat at constant volume per mole of these particles?

(*Suggestion:* There is no need to evaluate explicitly any integral to answer these questions.)

6.20 *Specific heat of a highly anisotropic solid*

Consider a solid which has a highly anisotropic crystalline layer structure. Each atom in this structure can be regarded as performing simple harmonic oscillations in three dimensions. The restoring forces in directions parallel to a layer are very large; hence the natural frequencies of oscillations in the x and y directions lying within the plane of a layer are both equal to a value $\omega_\parallel$ which is so large that $\hbar\omega_\parallel \gg 300k$, the thermal energy kT at room temperature. On the other hand, the restoring force perpendicular to a layer is quite small; hence the frequency of oscillation $\omega_\perp$ of an atom in the z direction perpendicular to a layer is so small that $\hbar\omega_\perp \ll 300k$. On the basis of this model, what is the molar specific heat (at constant volume) of this solid at 300°K?

6.21 *Quantum theory of the specific heat of solids*

To treat the atomic vibrations in a solid by quantum mechanics, use as a simplifying approximation a model which assumes that each atom of the solid vibrates independently of the other atoms with the same angular frequency ω in each of its three directions. The solid consisting of N atoms is then equivalent to an assembly of $3N$ independent one-dimensional oscillators vibrating with the frequency ω. The possible quantum states of every such oscillator have discrete energies given by

$$\epsilon_n = (n + \tfrac{1}{2})\hbar\omega \qquad \text{(i)}$$

where the quantum number n can assume the possible values $n = 0, 1, 2, 3, \ldots$.

(*a*) Suppose that the solid is in equilibrium at the absolute temperature T. By using the energy levels (i) and the canonical distribution, proceed as in Prob. 4.22 to calculate the mean energy $\bar{\epsilon}$ of an oscillator and thus also the total mean energy $\bar{E} = N\bar{\epsilon}$ of the vibrating atoms in the solid.

(*b*) Using the result of part (*a*), proceed as in Prob. 5.20 to calculate the molar specific heat c_V of the solid.

(*c*) Show that the result of part (*b*) can be expressed in the form

$$c_V = 3R\frac{w^2 e^w}{(e^w - 1)^2} \tag{ii}$$

where

$$w \equiv \frac{\hbar\omega}{kT} = \frac{\Theta}{T} \tag{iii}$$

and where $\Theta \equiv \hbar\omega/k$ is the temperature parameter previously defined in (70).

(*d*) Show that, when $T \gg \Theta$, the result (ii) approaches properly the classical value $c_V = 3R$.

(*e*) Show that the expression (ii) for c_V approaches properly the value zero as $T \to 0$.

(*f*) Find an approximate expression for the result (ii) in the limit when $T \ll \Theta$.

(*g*) Make a rough sketch of c_V as a function of the absolute temperature T.

(*h*) Apply the criterion (1) to find below what temperature the classical approximation is *not* expected to be applicable. Compare your result with the condition (69) for the applicability of the classical theory of specific heats.

[Using the approximations made in this problem, Einstein first derived the expression (ii) in 1907. Using the novel quantum ideas, he was thus able to account for the experimentally observed specific heat behavior which had been inexplicable on the basis of the classical theory.]

Chapter 7

General Thermodynamic Interaction

Chapter 7 General Thermodynamic Interaction

Our discussion up to now has dealt mainly with thermal interaction. In order to achieve full generality, we must extend our considerations slightly so as to encompass the case of arbitrary interaction between macroscopic systems. In the next two sections, accordingly, we shall try to generalize the discussion of Chap. 4 by considering what happens when the external parameters of interacting systems are allowed to vary. The interaction may then involve the performance of work as well as the exchange of heat. By understanding this general case, we shall complete the last missing link in the development of our ideas and thus shall have obtained all the basic results of the theory of *statistical thermodynamics.* The power of this theory is attested by its many diverse applications in physics, chemistry, biology, and engineering. We shall be content to mention only a few important illustrations.

7.1 Dependence of the Number of States on the External Parameters

Consider any macroscopic system which is characterized by one or more external parameters, such as its volume V or the applied magnetic field B in which it is located. For the sake of simplicity, we shall consider the case where only one of these external parameters, call it x, is free to vary; the generalization to the case where there are several such parameters will then be immediate. The number Ω of quantum states of this system in the fixed energy interval between E and $E + \delta E$ will depend not only on the energy E, but also on the particular value assumed by the external parameter x. Thus we can write the functional relationship $\Omega = \Omega(E,x)$. We are interested specifically in examining how Ω depends on x.

The energy E_r of each quantum state r depends on the value assumed by the external parameter x; that is, $E_r = E_r(x)$. When the value of the external parameter x changes by an infinitesimal amount dx, the energy E_r of the state r changes accordingly by an amount

$$dE_r = \frac{\partial E_r}{\partial x}\,dx = X_r\,dx \tag{1}$$

where we have introduced the convenient abbreviation

$$X_r \equiv \frac{\partial E_r}{\partial x}. \tag{2}$$

A given change dx of the external parameter ordinarily changes the energies of different states by different amounts. The value of $\partial E_r/\partial x$ depends therefore on the particular state r under consideration, and X_r may accordingly assume different values for different states.

To facilitate our thinking, let us subdivide the possible values of X_r into small intervals of fixed size δX. Consider then the total number $\Omega(E,x)$ of states having an energy between E and $E + \delta E$ when the external parameter has the value x. Among these states we shall focus attention first on the particular subset i of states for which X_r has a value lying in a particular interval between $X^{(i)}$ and $X^{(i)} + \delta X$. We shall denote the number of states in this subset by $\Omega^{(i)}(E,x)$. These states have the simple property that the energy of each one of them is changed by nearly the same amount $X^{(i)}\,dx$ when the external parameter changes by dx. If $X^{(i)}$ is positive, each of these states lying in the energy range $X^{(i)}\,dx$ below E will then have its energy changed from a value less than E to one greater than E. (See Fig. 7.2.) Since there are $\Omega^{(i)}/\delta E$ such states per unit energy range, there are $(\Omega^{(i)}/\delta E)(X^{(i)}\,dx)$ such states in the energy range $X^{(i)}\,dx$. Hence we can say that the quantity

$\Gamma^{(i)}(E) \equiv$ the number of states, among the $\Omega^{(i)}(E,x)$ states of the ith subset, whose energy is changed from a value *less* than E to a value *greater* than E when the external parameter is changed infinitesimally from x to $x + dx$ (3)

is simply equal to

$$\Gamma^{(i)}(E) = \frac{\Omega^{(i)}(E,x)}{\delta E} X^{(i)}\,dx. \tag{4}$$

If $X^{(i)}$ is negative, the relation (4) is still valid, but $\Gamma^{(i)}$ is negative; i.e., in this case a positive number $-\Gamma^{(i)}$ of states have their energy changed from a value *greater* than E to one *less* than E.†

Let us now look at *all* of the $\Omega(E,x)$ states which have an energy between E and $E + \delta E$ when the external parameter has the value x. To find the quantity

$\Gamma(E) \equiv$ the *total* number of states, among *all* the $\Omega(E,x)$ states, whose energy is changed from a value less than E to a value greater than E when the external parameter is changed from x to $x + dx$, (5)

† Note that (3) represents simply the number of energy levels crossing the energy E from below. The argument leading to (4) is thus similar to that used in Sec. 1.6 for finding the number of molecules striking a surface in a gas.

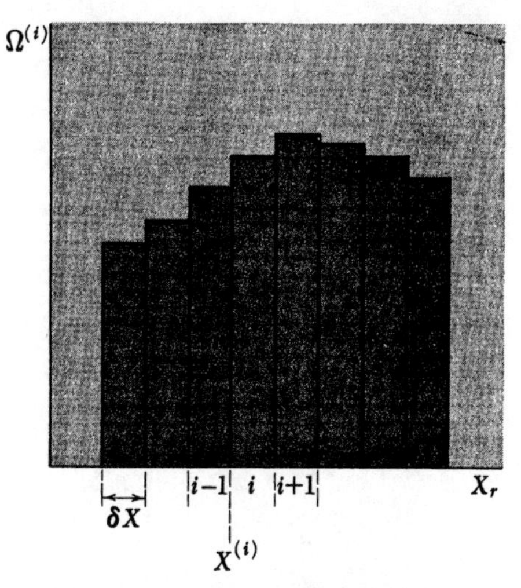

Fig. 7.1 The number $\Omega^{(i)}$ of those states for which $X_r \equiv \partial E_r/\partial x$ has a value in the interval between $X^{(i)}$ and $X^{(i)} + \delta X$ is here shown schematically as a function of the index i labeling these possible intervals. By summing $\Omega^{(i)}$ over all these possible intervals, one obtains the total number $\Omega(E,x)$ of states of interest, i.e., those which have an energy between E and $E + \delta E$ when the external parameter has the value x.

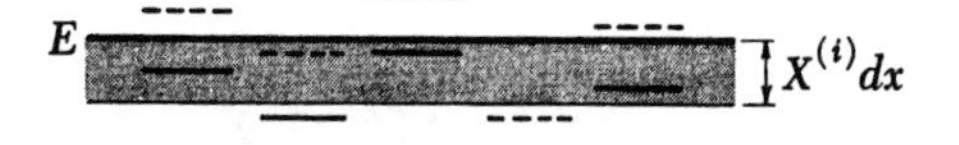

Fig. 7.2 Energy level diagram showing what happens when a change dx of the external parameter causes the energy E_r of each state r to change by an amount $X^{(i)}\,dx$ from an old value (indicated by a solid line) to a new value (indicated by a dotted line). As a result all those states with initial energies lying in the range $X^{(i)}\,dx$ below the energy E have their energy changed from a value less than E to one greater than E.

we need then only sum (4) over all the possible subsets i of states (i.e., over the states with all possible values of $\partial E_r/\partial x$). Thus we have

$$\Gamma(E) = \sum_i \Gamma^{(i)}(E) = \left[\sum_i \Omega^{(i)}(E,x)X^{(i)}\right]\frac{dx}{\delta E}$$

or

$$\boxed{\Gamma(E) = \frac{\Omega(E,x)}{\delta E}\bar{X}\,dx} \tag{6}$$

where we have used the definition

$$\bar{X} \equiv \frac{1}{\Omega(E,x)}\sum_i \Omega^{(i)}(E,x)X^{(i)}. \tag{7}$$

This is simply the mean value of X_r over all the states r lying in the interval between E and $E + \delta E$, each such state being considered equally probable as appropriate for an equilibrium situation. The mean value $\bar{X}$ defined in (7) is, of course, a function of E and x. Note that, by virtue of the definition (2),

$$\bar{X}\,dx = \overline{\frac{\partial E_r}{\partial x}}\,dx = đW \tag{8}$$

is simply the mean increase in energy of the system when the system is equally likely to be in any one of the states in its original accessible energy range. In other words, it is simply the macroscopic work $đW$ done on the system while the system remains in equilibrium, i.e., when the external parameter is changed quasi-statically.

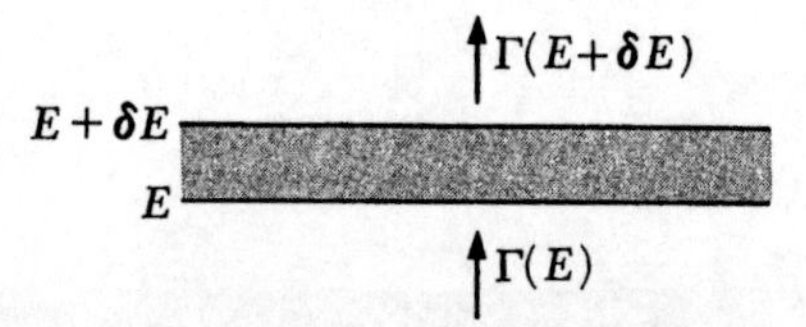

Fig. 7.3 When the external parameter is changed, the number of states in the indicated energy range between E and $E + \delta E$ changes because the energies of various states enter and leave this range.

Having found $\Gamma(E)$, it is now easy to consider some fixed energy E and to ask how $\Omega(E,x)$ changes when the external parameter x changes by an infinitesimal amount. Let us thus consider the total number $\Omega(E,x)$ of states lying in the particular energy range between E and $E + \delta E$. When the external parameter changes infinitesimally from x to $x + dx$, the number of states in this energy range changes by an amount $[\partial\Omega(E,x)/\partial x]\,dx$ which must be due to {the net number of states which enter this range by having their energy changed from a value less than E to one greater than E} minus {the net number of states which leave this range by having their energy changed from a value less than $E + \delta E$ to one greater than $E + \delta E$}. In symbols this can be written

$$\frac{\partial\Omega(E,x)}{\partial x}\,dx = \Gamma(E) - \Gamma(E + \delta E) = -\frac{\partial\Gamma}{\partial E}\delta E. \tag{9}$$

Substituting the expression (6) in (9), the fixed small quantities δE and dx cancel and we are left with

$$\frac{\partial \Omega}{\partial x} = -\frac{\partial}{\partial E}(\Omega \bar{X}) \tag{10}$$

or

$$\frac{\partial \Omega}{\partial x} = -\frac{\partial \Omega}{\partial E}\bar{X} - \Omega \frac{\partial \bar{X}}{\partial E}.$$

By dividing both sides by Ω, this can be written in the form

$$\frac{\partial \ln \Omega}{\partial x} = -\frac{\partial \ln \Omega}{\partial E}\bar{X} - \frac{\partial \bar{X}}{\partial E}. \tag{11}$$

For a macroscopic system, the first term on the right is, by (4.29), of the order of $f\bar{X}/(E - E_0)$ where f is the number of degrees of freedom of the system whose ground state energy is E_0. The second term on the right is roughly of the order $\bar{X}/(E - E_0)$. Since f itself is of the order of Avogadro's number so that $f \sim 10^{24}$, the second term on the right of (11) is thus utterly negligible compared to the first. Hence (11) reduces simply to

$$\frac{\partial \ln \Omega}{\partial x} = -\frac{\partial \ln \Omega}{\partial E}\bar{X} \tag{12}$$

or

$$\boxed{\left(\frac{\partial \ln \Omega}{\partial x}\right)_E = -\beta \bar{X}} \tag{13}$$

where we have used the definition (4.9) of the absolute temperature parameter β. Here we have embellished the partial derivative with the subscript E to indicate explicitly that the energy E is to be regarded as fixed in taking this derivative. In accordance with the definition (2),

$$\bar{X} \equiv \overline{\frac{\partial E_r}{\partial x}}. \tag{14}$$

In the special case where the external parameter x denotes a distance, the quantity $\bar{X}$ has the dimensions of a force. In general, $\bar{X}$ may have any dimensions and is called the *mean generalized force on the system, conjugate to the external parameter x.*

As a specific example, suppose that $x = V$, the volume of the system. Then the work $đW$ done *on* the system when its volume is increased quasi-statically by dV is given by $đW = -\bar{p}\, dV$ where $\bar{p}$ is the mean pressure exerted *by* the system. This work is thus properly of the form (8), i.e.,

$$đW = \bar{X}\, dV = -\bar{p}\, dV$$

so that

$$\bar{X} = -\bar{p}.$$

The mean generalized force $\bar{X}$ is in this case merely the mean pressure $-\bar{p}$ acting *on* the system. Hence (13) becomes

$$\left(\frac{\partial \ln \Omega}{\partial V}\right)_E = \beta \bar{p} = \frac{\bar{p}}{kT} \tag{15}$$

or

$$\left(\frac{\partial S}{\partial V}\right)_E = \frac{\bar{p}}{T} \tag{15a}$$

where $S = k \ln \Omega$ is the entropy of the system. Note that this relation allows us to calculate the mean pressure exerted by a system if its entropy is known as a function of its volume.

We derived the relation (13) by considering how the energy levels of the system move in or out of a given energy range when an external parameter is changed. The essential physics contained in this argument is very important. This becomes apparent if we look at the gist of what we have said. We only need to note that the relation (12) is equivalent to

$$\frac{\partial \ln \Omega}{\partial x}\, dx + \frac{\partial \ln \Omega}{\partial E}\, đW = 0 \tag{16}$$

where we have used (8) to write $\bar{X}\, dx = đW$ for the quasi-static work done on the system. The relation (16) represents merely the infinitesimal change of $\ln \Omega$ under a simultaneous change of the energy E and the external parameter x of the system. Thus (16) is equivalent to the statement

$$\ln \Omega(E + đW, x + dx) - \ln \Omega(E,x) = \frac{\partial \ln \Omega}{\partial E}\, đW + \frac{\partial \ln \Omega}{\partial x}\, dx = 0$$

or

$$\ln \Omega(E + đW, x + dx) = \ln \Omega(E,x). \tag{17}$$

Translated into words, this statement says the following: Suppose that an external parameter of an adiabatically isolated system is changed by a small amount. Then the energies of the various quantum states of the system are changed; correspondingly, the total energy of the system is changed by some amount $đW$ equal to the work done on the system. If the parameter change is quasi-static, the system merely tends to remain distributed over those states in which it was originally, while the energies of these states are changed. Hence one finds the system distributed over the *same* number of states at the end of the process (when its external parameter is $x + dx$ and its energy is $E + đW$) as it was at the beginning of the process (when its external parameter was x and its energy was E). This statement is the essential content of (17); it asserts that the entropy $S = k \ln \Omega$ of an adiabatically isolated system remains unchanged when its external parameters are changed quasi-statically by an infinitesimal amount. If one keeps on changing the external parameters quasi-statically until they have changed appreciably, this succession of infinitesimal processes must then also result in a zero entropy change. Hence we arrive at the important conclusion that no entropy change results if the external parameters of an adiabatically isolated system are changed quasi-statically by arbitrary amounts; in short,

$$\boxed{\begin{array}{c}\text{in a quasi-static adiabatic process,}\\ \Delta S = 0.\end{array}} \qquad (18)$$

Thus, although the performance of quasi-static work changes the energy of an adiabatically isolated system, the entropy of the system remains unaffected.

We should stress that the statement (18) is only true if the external parameters are changed *quasi-statically*. Otherwise, as illustrated by the discussion of Sec. 3.6, the entropy of an adiabatically isolated system will tend to *increase*. [Consider, for instance, the process described in Example (ii) at the end of that section.]

7.2 *General Relations Valid in Equilibrium*

We are now ready to consider the most general interaction between systems, i.e., the case where two macroscopic systems A and A' can interact both by exchanging heat and by doing work on each other. (A specific example is illustrated in Fig. 7.4 where two gases A and A' are separated by a piston which is not thermally insulating and which is free to move.) The analysis of the situation is merely a generalization of that of Sec. 4.1. If the energy E of system A is specified, the energy E' of the system A' is determined since the total energy E^* of the combined isolated system A^* consisting of A and A' is constant. The number Ω^* of states accessible to A^*, or equivalently its entropy $S^* = k \ln \Omega^*$, is then a function of the energy E of system A and of

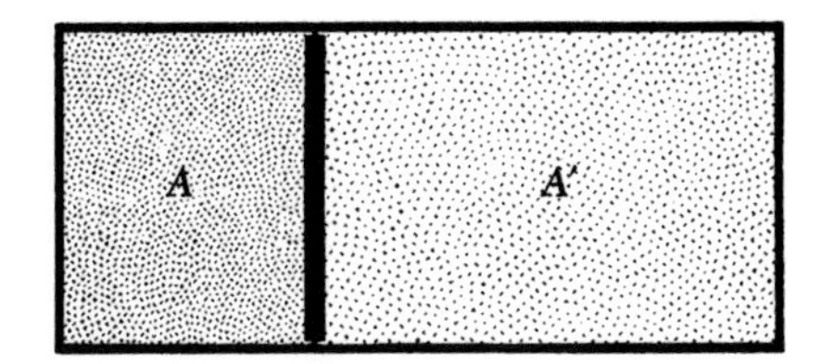

Fig. 7.4 Two gases A and A' separated by a thermally conducting piston which is free to move.

some external parameters $x_1, x_2, \ldots, x_n$; that is, $\Omega^* = \Omega^*(E;x_1, \ldots, x_n)$. This number Ω^* of states ordinarily has some extremely sharp maximum for some particular values $E = \tilde{E}$ of the energy and $x_\alpha = \tilde{x}_\alpha$ of each of the external parameters (where $\alpha = 1, 2, \ldots, n$). In equilibrium the combined system A^* is then with overwhelming probability found to be in a situation where the energy of A has the value $\tilde{E}$ and the external parameters have values $\tilde{x}_\alpha$. Correspondingly, the mean value of E is then given by $\bar{E} = \tilde{E}$ and the mean value of each external parameter by $\bar{x}_\alpha = \tilde{x}_\alpha$.

Equilibrium conditions

To make the discussion specific, consider two arbitrary systems A and A' (such as those shown in Fig. 7.4) each of which is characterized by a single external parameter, namely its volume. The conservation of energy for the combined isolated system implies that

$$E + E' = E^* = \text{constant}. \tag{19}$$

As the piston moves, a change in the volume V of the system A must be accompanied by a corresponding change in the volume V' of A' so that the total volume remains unchanged. Thus

$$V + V' = V^* = \text{constant}. \tag{20}$$

Let $\Omega(E,V)$ denote the number of states accessible to the system A in the energy interval between E and $E + \delta E$ when its volume lies between V and $V + \delta V$. Let $\Omega'(E',V')$ denote the corresponding number of states accessible to A'. Then the total number Ω^* of states accessible to the combined system A^* is, as in (4.4), given by the simple product

$$\Omega^* = \Omega(E,V)\,\Omega'(E',V') \tag{21}$$

where E' and V' are related to E and V by (19) and (20). Hence Ω^* is a function of the two independent variables E and V. Taking the logarithm of (21), we obtain

$$\ln \Omega^* = \ln \Omega + \ln \Omega' \tag{22}$$

or

$$S^* = S + S'$$

where we have used the definition $S \equiv k \ln \Omega$ for each system. Our basic statistical postulate (3.19) then leads to the following conclusion: In equilibrium, the most probable situation corresponds to those values of the parameters E and V where Ω^*, or equivalently S^*, is maximum.

The location of this maximum is determined by the condition that

$$d \ln \Omega^* = d \ln \Omega + d \ln \Omega' = 0 \tag{23}$$

for arbitrary small changes dE and dV of the energy E or volume V. But we can write the purely mathematical equation

$$d \ln \Omega = \frac{\partial \ln \Omega}{\partial E} dE + \frac{\partial \ln \Omega}{\partial V} dV.$$

Using the definition of β and the relation (15), this equation becomes

$$d \ln \Omega = \beta \, dE + \beta \bar{p} \, dV \tag{24}$$

where $\bar{p}$ is the mean pressure exerted by the system A. Similarly, we obtain for the system A'

$$d \ln \Omega' = \beta' \, dE' + \beta' \bar{p}' \, dV'$$

or

$$d \ln \Omega' = -\beta' \, dE - \beta' \bar{p}' \, dV \tag{25}$$

if we make use of the conditions (19) and (20) which imply that $dE' = -dE$ and $dV' = -dV$. The condition (23) of maximum probability in the equilibrium situation then becomes

$$(\beta - \beta') \, dE + (\beta \bar{p} - \beta' \bar{p}') \, dV = 0. \tag{26}$$

Since this relation must be satisfied for any *arbitrary* infinitesimal values of dE and dV, it follows that the coefficients of both these differentials must vanish separately. Hence we obtain at equilibrium

$$\beta - \beta' = 0$$

and

$$\beta \bar{p} - \beta' \bar{p}' = 0,$$

or

$$\boxed{\begin{aligned} \beta &= \beta' \\ \bar{p} &= \bar{p}'. \end{aligned}} \tag{27}$$

and

In equilibrium, the energies and volumes of the systems thus adjust themselves so that the conditions (27) are satisfied. These conditions merely assert that the temperatures of the systems must then be equal to guarantee their thermal equilibrium, and that their mean pressures must then be equal to guarantee their mechanical equilibrium. Although these equilibrium conditions are so obvious that we could have guessed them, it is satisfying to see them appear as automatic consequences of the general requirement that the total entropy S^* be maximum.

Infinitesimal quasi-static process

Consider a completely general quasi-static process in which the system A, by virtue of its interaction with some other system A', is brought from an equilibrium situation described by a mean energy $\bar{E}$ and external parameter values $\bar{x}_\alpha$ (for $\alpha = 1, 2, \ldots, n$) to an infinitesimally different equilibrium situation described by $\bar{E} + d\bar{E}$ and $\bar{x}_\alpha + d\bar{x}_\alpha$. In this infinitesimal process the system A can, in general, absorb heat and do work. What is the change of entropy of the system A as a result of this process?

Since $\Omega = \Omega(E; x_1, \ldots, x_n)$, one can write for the resultant change in $\ln \Omega$ the purely mathematical equation

$$d \ln \Omega = \frac{\partial \ln \Omega}{\partial E} d\bar{E} + \sum_{\alpha=1}^{n} \frac{\partial \ln \Omega}{\partial x_\alpha} d\bar{x}_\alpha. \tag{28}$$

The relation (13) was derived by considering a change of one external parameter, while considering all the others to be fixed. Hence this relation can be applied to each partial derivative in (28) to give

$$\frac{\partial \ln \Omega}{\partial x_\alpha} = -\beta \bar{X}_\alpha \equiv -\beta \frac{\overline{\partial E_r}}{\partial x_\alpha} \tag{29}$$

and (28) becomes

$$d \ln \Omega = \beta\, d\bar{E} - \beta \sum_{\alpha=1}^{n} \bar{X}_\alpha\, d\bar{x}_\alpha. \tag{30}$$

But the sum over all external parameter changes is merely

$$\sum_{\alpha=1}^{n} \bar{X}_\alpha\, d\bar{x}_\alpha = \sum_{\alpha=1}^{n} \frac{\overline{\partial E_r}}{\partial x_\alpha} d\bar{x}_\alpha = đW,$$

the mean increase in energy of the system caused by the external parameter changes, i.e., the work $đW$ done on the system in the infinitesimal process. Hence (30) becomes

$$d \ln \Omega = \beta(d\bar{E} - đW) = \beta\, đQ \tag{31}$$

since $(d\bar{E} - đW)$ is simply the infinitesimal heat $đQ$ absorbed by the system. Using $\beta = (kT)^{-1}$ and $S = k \ln \Omega$, Eq. (31) then asserts that,

in any infinitesimal quasi-static process,

$$dS = \frac{đQ}{T}. \tag{32}$$

This same relation was already derived in (4.42) for the special case where all external parameters of the system remain fixed. What we have now done is to generalize this result to show that it is applicable in *any* quasi-static process, even if work is done. Note also that when no heat is absorbed so that $đQ = 0$ (i.e., when the increase in mean energy of the system is simply due to the work done on the system) the entropy change $dS = 0$ in accordance with our previous statement (18).

We shall call (32) the *fundamental thermodynamic relation*. It is a very important and useful statement which can be written in many equivalent forms, such as

$$T\,dS = đQ = d\bar{E} - đW. \tag{33}$$

If the only external parameter of relevance is the volume V of the system, then the work done on the system is $đW = -\bar{p}\,dV$ if its mean pressure is $\bar{p}$. In this case (33) becomes

$$\boxed{T\,dS = d\bar{E} + \bar{p}\,dV.} \tag{34}$$

The relation (32) makes it possible to generalize the discussion of Sec. 5.5 since it permits us to calculate the entropy difference between *any* two macrostates of a system from measurements of the heat absorbed by it.† Consider thus any two macrostates a and b of the system. The entropy of the system then has a definite value S_a in macrostate a and a definite value S_b in macrostate b. The entropy difference can be calculated in any convenient way and will always yield the same value $S_b - S_a$. In particular, if one goes from the macrostate a to the macrostate b by *any* process whatever which is *quasi-static*, then the system remains always arbitrarily close to equilibrium and (32) applies at every stage of the process. Thus we can write the total entropy change of interest as the sum (or integral)

$$S_b - S_a \equiv \int_a^b \frac{đQ}{T} \qquad \text{(quasi-static)}. \tag{35}$$

Here the statement in parentheses is an explicit reminder that the integral must be evaluated for a *quasi-static* process leading from a to b. Since the absolute temperature T has then a well-defined measurable value at any stage of the process and since the absorbed heat $đQ$ can

† In Sec. 5.5 we could show how this can be done only in the special case where the macrostates under consideration are characterized by the *same* values of the external parameters of the system.

also be measured, (35) permits us to determine entropy differences by suitable measurements of heat.

Since the left side of (35) depends only on the initial and final macrostates, the value of the integral on the right side of (35) must be independent of the particular quasi-static process chosen to go from macrostate a to macrostate b. Thus

$$\int_a^b \frac{đQ}{T} \text{ has the } \textit{same} \text{ value for any quasi-static process } a \to b. \tag{36}$$

Note that the values of other integrals involved in the process *do* ordinarily depend on the nature of the process. For example, the total heat Q absorbed by the system in a quasi-static process going from macrostate a to macrostate b is given by

$$Q = \int_a^b đQ$$

and the value of this heat *does* ordinarily depend in an essential way on the particular process used in going from a to b. We shall illustrate these comments in the next section.

7.3 *Applications to an Ideal Gas*

In order to gain a better understanding of the results of the preceding section, we shall apply them to the simple case of an ideal gas. Macroscopically any such gas, whether monatomic or not, is characterized by the following two properties:

(i) The equation of state which relates the mean pressure $\bar{p}$ of ν moles of the gas to its volume V and absolute temperature T is given by (4.93), i.e.,

$$\bar{p}V = \nu RT. \tag{37}$$

(ii) At a fixed temperature, the mean internal energy $\bar{E}$ of such a gas is, by (4.86), independent of its volume; i.e.,

$$\bar{E} = \bar{E}(T), \qquad \text{independent of } V. \tag{38}$$

The mean internal energy $\bar{E}$ can be readily related to the specific heat c_V (at constant volume) per mole of the gas. Indeed, it follows from (5.23) that

$$c_V = \frac{1}{\nu}\left(\frac{\partial \bar{E}}{\partial T}\right)_V \tag{39}$$

where the subscript V indicates that the volume V is kept constant in taking the derivative. By virtue of (38), the specific heat c_V is thus also independent of the volume V of the gas, although it may depend on its temperature T. If the volume V is kept constant, (39) allows us to write the following relation for the mean energy change $d\bar{E}$ resulting from an absolute temperature change dT:

$$d\bar{E} = \nu c_V \, dT. \tag{40}$$

But the property (38) implies that an energy difference of the gas can result only from a difference of its temperature and does not depend on what may happen to its volume. Hence the relation (40) must be generally valid, irrespective of what volume change dV may accompany the temperature change dT. As a special case, (40) implies that

$$\text{if } c_V \text{ is independent of } T, \qquad \bar{E} = \nu c_V T + \text{constant}. \tag{41}$$

The previous comments make it easy to write down a general expression for the heat $đQ$ absorbed by an ideal gas in an infinitesimal quasi-static process in which the temperature of the gas changes by dT and its volume by dV. Using the expression (5.14) for the work done on the gas, we have

$$đQ = d\bar{E} - đW = d\bar{E} + \bar{p}\, dV. \tag{42}$$

By (40) and (37) this becomes then

$$\boxed{đQ = \nu c_V \, dT + \frac{\nu R T}{V} \, dV.} \tag{43}$$

The entropy change of the gas in this infinitesimal process is then, according to (32), given by

$$dS = \frac{đQ}{T} = \nu c_V \frac{dT}{T} + \nu R \frac{dV}{V}. \tag{44}$$

Entropy of an ideal gas

What is the entropy $S(T,V)$ of the gas in a macrostate where its temperature is T and its volume is V, compared to its entropy $S(T_0,V_0)$ in some other macrostate where its temperature is T_0 and its volume is V_0? To answer this question it is only necessary to go quasi-statically from the initial macrostate (T_0,V_0) to the final macrostate (T,V) through a succession of near-equilibrium macrostates where the gas has temperatures T' and volumes V'. For example, we can first keep the volume fixed at the initial value V_0 and change the temperature quasi-statically from T_0 to T by bringing the gas in contact with a succession of heat reservoirs differing in temperature progressively by infinitesimal amounts. In this process (44) shows that the entropy of the gas changes by an amount

$$S(T,V_0) - S(T_0,V_0) = \nu \int_{T_0}^{T} \frac{c_V(T')}{T'}\, dT'. \tag{45}$$

We can then keep the temperature at the value T and very slowly change the volume of the gas (say, by moving a piston) from its initial value V_0 to its final value V. In this process (44) shows that the entropy of the gas changes by an amount

$$S(T,V) - S(T,V_0) = \nu R \int_{V_0}^{V} \frac{dV'}{V'} = \nu R(\ln V - \ln V_0). \tag{46}$$

Adding (45) and (46), the total entropy change is then given by

$$\boxed{S(T,V) - S(T_0,V_0) = \nu \left[\int_{T_0}^{T} \frac{c_V(T')}{T'}\, dT' + R \ln \frac{V}{V_0} \right].} \tag{47}$$

The macrostate (T_0,V_0) can be regarded as some standard macrostate of the gas. The expression (47) then gives the dependence of the entropy S on the temperature T and volume V of any other macrostate of the gas. It can be written in the simple form

$$S(T,V) = \nu \left[\int \frac{c_V(T)}{T}\, dT + R \ln V + \text{constant} \right] \tag{48}$$

where the constant includes the fixed parameters T_0 and V_0 of the standard macrostate and where the indefinite integral is a function of T. The expression (48) is, of course, merely the integrated form of Eq. (44). The results (47) and (48) show properly that the number of states accessible to the gas increases as its absolute temperature (or energy) increases and as the volume available to its molecules increases.

A special simple case is that where the specific heat c_V is constant (i.e., independent of temperature) in the temperature range of interest. [For example, for a monatomic gas we showed in (5.26) that $c_V = \frac{3}{2}R$.] In this case c_V can be taken outside the integral. Since $dT'/T' = d(\ln T')$, the relations (47) and (48) then become if c_V is independent of T,

$$S(T,V) - S(T_0,V_0) = \nu\left[c_V \ln \frac{T}{T_0} + R \ln \frac{V}{V_0}\right] \tag{49}$$

or

$$S(T,V) = \nu\left[c_V \ln T + R \ln V + \text{constant}\right]. \tag{50}$$

Remark:

Note that the expressions (47) or (48) for the entropy change depend properly *only* on the temperatures and volumes of the initial macrostate a specified by (T_0,V_0) and of the final macrostate b specified by (T,V). On the other hand, the total heat Q absorbed depends on the particular process used in going from a to b. Consider, for example, the following two processes, (i) and (ii), both of which take the system from its macrostate a to its macrostate b.

(i) Keeping the volume constant at its value V_0, we first proceed quasi-statically from the initial macrostate a specified by (T_0,V_0) to the macrostate a' specified by (T,V_0). Keeping the temperature constant now at its value T, we then proceed quasi-statically from this macrostate a' to the final microstate b specified by (T,V). Using (43) with an assumed constant specific heat c_V, we find for the total heat $Q_{(i)}$ absorbed in this process $a \to a' \to b$ the result

$$Q_{(i)} = \nu c_V(T - T_0) + \nu RT \ln \frac{V}{V_0} \tag{51}$$

where the first term on the right represents the heat absorbed in going from a to a', and the second term the heat absorbed in going from a' to b.

(ii) Keeping the temperature constant at its value T_0, we first proceed quasi-statically from the initial macrostate a specified by (T_0,V_0) to the macrostate b' specified by (T_0,V). Keeping the volume constant now at its value V, we then proceed quasi-statically from this macrostate b' to the final macrostate b specified by (T,V). By (43) we then find for the total heat $Q_{(ii)}$ absorbed in this process $a \to b' \to b$ the result

$$Q_{(ii)} = \nu RT_0 \ln \frac{V}{V_0} + \nu c_V(T - T_0) \tag{52}$$

where the first term on the right represents the heat absorbed in going from a to b' and the second term the heat absorbed in going from b' to b. Note that the heats (51) and (52) absorbed in the two processes are *not* the same, since the coefficient of $\ln (V/V_0)$ involves T in the first process and T_0 in the second process. On the other hand, the entropy change (49) is, of course, the same for both processes in accordance with the general conclusion (36).

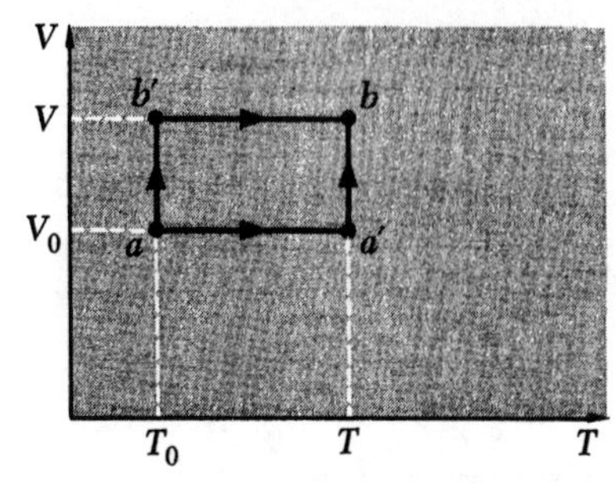

Fig. 7.5 Alternative quasi-static processes leading from an initial macrostate a, specified by a temperature T_0 and volume V_0, to a final macrostate b, specified by a temperature T and volume V.

Adiabatic compression or expansion

Consider an ideal gas which is adiabatically isolated so that it cannot absorb any heat. Suppose now that the volume of the gas is changed quasi-statically; then the temperature and pressure of the gas must change correspondingly. Indeed, the relation (43) must be valid at all stages of the quasi-static process if we put $đQ = 0$ since no heat is absorbed. Thus

$$c_V \, dT + \frac{RT}{V} \, dV = 0$$

Division of both sides by RT then yields

$$\frac{c_V}{R} \frac{dT}{T} + \frac{dV}{V} = 0. \qquad (53)$$

Let us assume that the specific heat c_V is temperature-independent, at least within the limited range corresponding to the temperature change occurring in the contemplated process. The relation (53) can then be integrated immediately to give

$$\frac{c_V}{R} \ln T + \ln V = \text{constant.}\dagger \qquad (54)$$

Thus

$$\ln T^{(c_V/R)} + \ln V = \text{constant,}$$

$$\ln \left[T^{(c_V/R)} \, V \right] = \text{constant,}$$

or

$$\boxed{T^{(c_V/R)} \, V = \text{constant.}} \qquad (55)$$

This relation predicts how the temperature of a thermally insulated ideal gas depends on its volume.

If one is interested in how the pressure of the gas depends on its volume, it is only necessary to use the fact that $T \propto \bar{p}V$ by virtue of the equation of state (37). Thus (55) becomes

$$(\bar{p}V)^{(c_V/R)} \, V = \text{constant.}$$

Raising both sides to the power (R/c_V), we then obtain

$$\boxed{\bar{p}V^{\gamma} = \text{constant}} \qquad (56)$$

where

$$\gamma \equiv 1 + \frac{R}{c_V} = \frac{c_V + R}{c_V}. \qquad (57)$$

The relation (56) is to be contrasted with the relation applicable in a quasi-static process where the gas is not thermally insulated, but is

† Note that (54) also follows immediately from (50) if one makes use of the general result (18) that the entropy of any adiabatically isolated system must remain unchanged in any quasi-static process.

maintained at a constant temperature T by being in contact with a heat reservoir at that temperature. In this case (37) gives

$$\bar{p}V = \text{constant.} \tag{58}$$

Comparison of (56) and (58) shows that the pressure of the gas falls off more rapidly with increasing volume when the gas is thermally insulated than when it is maintained at a constant temperature.

An interesting application of (56) is to the propagation of sound in a gas. If the frequency of vibration of a sound wave is ω, an alternate compression and expansion of any small amount of gas occurs in a time $\tau \sim 1/\omega$. The frequency ω of ordinary sound is sufficiently high so that τ is too short to allow an appreciable amount of heat to flow during this time τ between any such small amount of gas and the gas that surrounds it. Any small amount of gas under consideration thus experiences compressions which are adiabatic; accordingly, its elastic properties are described by (56). As a result, the velocity of sound in a gas depends on the specific heat of this gas through the constant γ. Conversely, measurements of the velocity of sound provide a direct method for determining the quantity γ defined in (57).

7.4 *Basic Statements of Statistical Thermodynamics*

Starting from the statistical postulates of Sec. 3.3, we have now essentially completed our basic investigation of the thermal and mechanical interaction between macroscopic systems. In particular, our discussion has yielded all the fundamental statements of the theory commonly called *statistical thermodynamics*. It seems, therefore, useful to pause and take stock by summarizing the fundamental statements which we have derived.

The first four of these statements are called the *laws of thermodynamics*. We shall list them in their conventional order and, so as to designate them by their traditional names, shall number them starting from zero.†

† The first of these laws is commonly called the "zeroth law" since its importance was appreciated only after the first and second laws had already been assigned their numerical designations.

Statement 0

The first of these statements is the following simple result derived in Sec. 4.3:

> *Zeroth law of thermodynamics*
> If two systems are in thermal equilibrium with a third system, they must be in thermal equilibrium with each other.

This statement is important because it permits one to introduce the notion of thermometers and the concept of a temperature parameter characterizing the macrostate of a system.

Statement 1

The discussion in Sec. 3.7 of the various kinds of interaction between macroscopic systems led us to the following statement about the energy of a system:

> *First law of thermodynamics*
> An equilibrium macrostate of a system can be characterized by a quantity $\bar{E}$ (called its *internal energy*) which has the property that
>
> $$\text{for an isolated system,} \qquad \bar{E} = \text{constant.} \tag{59}$$
>
> If the system *is* allowed to interact and thus goes from one macrostate to another, the resulting change in $\bar{E}$ can be written in the form
>
> $$\Delta\bar{E} = W + Q \tag{60}$$
>
> where W is the macroscopic work done on the system as a result of changes of the external parameters of the system. The quantity Q, defined by (60), is called the *heat absorbed by the system.*

The statement (60) is an expression of the conservation of energy which recognizes heat as a form of energy transfer unaccompanied by any change of external parameters. The relation (60) is important because it introduces the concept of another parameter, the internal energy $\bar{E}$, which characterizes the macrostate of a system. Furthermore, it provides a method for determining this internal energy and for determining absorbed heats in terms of measurements of macroscopic work (as discussed in Sec. 5.3).

Statement 2

We have seen that the number of states accessible to a system (or, equivalently, its entropy) is a quantity of fundamental importance in describing the macrostate of a system. In Sec. 7.2 we showed that changes in the entropy of a system can be related by (32) to the heat absorbed by this system. In Sec. 3.6 we also showed that an *isolated* system tends to approach a situation of greater probability where the number of states accessible to it (or equivalently, its entropy) is larger than initially. (As a special case, when the system is initially already in its most probable situation, it remains in equilibrium and its entropy remains unchanged.) Thus we arrive at the following statement:

> *Second law of thermodynamics*
>
> An equilibrium macrostate of a system can be characterized by a quantity S (called *entropy*) which has the following properties:
>
> (i) In any infinitesimal quasi-static process in which the system absorbs heat $đQ$, its entropy changes by an amount
>
> $$dS = \frac{đQ}{T} \tag{61}$$
>
> where T is a parameter characteristic of the macrostate of the system and is called its *absolute temperature.*
>
> (ii) In any process in which a thermally isolated system changes from one macrostate to another, its entropy tends to increase, i.e.,
>
> $$\Delta S \geq 0. \tag{62}$$

The relation (61) is important because it allows one to determine entropy *differences* by measurements of absorbed heat and because it serves to characterize the absolute temperature T of a system. The relation (62) is significant because it specifies the direction in which nonequilibrium situations tend to proceed.

Statement 3

In Sec. 5.2 we established the fact that the entropy of a system approaches a definite limiting value as its absolute temperature approaches zero. The statement, given by Eq. (5.12), is the following:

> *Third law of thermodynamics*
>
> The entropy S of a system has the limiting property that
>
> $$\text{as } T \to 0_+, \qquad S \to S_0, \tag{63}$$
>
> where S_0 is a constant independent of the structure of the system.

This statement is important because it asserts that, for a system consisting of a given number of particles of a particular kind, there exists near $T = 0$ a standard macrostate which has a unique value of the entropy with respect to which all other entropies of the system can be measured. The entropy *differences* determined by (61) can thus be converted into absolute measurements of actual values of the entropy of the system.

Statement 4

The number Ω of states accessible to a system, or its entropy $S = k \ln \Omega$, can be considered as a function of some set of macroscopic parameters $(y_1, y_2, \ldots, y_n)$. If the system is isolated and in equilibrium, our fundamental statistical postulate allows us then to calculate probabilities by the relation (3.20). The probability P of finding the system in a situation characterized by particular values of its parameters is thus simply proportional to the number of states Ω accessible to the system under these conditions. Since $S = k \ln \Omega$, or $\Omega = e^{S/k}$, this yields the following statement:

> *Statistical relation*
>
> If an isolated system is in equilibrium, the probability of finding it in a macrostate characterized by an entropy S is given by
>
> $$P \propto e^{S/k}. \tag{64}$$

This statement is important because it allows one to calculate the probability of occurrence of various situations, in particular, to calculate the statistical fluctuations occurring in any equilibrium situation.

Statement 5

The statistical definition of the entropy is of fundamental importance. It can be expressed in the following way:

> *Connection with microscopic physics*
> The entropy S of a system is related to the number Ω of states accessible to the system by
>
> $$S = k \ln \Omega. \tag{65}$$

This statement is important because it allows one to calculate the entropy from a microscopic knowledge about the quantum states of the system.

Discussion

Note that the statements 0 through 4, i.e., the four laws of thermodynamics and the statistical relation, are very general statements which are completely *macro*scopic in content. They make *no* explicit references whatever to the atoms composing the systems under consideration. They are, therefore, completely independent of any detailed microscopic models which might be assumed about the atoms or molecules in the systems. These statements thus have the virtue of very great generality and can be used even in the absence of any knowledge about the atomic constitution of the systems of interest. Historically, the laws of thermodynamics were introduced as purely macroscopic postulates before the atomic theory of matter had become established. A completely macroscopic discussion of these laws leads to a large body of consequences and constitutes the subject of *thermodynamics.* This approach is sufficiently fruitful to give rise to a discipline of major importance. The discipline can be enlarged, without changing its generality or completely macroscopic content, by adding the statistical relation (64); it thereupon becomes *statistical thermodynamics.*

Of course, *if* one does combine statistical concepts with *micro*scopic knowledge about the atoms or molecules in a system, one's powers of understanding and prediction are greatly enhanced. One obtains thus the discipline of *statistical mechanics* which includes also the relation (65). It is then possible to calculate the entropy of a system from first principles and to make detailed probability statements based on (64) or its consequences (such as the canonical distribution). Thus one is in a position to calculate the properties of macroscopic systems on the basis of microscopic information. The subject of statistical mechanics, upon which we have based the entire discussion of this book, is thus the all-inclusive discipline. It encompasses, as a special case, the thermodynamic laws which are independent of any models assumed about the atomic constitution of the systems under consideration.

7.5 *Equilibrium Conditions*

Our fundamental statistical postulates of Sec. 3.3 deal specifically with the equilibrium situation of an isolated system and with its approach to equilibrium. These postulates, which formed the basis of all our considerations, were formulated in terms of the number of states accessible to the system, or equivalently, in terms of its entropy. We now want to return to these fundamental ideas in order to express them in some alternative forms useful for many practical applications.

Isolated system

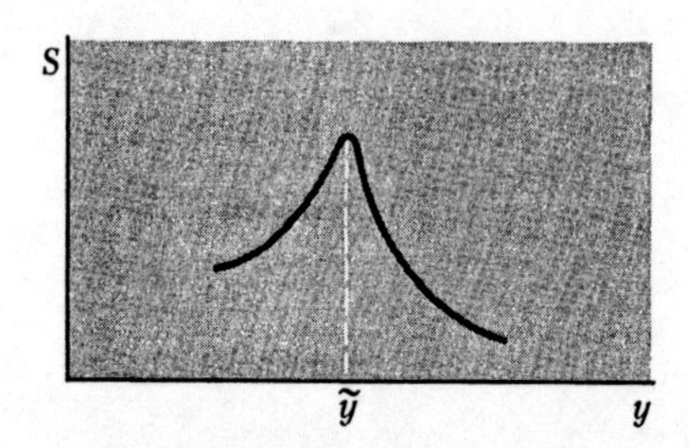

Fig. 7.6 Schematic diagram illustrating the dependence of the entropy S on some macroscopic parameter y.

Let us begin by reviewing the implications of the postulates for an isolated system. The total energy of the system then remains constant. Let us suppose that the system can be described macroscopically by a parameter y, or by several such parameters. (For example, y might denote the energy of the subsystem A in Fig. 3.9, or the position of the piston in Fig. 3.10.) The number of states accessible to the system is then some function of y. We shall subdivide the possible values of y into small equal intervals of fixed size δy. We shall then denote by $\Omega(y)$ the number of states accessible to the system when the parameter assumes a value in the interval between y and $y + \delta y$. The corresponding entropy of the system is, by definition, $S = k \ln \Omega$. Our fundamental postulate (3.19) asserts that, when the system is in equilibrium, it is equally likely to be found in each one of its accessible states. If the parameter y is free to vary, the probability $P(y)$ of finding the system in a situation where its parameter lies between y and $y + \delta y$ is thus,

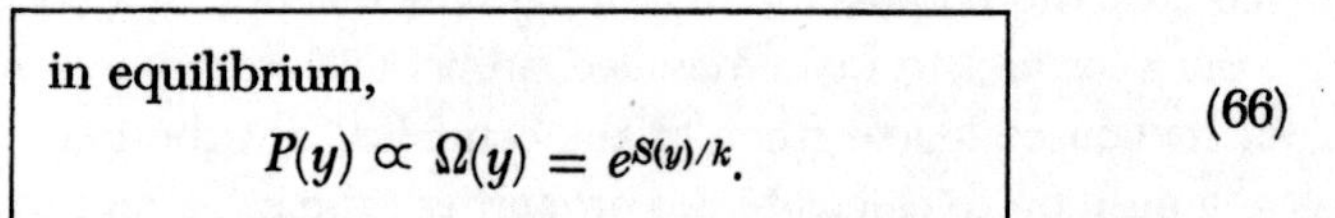

in equilibrium,

$$P(y) \propto \Omega(y) = e^{S(y)/k}. \tag{66}$$

If the parameter y assumes the value y_0 in some standard macrostate of the system, (66) implies the proportionality

$$\frac{P(y)}{P(y_0)} = \frac{e^{S(y)/k}}{e^{S(y_0)/k}}$$

or

$$P(y) = P_0\, e^{\Delta S/k} \tag{67}$$

where

$$\Delta S \equiv S(y) - S(y_0)$$

and $P_0 \equiv P(y_0)$. Probability ratios can thus be obtained immediately from entropy *differences*.

According to (66), it is most probable that the parameter y of the system in equilibrium assumes values where the entropy $S(y)$ is largest. Even a gentle maximum of $S = k \ln \Omega$ corresponds to a very sharp maximum of Ω itself and thus of the probability P. Hence it follows that y ordinarily assumes with overwhelming probability values very close to the particular value $\tilde{y}$ where the entropy S has its maximum value. In short,

> The equilibrium situation of an isolated system is characterized by values of its parameters such that
>
> $$S = \text{maximum}. \tag{68}$$

In equilibrium the probability $P(y)$ of observing in an ensemble of systems a value of y significantly different from $\tilde{y}$ is thus very small. On the other hand, as a result of prior external intervention or special preparation, the system may at some particular time t_0 have a large probability of being found in a macrostate where y has a value quite different from $\tilde{y}$. If the system after the time t_0 is now left isolated while its parameter y is free to vary, it is *not* in equilibrium. In accordance with the postulate (3.18), the situation will then change in time until the equilibrium probability distribution (66) is attained. In other words, the situation changes in such a direction that values of y corresponding to larger entropy become more probable, i.e., so that the entropy S tends to increase and the resulting entropy change ΔS satisfies the inequality

$$\Delta S \geq 0. \tag{69}$$

This change continues until the final equilibrium condition is reached where there is overwhelming probability that the parameter y has a value corresponding to the maximum of the entropy S.

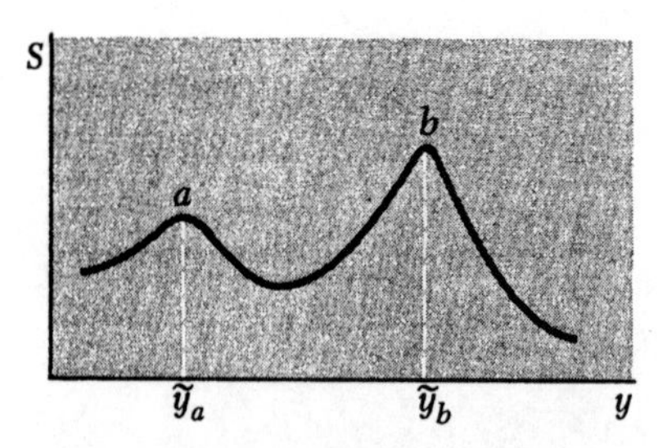

Fig. 7.7 Schematic diagram illustrating an entropy function S exhibiting maxima at two different values of a macroscopic parameter y.

Remark on metastable equilibrium

It is possible that the entropy S may exhibit more than one maximum, as illustrated in Fig. 7.7. If the maximum of the entropy S at $\tilde{y}_b$ is higher than its maximum at $\tilde{y}_a$, the corresponding probability $P(y)$ of (66) is, because of its exponential dependence on S, very much higher at $\tilde{y}_b$ than at $\tilde{y}_a$. In a genuine equilibrium situation the system would thus almost always be found with its parameter near $\tilde{y}_b$.

Suppose now that the system has been prepared in such a way that its parameter has at some initial time a value not too far removed from $\tilde{y}_a$. The system can then readily change so as to attain the very probable situation where its parameter is nearly equal to $\tilde{y}_a$. Although the situation where y is close to $\tilde{y}_b$ is much *more* probable, it can only be reached if the system passes through the very improbable situations where

$$\tilde{y}_a < y < \tilde{y}_b.$$

Without any outside assistance, the probability of passing through these intermediate situations may, however, be so low that it takes a very long time before the system can attain its ultimate equilibrium situation where y is near $\tilde{y}_b$. During times of experimental interest, the states near $\tilde{y}_b$ may thus effectively remain inaccessible to the system. The system can, however, readily attain an equilibrium situation where it is distributed with equal probability over all its accessible states where y is near $\tilde{y}_a$. Such a situation is called *metastable equilibrium.* If some means is found to facilitate the transition of the system from states where the parameter is near $\tilde{y}_a$ to states where the parameter is near $\tilde{y}_b$, then the system may quickly leave its situation of metastable equilibrium to attain its genuine equilibrium situation where y is close to $\tilde{y}_b$.

Examples of this sort can be quite striking. For instance, water in genuine equilibrium becomes ice at a temperature below 0°C. Very pure water cooled very gently below 0°C can, however, remain liquid in a state of metastable equilibrium down to −20°C, or less. But if one drops a grain of dust into the water, thus helping to initiate the growth of ice crystals, the liquid will suddenly freeze to attain its genuine equilibrium form of ice.

System in contact with a reservoir

Suppose that a system of interest (call it A) is not isolated but is free to interact with one or more other systems (which we shall denote collectively by A'). The combined system A^* consisting of A and A' is isolated. The equilibrium conditions which must be satisfied by A can then be obtained by reducing the discussion to the familiar case of the isolated system A^*.

Most experiments in the laboratory are carried out under conditions of constant temperature and pressure. Thus the system A under consideration is usually in thermal contact with a heat reservoir (which might be the surrounding atmosphere or a more carefully controlled water bath) whose temperature remains essentially constant. In addition, usually no attempt is made to keep the volume of the system A fixed. Instead, the system is kept at a constant pressure (ordinarily the pressure of the surrounding atmosphere). Let us, therefore, investigate the equilibrium conditions of a system A in contact with a reservoir A' which remains at a constant temperature T' and at a constant pressure p'. The system A can exchange heat with the reservoir A', but the latter is so large that its temperature T' remains essentially unchanged. Similarly, the system A can change its volume V at the expense of the reservoir A', doing work on the reservoir in the process; but again A' is so large that its pressure p' remains unaffected by this relatively small volume change.†

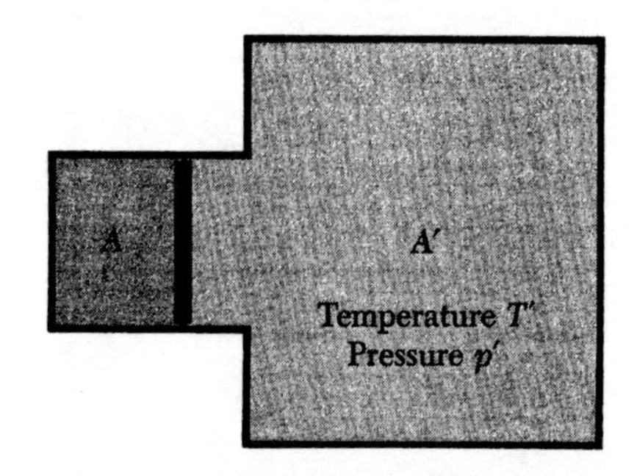

Fig. 7.8 A system A in contact with a reservoir A' at a constant temperature T' and a constant pressure p'.

Suppose that the system A is described by some macroscopic parameter y (or by several such parameters). When this parameter has a particular value y, the number $\Omega^*(y)$ of states accessible to the combined system A^* is then given by the product of the number of states $\Omega(y)$ accessible to A and the number of states $\Omega'(y)$ accessible to the reservoir A' under these circumstances. Thus

$$\Omega^* = \Omega\Omega'.$$

By virtue of the definition $S \equiv k \ln \Omega$, it thus follows that

$$S^* = S + S' \tag{70}$$

where S^* is the entropy of the combined system A^*, while S and S' are the entropies of A and A', respectively. Let us consider some standard macrostate where the parameter y has the value y_0. By applying (67) to the combined system A^*, which is isolated, we then obtain for the probability $P(y)$ that the parameter assumes in this system a value between y and $y + \delta y$

$$P(y) = P_0 e^{\Delta S^*/k} \tag{71}$$

where $$\Delta S^* = S^*(y) - S^*(y_0).$$

But by virtue of (70)

$$\Delta S^* = \Delta S + \Delta S' \tag{72}$$

where ΔS denotes the entropy change of A and $\Delta S'$ that of A' when the parameter is changed from y_0 to y. We shall now try to simplify (72) by expressing the entropy change $\Delta S'$ of the reservoir in terms of quantities referring to the system A of interest.

† The system A' may be a single reservoir with which A can interact by both heat transfer and pressure work. Alternatively, A' may be a combination of two reservoirs, one having a temperature T' and interacting with A by heat transfer only, the other having a pressure p' and interacting with A by pressure work only.

Since the reservoir is so large that it always remains in equilibrium at the constant temperature T' and constant pressure p' while absorbing a relatively small amount of heat Q' from the system A, its change of entropy in this quasi-static process is given by (32) so that

$$\Delta S' = \frac{Q'}{T'}. \tag{73}$$

But the heat Q' absorbed by the reservoir when the parameter changes from y_0 to y is equal to

$$Q' = \Delta \bar{E}' - W'. \tag{74}$$

Here $\Delta \bar{E}'$ is the change in mean energy of A', while W' is the work done on A' when the volume of A is changed by an amount $\Delta V \equiv V(y) - V(y_0)$ against the constant pressure p' of the reservoir. Since the volume of the reservoir is then changed by an amount $-\Delta V$, it follows by (5.14) that $W' = p' \Delta V$. In addition, the conservation of energy applied to the combined isolated system A^* implies that $\Delta \bar{E}' = -\Delta \bar{E}$, where $\Delta \bar{E} \equiv \bar{E}(y) - \bar{E}(y_0)$ is the change in mean energy of A. Thus (74) becomes

$$Q' = -\Delta \bar{E} - p' \Delta V.$$

Using (72) and (73), we then obtain

$$\Delta S^* = \Delta S - \frac{\Delta \bar{E} + p' \Delta V}{T'} = -\frac{-T' \Delta S + \Delta \bar{E} + p' \Delta V}{T'}. \tag{75}$$

To simplify the expression on the right, let us define the function

$$\boxed{G \equiv \bar{E} - T'S + p'V} \tag{76}$$

which involves, besides the constant specified temperature T' and pressure p' of the reservoir, only the functions $\bar{E}$, S, and V of the system A. Since T' and p' are constants, we have

$$\Delta G = \Delta \bar{E} - T' \Delta S + p' \Delta V.$$

Hence (75) can then be written in the simple form

$$\boxed{\Delta S^* = -\frac{\Delta G}{T'}} \tag{77}$$

where $\Delta G \equiv G(y) - G(y_0)$. The function G defined by (76) is seen to have the dimensions of energy and is called the *Gibbs free energy* of the system A at the specified constant temperature T' and constant pressure p'.

The result (77) shows that the entropy S^* of the total system A^* *in*creases as the Gibbs free energy of the subsystem A *de*creases. The situation of *maximum* probability (71) or *maximum* entropy S^* of the total isolated system A^* corresponds therefore to a *minimum* value of the Gibbs free energy G of the subsystem A. In accordance with the statement (68) for an isolated system, we thus arrive at the following conclusion:

> The equilibrium situation of a system in contact with a reservoir at constant temperature and constant pressure is characterized by values of its parameters such that
>
> $$G = \text{minimum}. \tag{78}$$

Suppose that the condition (78) is not satisfied so that the system A is not in equilibrium. Then the situation will change in such a direction that the entropy S^* of the total system A^* tends to increase until the final equilibrium situation is reached where the parameters of A have, with overwhelming probability, values corresponding to the maximum of S^*. Equivalently, this statement can be expressed more conveniently in terms of the Gibbs free energy of A. Thus the situation will change in such a direction that the Gibbs free energy G of A tends to *de*crease, i.e.,

$$\Delta G \leq 0, \tag{79}$$

until a final equilibrium situation is reached where the parameters of A have with overwhelming probability values corresponding to the minimum value of G.

By using (77) in (71), we obtain the explicit probability statement

$$P = P_0\, e^{-\Delta G/kT'}. \tag{80}$$

Equivalently, since $\Delta G = G(y) - G(y_0)$, where $G(y_0)$ is merely some constant referring to the standard macrostate, the result (80) can be expressed by the following proportionality:

> in equilibrium,
>
> $$P(y) \propto e^{-G(y)/kT}. \tag{81}$$

This result is analogous to the result (66) for an isolated system and shows explicitly that the probability $P(y)$ is maximum when $G(y)$ is minimum.

Since most systems of physical or chemical interest are studied under conditions of specified constant temperature and pressure, the statements (78) or (81) represent very convenient formulations of the equilibrium conditions. They are, therefore, the usual starting point in any discussion dealing with physical or chemical systems. We shall give a specific illustration in the next section.

7.6 *Equilibrium between Phases*

Every substance can exist in distinctly different forms, called *phases,* which correspond to different types of aggregation of the same molecules. Thus one finds that a substance can exist in the form of a solid, a liquid, or a gas.† (The gaseous form is sometimes also called a *vapor.*) For example, water can exist in the form of ice, liquid water, or water vapor. Different phases are found to exist in different ranges of pressure and temperature. In addition, one phase may change into another phase at particular temperatures and pressures. Thus a solid can *melt* to become a liquid, a liquid can *vaporize* to become a gas, or a solid can *sublime* to become a gas. In this section we shall try to apply our general theory to gain a greater understanding of such phase transformations.

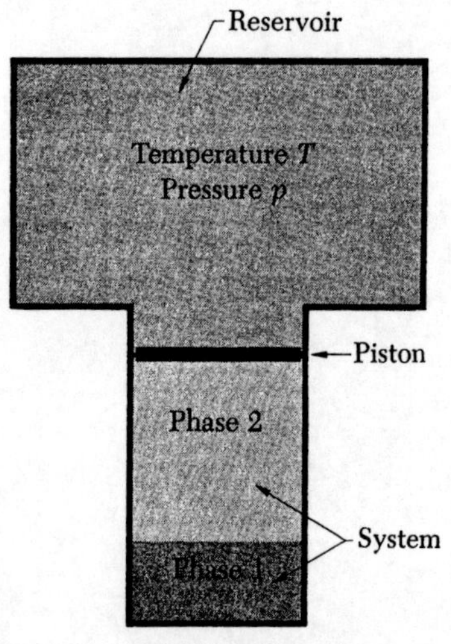

Fig. 7.9 A system which consists of two phases and which is maintained at a constant temperature T and constant pressure p by being kept in contact with a suitable reservoir.

We shall consider a system consisting of two spatially distinct phases of a substance consisting of a single type of molecule. For example, these phases might be a solid and a liquid, or they might be a liquid and a gas. In general, we shall call them phase 1 and phase 2, respectively. We shall examine this system at any specified constant temperature T and pressure p provided by placing the system in contact with a suitable reservoir having these values of temperature and pressure. Except for small fluctuations (which are not of interest to us in the present context), both phases of the substance in equilibrium will then always have a temperature T and a pressure p. We shall denote by N_1 the number of molecules of the substance present in the form of phase 1, and by N_2 the number of molecules of the substance present in the form of phase 2. By virtue of the conservation of matter, the total number N of molecules must, of course, remain constant irrespective of the way in which they may be distributed between the two phases. Thus

$$N_1 + N_2 = N = \text{constant}. \tag{82}$$

---

† There may also exist different forms of the solid, corresponding to different crystal structures.

The questions of interest are then the following: In an equilibrium situation at the specified temperature T and pressure p, will phase 1 alone be present, or will phase 2 alone be present, or will both phases be present simultaneously?

Since the temperature T and pressure p are kept constant, all these questions can be reduced to an examination of the total Gibbs free energy G of the system. This free energy can be regarded as a function of N_1 and N_2. Our general formulation (78) of the equilibrium conditions asserts then that the parameters N_1 and N_2 must assume values such that G is minimum, where G is defined by the relation (76). Thus†

$$G = \bar{E} - TS + pV = \text{minimum.} \tag{83}$$

Here the total mean energy $\bar{E}$ of the system is simply equal to the sum of the mean energies of the two phases, the total entropy S of the system is simply equal to the sum of the entropies of the two phases,‡ and the total volume V of the system is simply equal to the sum of the volumes of the two phases. Hence it follows that

$$G = G_1 + G_2 \tag{84}$$

where G_1 is the Gibbs free energy of phase 1 and G_2 is that of phase 2. But at a given temperature and pressure the mean energy, entropy, and volume of any particular phase are each simply proportional to the amount of phase present (i.e., each is an extensive quantity as discussed in Sec. 5.6). Thus one can write $G_1 = N_1 g_1$ and $G_2 = N_2 g_2$ where

$$g_i(T,p) \equiv \text{the Gibbs free energy per molecule of phase } i \text{ at the given temperature } T \text{ and pressure } p \tag{85}$$

characterizes the intrinsic properties of the ith phase irrespective of its amount present. Hence (84) becomes

$$G = N_1 g_1 + N_2 g_2 \tag{86}$$

where g_1 and g_2 depend on T and p, but not on the numbers N_1 or N_2.

† The primes occurring in (76) have disappeared in (83) since we have now denoted the temperature and pressure of the reservoir simply by T and p, respectively. The temperature and pressure of the system are also supposed to be equal to T and p since we disregard small temperature and pressure fluctuations of our system.

‡ This corresponds merely to the relation (70), i.e., the number of states accessible to the total system is the product of the numbers of states accessible to each phase.

If both phases coexist in equilibrium, N_1 and N_2 must be such that G is a minimum in accordance with (83). Thus G must be unchanged for infinitesimal changes of N_1 and N_2 so that

$$dG = g_1\, dN_1 + g_2\, dN_2 = 0$$

or

$$(g_1 - g_2)\, dN_1 = 0$$

since the conservation of matter expressed by (82) implies that $dN_2 = -dN_1$. Thus we obtain as a necessary condition for the coexistence of two phases in equilibrium the following condition:

$$\boxed{\text{for coexistence in equilibrium,} \quad g_1 = g_2.} \tag{87}$$

When this condition is satisfied, the transfer of one molecule of substance from one phase to another clearly leaves the value of G in (86) unchanged so that G has an extremum as required.†

Let us now examine the Gibbs free energy (86) somewhat more closely. Remembering that the Gibbs free energy $g_i(T,p)$ per molecule of each phase is a well-defined function characterizing the particular phase i at the given temperature and pressure, we can then make the following statements:

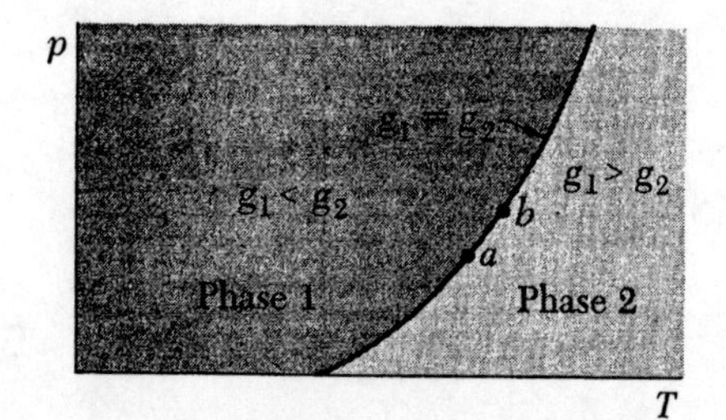

Fig. 7.10 A plot of pressure p versus temperature T showing the regions where each of two phases exists separately in equilibrium and the phase-equilibrium curve where both phases can coexist in equilibrium.

If T and p are such that $g_1 < g_2$, then the minimum value of G in (86) is achieved if all the N molecules of the substance transform into phase 1 so that $G = Ng_1$. Only phase 1 alone can then exist in stable equilibrium.

If T and p are such that $g_1 > g_2$, then the minimum value of G is achieved if all the N molecules of the substance transform into phase 2 so that $G = Ng_2$. Only phase 2 alone can then exist in stable equilibrium.

If T and p are such that $g_1 = g_2$, the condition (87) is satisfied and any number N_1 of molecules in phase 1 can coexist in equilibrium with the remaining number $N_2 = N - N_1$ of molecules in phase 2. The value G remains then unchanged when N_1 is varied. The locus of points where T and p are such that the condition (87) is fulfilled then represents the *phase-equilibrium curve* along which the two phases can coexist in equilibrium. This curve, along which $g_1 = g_2$, divides

† Although the condition (87) is only a necessary condition for the existence of a minimum, the sufficient conditions which guarantee that G is actually a minimum rather than a maximum are not of sufficient interest to merit investigation on our part.

the pT plane into two regions: one where $g_1 < g_2$ so that phase 1 is the stable one, and the other where $g_1 > g_2$ so that phase 2 is the stable one.

It is possible to characterize the phase-equilibrium curve by a differential equation. In Fig. 7.10 consider any point, such as a, which lies on the phase-equilibrium curve and corresponds to a temperature T and pressure p. Then the condition (87) implies that

$$g_1(T,p) = g_2(T,p). \tag{88}$$

Consider now a neighboring point, such as b, which also lies on the phase-equilibrium curve and corresponds to a temperature $T + dT$ and pressure $p + dp$. Then the condition (87) implies that

$$g_1(T + dT, p + dp) = g_2(T + dT, p + dp). \tag{89}$$

Subtraction of (88) from (89) thus yields the condition

$$dg_1 = dg_2 \tag{90}$$

where dg_i is the change in the free energy per molecule of phase i if this phase is brought from the temperature T and pressure p of point a to the temperature $T + dT$ and pressure $p + dp$ of point b.

But, by virtue of the definition (83), the free energy per molecule of phase i is merely

$$g_i = \frac{G_i}{N_i} = \frac{\bar{E}_i - TS_i + pV_i}{N_i}$$

or

$$g_i = \bar{\epsilon}_i - Ts_i + pv_i$$

where $\bar{\epsilon}_i = \bar{E}_i/N_i$ is the mean energy, $s_i \equiv S_i/N_i$ is the entropy, and $v_i \equiv V/N_i$ is the volume per molecule of phase i. Hence

$$dg_i = d\bar{\epsilon}_i - T\,ds_i - s_i\,dT + p\,dv_i + v_i\,dp.$$

But the fundamental thermodynamic relation (34) allows us to relate the entropy change ds_i to the heat absorbed by the phase in this change, i.e.,

$$T\,ds_i = d\bar{\epsilon}_i + p\,dv_i.$$

Hence we get simply

$$dg_i = -s_i\,dT + v_i\,dp. \tag{91}$$

Applying this result to each phase, (90) becomes

$$-s_1\,dT + v_1\,dp = -s_2\,dT + v_2\,dp$$
$$(s_2 - s_1)\,dT = (v_2 - v_1)\,dp$$

or

$$\boxed{\frac{dp}{dT} = \frac{\Delta s}{\Delta v}} \tag{92}$$

where $\Delta s \equiv s_2 - s_1$ and $\Delta v \equiv v_2 - v_1$.

The relation (92) is called the *Clausius-Clapeyron equation.* Consider any point on the phase-equilibrium curve at a temperature T and corresponding pressure p. Equation (92) then relates the slope of the phase-equilibrium curve at this point to the entropy change Δs and volume change Δv per molecule when the curve is crossed at this point, i.e., when a change of phase occurs at this temperature and pressure. Note that if one deals with any *arbitrary* amount consisting of N molecules of the substance, its entropy change and volume change in the transformation are simply given by $\Delta S = N\,\Delta s$ and $\Delta V = N\,\Delta v$; hence (92) can equally well be written as

$$\boxed{\frac{dp}{dT} = \frac{\Delta S}{\Delta V}.} \tag{93}$$

Since there is an entropy change associated with the phase transformation, heat must also be absorbed. The *latent heat of transformation* L_{12} is defined as the heat absorbed when a given amount of phase 1 is transformed to phase 2 when the phases coexist in equilibrium. Since the process takes place at the constant temperature T, the corresponding entropy change is related to L_{12} by (32) so that

$$\Delta S = S_2 - S_1 = \frac{L_{12}}{T}. \tag{94}$$

where L_{12} is the latent heat at this temperature. Thus the Clausius-Clapeyron equation (93) can also be written in the form

$$\boxed{\frac{dp}{dT} = \frac{L_{12}}{T\,\Delta V}.} \tag{95}$$

If V refers to the volume per mole, then L_{12} is the latent heat per mole; if V refers to the volume per gram, then L_{12} is the latent heat per gram.

Let us discuss a few important applications of these results.

Phase transformations of a simple substance

As already mentioned, simple substances are capable of existing in phases of three types: solid, liquid, and gas. (There may also be several solid phases with different crystal structures.) The phase-equilibrium curves between these three phases can be represented on a diagram of pressure versus temperature in the general manner indicated in Fig. 7.11. These curves separate on this diagram the region of the solid from that of the liquid, the region of the solid from that of the gas, and the region of the liquid from that of the gas. Since the three curves can only intersect in such a way that they divide the plane into no more than three distinct regions, they must meet at one common point t, called the *triple point*. At this unique temperature and pressure arbitrary amounts of all three phases can, therefore, coexist in equilibrium with each other. (This is the property which makes the triple point of water so suitable as a readily reproducible temperature standard.) At point c, the so-called *critical point*, the liquid-gas equilibrium curve ends. The volume change ΔV between a given amount of liquid and gas has then approached zero. Beyond c there is no further phase transformation, since there exists then only one "fluid" phase. (The pressure has then become so large that the dense gas is indistinguishable from the liquid.)

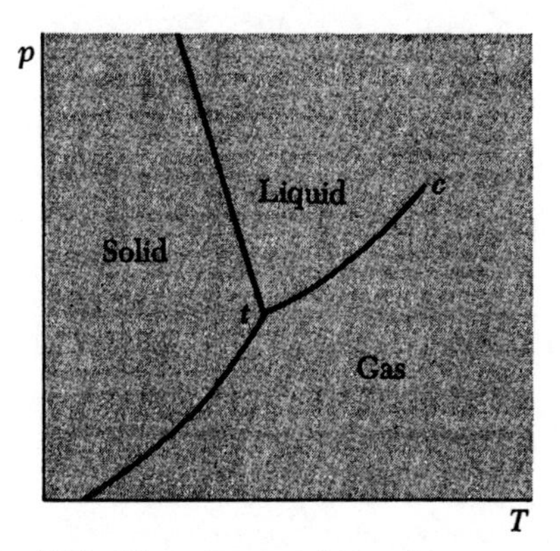

Fig. 7.11 Phase diagram for a substance such as water. Point t is the triple point, point c the critical point.

As a substance goes from its solid form (s) to its liquid form (l), its entropy (or degree of disorder) almost always increases.† Thus the corresponding latent heat L_{sl} is positive and heat is absorbed in the transformation. In most cases the solid expands upon melting, so that $\Delta V > 0$. In this case the Clausius-Clapeyron equation (93) asserts that the slope of the solid-liquid equilibrium line (i.e., of the melting curve) is positive. There are some substances, like water, which contract upon melting so that $\Delta V < 0$. For these, therefore, the slope of the melting curve must be negative (as drawn in Fig. 7.11).

Approximate calculation of the vapor pressure

The Clausius-Clapeyron equation can be used to derive an approximate expression for the pressure of a vapor in equilibrium with a liquid (or solid) at a temperature T. This pressure is called the *vapor pressure*

† An exceptional case arises in the case of solid ^{3}He at very low temperatures. Here quantum-mechanical effects give rise to an antiparallel alignment of nuclear spins in the liquid, while these spins remain randomly oriented in the solid.

of the liquid (or solid) at this temperature. Equation (95) applied to a mole of the substance gives

$$\frac{dp}{dT} = \frac{L}{T\,\Delta V} \tag{96}$$

where $L \equiv L_{12}$ is the latent heat of vaporization per mole and V is the molar volume. Let 1 refer to the liquid (or solid) phase and 2 to the vapor. Then

$$\Delta V = V_2 - V_1 \approx V_2$$

since the vapor is much less dense than the liquid (or solid), so that $V_2 \gg V_1$. Let us also assume that the vapor can be adequately treated as an ideal gas, so that its equation of state per mole is simply

$$pV_2 = RT.$$

Then

$$\Delta V \approx V_2 = \frac{RT}{p}.$$

With these approximations (96) becomes

$$\frac{1}{p}\frac{dp}{dT} \approx \frac{L}{RT^2}. \tag{97}$$

Ordinarily, L is approximately temperature-independent. Then (97) can be immediately integrated to give

$$\ln p \approx -\frac{L}{RT} + \text{constant}$$

or

$$\boxed{p \approx p_0 e^{-L/RT}} \tag{98}$$

where p_0 is some constant. Thus we see that the temperature dependence of the vapor pressure is determined by the magnitude of the latent heat. This latent heat is approximately the energy required to dissociate a mole of the liquid (or solid) into individual widely separated molecules. It must thus be much greater than the thermal energy RT per mole if the liquid (or solid) is to exist as an undissociated phase. Since $L \gg RT$, the vapor pressure given by (98) is thus a very rapidly increasing function of the temperature T.

Note that it should be possible to calculate the vapor pressure from first principles. Indeed, a knowledge of the microscopic constitution of each phase allows one to calculate the number of accessible states of the phase. Hence its entropy as well as mean energy can be determined so as to calculate its Gibbs free energy per molecule as a function of T and p. The fundamental equilibrium condition

$$g_1(T,p) = g_2(T,p)$$

then provides an equation which permits one to solve for p in terms of T. Thus one can find an expression which relates the vapor pressure to the temperature and which involves no unknown constants of proportionality (such as p_0). Microscopic calculations of this kind can actually be carried out in simple cases.

7.7 *The Transformation of Randomness into Order*

Any isolated system tends to approach a situation of maximum randomness, i.e., one where its entropy is maximum. This has been the key principle incorporated in our fundamental statistical postulates and pervading the entire discussion of this book. Situations illustrating this principle are extremely common. Let us just single out two specific examples:

(i) Consider a system consisting of a container full of water, a paddle wheel, and a weight connected to this wheel by a string (see Fig. 5.7). Suppose that this isolated system is left to itself. The weight can then move either up or down, turning the paddle wheel and thus exchanging energy with the water. If the weight descends, the gravitational potential energy associated with its single degree of freedom (i.e., with its height above the floor) becomes transformed into an equivalent amount of internal energy randomly distributed over the many molecules of the water. If the weight moves up, the energy randomly distributed over the molecules of the water becomes converted into potential energy associated with the nonrandom upward displacement of the weight. Since the entropy tends to increase, the process which actually occurs with overwhelming probability is the first one. Thus the weight descends so that the system attains a situation which is less orderly or more random.

(ii) Consider an animal or any other biological organism. Although it consists merely of simple atoms (such as carbon, hydrogen, oxygen, and nitrogen), these are not simply mixed together at random. Indeed, they are arranged in an exquisitely organized way to give rise to a highly ordered system. Typically, the atoms are first used to form some special organic molecules (e.g., some twenty different kinds of amino acids). These organic molecules are then used as building blocks which are joined end-to-end in a very precise sequence to give rise to different kinds of large molecules, called *macromolecules,* with very special properties. (For example, the amino acids are used thus as building blocks to form different proteins.) Now suppose that an animal is enclosed in a box so that it is completely isolated. Then its highly ordered structure would not be maintained. In accordance with the principle of increasing entropy, the animal would not survive and its elaborate organization of complex macromolecules would gradually disintegrate into a much more random mixture of simple organic molecules.

The principle of increasing entropy thus gives the impression of a world approaching a situation of ever greater randomness. Even without making statements about the whole universe (which, perhaps, might not legitimately be considered an isolated system), we can certainly say that every process occurring spontaneously in an isolated system has a preferred direction from a more orderly to a more random situation. We can then pose the following interesting question: *To what extent is it possible to reverse the direction of such processes so as to bring a system from a more random to a more orderly situation?* To show the significance of this question, we shall phrase it in more concrete terms related to our previous two examples.

(i) To what extent is it possible to convert the internal energy distributed randomly over many molecules of a substance (such as water, oil, or coal) into the energy associated with the systematic change of an external parameter (such as the motion of a piston or the rotation of a shaft), i.e., into work useful for lifting weights or driving trucks? In other words, to what extent is it possible to construct the various engines and machines which are responsible for the industrial revolution?

(ii) To what extent is it possible to convert a random mixture of simple molecules into the complex and highly organized macromolecules which constitute an animal or a plant? In other words, to what extent is it possible for living organisms to exist?

The question which we have posed is thus far from trivial; as we have seen, it is directly relevant to issues as profound as the possibility of life or the possibility of the industrial revolution. Let us then formulate the question in the following very general terms: To what extent is it possible to take a system A from a more random to a less random situation? Or more quantitatively, to what extent is it possible to take a system A from a macrostate a where its entropy is S_a to another macrostate b of lower entropy S_b such that $\Delta S \equiv S_b - S_a < 0$?

We shall now try to answer this question with similar generality. *If* the system A is isolated, it is overwhelmingly probable that its entropy will increase (or, at most, remain unchanged) so that $\Delta S \geq 0$. The answer to our question is then simply that the desired decrease of randomness cannot be achieved. Suppose, however, that the system A is *not* isolated but is free to interact with some other system A'. Then it is similarly true that the entropy S^* of the isolated combined system A^* consisting of A and A' must tend to increase so that $\Delta S^* \geq 0$. But

$$S^* = S + S'$$

if S' denotes the entropy of the system A'. The statement of increasing entropy applied to the isolated system A^* thus asserts that

$$\boxed{\Delta S^* = \Delta S + \Delta S' \geq 0.} \tag{99}$$

This condition does *not*, however, imply necessarily that $\Delta S \geq 0$. Indeed, it is quite possible that the entropy S of A may decrease, provided that the entropy S' of A' increases by at least a compensating amount to satisfy the condition (99) for the *total* system. The randomness of the system A of interest is then decreased at the expense of the other system A' with which it is allowed to interact. Thus we arrive at the following conclusion, which we might call the "principle of entropy compensation":

> The entropy of a system can be reduced only if it is made to interact with one or more auxiliary systems in a process which imparts to these at least a compensating amount of entropy. (100)

The statement (100), which merely expresses the content of (99) in verbal form, provides the general answer to our question. Suppose that one is faced with the problem of how the entropy of some system might be reduced. The task might be accomplished by various procedures involving different auxiliary systems and different processes. The statement (100) is then useful in the following ways:

(i) It can immediately rule out certain suggested procedures by showing that they cannot be feasible (if they are such that $\Delta S^* < 0$).

(ii) It can suggest that, among possible alternative procedures, some might be more efficient than others in achieving desired ends.

The statement (100) provides, however, *no* information about the detailed procedures or mechanisms which can actually be used to reduce the entropy of a system. Thus there have been some very ingenious people who invented the specific steam engines, gasoline engines, or Diesel engines useful for converting internal energy into work. Similarly, biological evolution over billions of years has resulted in the selection of those particular biochemical reactions well adapted to accomplish synthesis of the macromolecules that make life possible.

Let us now illustrate the wide applicability of the general principle (100) to some concrete situations.

Engines

An engine is a device used to convert some of the internal energy of a system into work. The engine mechanism M (which may consist of various pistons, cylinders, etc.) should itself remain unchanged in the process. This is achieved by letting the mechanism M go through a cycle of steps so that it returns at the end of the cycle to the same macrostate from which it started at the beginning. The engine can then be made to operate continuously by going through a succession of such repeated cycles. The entropy of the mechanism M itself does not change in a cycle since it returns to its initial macrostate. The work w done by the engine is merely supposed to change the external parameter of some other system B (e.g., to lift a weight or move a piston), while leaving the entropy of B unchanged. As the engine goes through a cycle, the only entropy change occurring then is that associated with the system A whose internal energy $\bar{E}$ is partly converted into macroscopic work.

The simplest case is that where the system A is merely a heat reservoir at some constant absolute temperature T. Ideally, we would want the engine to extract from the heat reservoir A, in one cycle, some amount of heat q (thus reducing the internal energy of the reservoir by q) and to use this heat to perform an amount of work w on the system B.† Here $w = q$ to satisfy the conservation of energy. We would then have the "perfect engine" indicated schematically in Fig. 7.12. But although it would be eminently desirable, it is clear that such a perfect engine cannot be realized. Indeed, since the reservoir A absorbs in a cycle an amount of heat $(-q)$, its corresponding entropy change is

$$\Delta S = -\frac{q}{T} \tag{101}$$

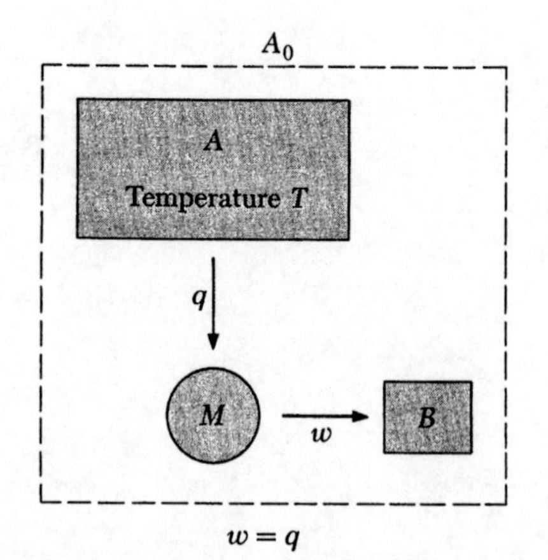

Fig. 7.12 A perfect engine A_0 consisting of the engine mechanism M, the system B upon which it does work, and the single heat reservoir A from which it extracts heat.

† We use the small letters q and w to denote quantities of heat and work which are intrinsically positive.

and thus is negative. This is also the entropy change per cycle of the entire system A_0 illustrated in Fig. 7.12. In accordance with our general discussion, such a perfect engine cannot be built precisely because its sole effect would be to reduce the randomness of the heat reservoir by extracting energy from it.

If we want to persist in our goal of converting some of the internal energy of the reservoir A into work, we must be prepared to cope with the entropy decrease (101) by using the principle (100) of entropy compensation. Thus we must introduce some auxiliary system A' with which the system A_0 of Fig. 7.12 can interact. Let us choose for A' some other heat reservoir at an absolute temperature T'. This reservoir can then interact with our previous system A_0 by absorbing from it in a cycle an amount of heat q'. Correspondingly, the entropy S' of A' will be increased by an amount

$$\Delta S' = \frac{q'}{T'}. \tag{102}$$

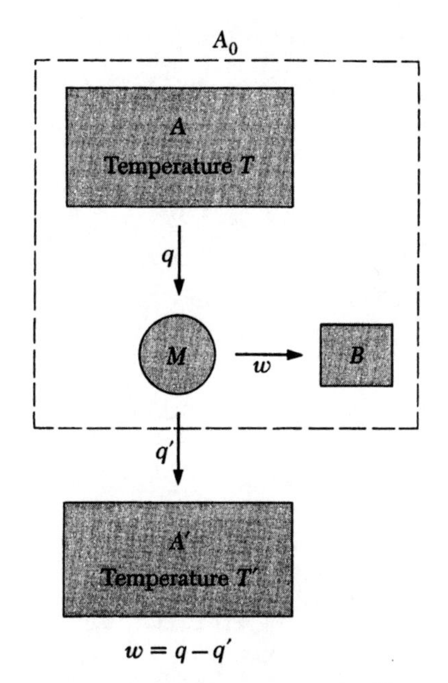

Fig. 7.13 A realizable engine assembly consisting of the system A_0 of Fig. 7.12 coupled to an auxiliary heat reservoir A' which has an absolute temperature lower than that of the heat reservoir A.

To achieve the desired entropy compensation, we require that the entropy S^* of the total isolated system A^* consisting of A_0 and A' satisfy the condition

$$\Delta S^* = \Delta S + \Delta S' \geq 0. \tag{103}$$

To make it easy to satisfy this condition, we should like the entropy of A to decrease as little as possible for a given amount of heat q extracted from it, i.e., we should like the absolute temperature of A to be as high as possible. Similarly, we should like to waste as little energy as possible in the form of heat q' given to A' in order to achieve the largest possible compensating entropy increase $\Delta S'$; i.e., we should like the absolute temperature T' of the auxiliary reservoir to be as low as possible.

We have now designed a realizable engine of the general type illustrated in Fig. 7.13. To investigate its properties, we note first that the condition (103) becomes, by virtue of (101) and (102),

$$\Delta S^* = -\frac{q}{T} + \frac{q'}{T'} \geq 0. \tag{104}$$

Furthermore, the conservation of energy implies that the work w done by the engine in a cycle must be equal to

$$w = q - q'. \tag{105}$$

To get the maximum amount of work w from the engine, the heat q' given to the auxiliary reservoir A' should thus be as small as possible consistent with the purpose of achieving the compensating entropy increase $\Delta S'$. By virtue of (105), $q' = q - w$ so that (104) becomes

$$-\frac{q}{T} + \frac{q-w}{T'} \geq 0$$

$$\frac{w}{T'} \leq q\left(\frac{1}{T'} - \frac{1}{T}\right)$$

or

$$\boxed{\frac{w}{q} \leq 1 - \frac{T'}{T} = \frac{T-T'}{T}.} \tag{106}$$

In the hypothetical case of a perfect engine, all the heat extracted from the reservoir A would be transformed into work so that $w = q$. For the realizable engine which we have just discussed, $w < q$ since some heat q' must be given to the auxiliary reservoir A'. The ratio

$$\eta \equiv \frac{w}{q} = \frac{q-q'}{q} \tag{107}$$

is accordingly called the *efficiency* of the engine. This efficiency would be unity in the case of a perfect engine, and is less than unity for all realizable engines. Equation (106) provides then an expression for the maximum possible efficiency of a heat engine operating between two heat reservoirs of given absolute temperatures, i.e.,

$$\eta \leq \frac{T-T'}{T}. \tag{108}$$

As expected from our previous comments, this efficiency is greater the larger the temperature difference between the reservoirs.

Our highly industrial society is, of course, full of engines of various kinds. But none of them is perfect, i.e., each rejects heat to some auxiliary low-temperature reservoir, usually the surrounding atmosphere. For example, steam engines have condensers and gasoline engines have exhausts. The expression (108) provides a theoretical upper limit for the possible efficiency of an engine. Although this maximum theoretical efficiency is, in practice, not attained by real engines, it does provide a useful guide in engineering applications. For example, the desirability of using superheated steam in steam engines, instead of ordinary steam near 100°C, is based on the fact that the larger temperature difference between superheated steam and room temperature does, in accordance with (108), lead to an engine of increased efficiency.

Fig. 7.14 N. L. Sadi Carnot (1796–1832). In 1824, before heat had become generally recognized as a form of energy, the young French engineer Carnot published a penetrating theoretical analysis of heat engines. The later development of his ideas by Kelvin and by Clausius led to the macroscopic formulation of the second law of thermodynamics. (*From Sadi Carnot, "Reflections on the Motive Power of Fire," edited by E. Mendoza, reprinted by Dover Publications, Inc., New York, 1960.*)

From a theoretical point of view, it is worth noting that the maximum efficiency of an engine operating between two reservoirs of fixed temperatures is obtained if the equals sign holds in (108). This is only true if the equals sign holds in (103), i.e., if the process is quasi-static so that it does not lead to an entropy change. The relation (108) then asserts that no engine operating between the two given heat reservoirs can have an efficiency greater than that of an engine which operates between the same two reservoirs in a quasi-static manner. Furthermore (108) implies that *any* engine which operates between these two reservoirs in a quasi-static manner has the *same* efficiency; i.e.,

$$\text{for any quasi-static engine,} \qquad \eta = \frac{T - T'}{T}. \tag{109}$$

Biochemical synthesis

Let us give a simple application representative of the biological processes involved in the synthesis of macromolecules. The sugar molecule *glucose* has a ring structure built of six carbon atoms and is very important in all metabolism. The sugar molecule *fructose* has a different ring structure built of six carbon atoms. These two molecules can be combined to form the more complicated sugar molecule *sucrose* which consists of the glucose and fructose carbon rings joined together. The corresponding chemical reaction can be written in the form

$$\text{glucose} + \text{fructose} \rightleftarrows \text{sucrose} + H_2O. \tag{110}$$

Since all chemical reactions of interest take place at constant temperature and pressure, entropy changes involving the *entire* isolated system (including the reservoir which maintains the constant temperature and pressure) can be most conveniently expressed in terms of the Gibbs free energy G of the system under consideration. Measurements performed on the system consisting of the molecules in (110) show that, under standard conditions (one mole per liter concentrations of each of the reactants) the reaction (110) proceeding in the direction from left to right is accompanied by a free energy change $\Delta G = +0.24$ ev (electron volts). But in accordance with our previous discussion of Sec. 7.5, the Gibbs free energy of a system at constant temperature and pressure tends to *de*crease. Correspondingly, the reaction (110) tends to proceed from right to left so as to produce more of the simple molecules glucose and fructose at the expense of the more complex sucrose molecules. Thus the reaction (110) by itself cannot achieve the synthesis of sucrose. Indeed, in a solution containing the molecules of (110) in equilibrium, most of these would be the simple molecules glucose and fructose, rather than the more complex sucrose molecules.

In order to achieve the desired synthesis of sucrose, our principle of entropy compensation (100) suggests that we couple the reaction (110) with some other reaction accompanied by a free energy change $\Delta G'$ negative and large enough so that the total free energy change of both reactions *combined* satisfies the condition

$$\Delta G + \Delta G' \leq 0. \tag{111}$$

The reaction used most extensively by biological organisms to achieve such a negative value of $\Delta G'$ is one involving the molecule ATP (adenosine triphosphate) which can easily lose one of its weakly bound phosphate groups to be transformed into the molecule ADP (adenosine diphosphate) according to the reaction

$$\text{ATP} + \text{H}_2\text{O} \rightarrow \text{ADP} + \text{phosphate}. \quad (112)$$

The free energy change (under standard conditions) of this particular reaction is $\Delta G' = -0.30$ ev. This is sufficient to overcome the positive free energy change ΔG of the reaction (110); indeed,

$$\Delta G + \Delta G' = 0.24 - 0.30 = -0.06 \text{ ev}. \quad (113)$$

It should thus be possible to make the reactions (110) and (112) proceed simultaneously in order to achieve the desired synthesis of sucrose, provided that these two reactions can be suitably coupled to each other. This can be achieved by means of a common intermediate, the molecule *glucose 1-phosphate* which consists of a phosphate group attached to a glucose molecule. In the presence of suitable catalysts (enzymes) to make the reaction rate appreciable, the actual mechanism by which the synthesis is achieved in biological organisms consists thus of the following two sequential reactions:

$$\text{ATP} + \text{glucose} \rightarrow \text{ADP} + (\text{glucose 1-phosphate}),$$

$$(\text{glucose 1-phosphate}) + \text{fructose} \rightarrow \text{sucrose} + \text{phosphate}.$$

The sum of these reaction yields the net reaction

$$\text{ATP} + \text{glucose} + \text{fructose} \rightarrow \text{sucrose} + \text{ADP} + \text{phosphate}.$$

This is equivalent, as far as the initial and final macrostates are concerned, to the reactions (110) and (112) proceeding simultaneously. The synthesis of the more complex sucrose molecule is thus compensated by the breakdown of the ATP molecule into the simpler ADP molecule.

The basic principles underlying the synthesis of proteins from amino acids (or of DNA molecules, which carry the genetic information, from nucleic acids) are similar to those sketched in our simple example. The interested reader is referred to A. L. Lehninger's book listed in the references at the end of this chapter.

Summary of Definitions

generalized force The generalized force X_r, conjugate to an external parameter x of a system in a state r of energy E_r, is defined as $X_r \equiv \partial E_r/\partial x$.

Gibbs free energy If a system is in contact with a reservoir at the constant temperature T' and constant pressure p', its Gibbs free energy G is defined by

$$G \equiv \bar{E} - T'S + p'V$$

where $\bar{E}$ is the mean energy, S the entropy, and V the volume of the system.

phase A particular form of aggregation of the molecules of a substance.

latent heat The heat which must be absorbed to transform a given amount of one phase into an equivalent amount of another phase when the two phases are in equilibrium with each other.

vapor pressure The pressure of the gaseous phase in equilibrium with a liquid (or solid) at a specified temperature.

phase-equilibrium curve The curve of corresponding values of temperature and pressure where two phases can coexist in equilibrium with each other.

Clausius-Clapeyron equation The equation $dp/dT = \Delta S/\Delta V$ which relates the slope of a phase equilibrium curve to the entropy change ΔS and volume change ΔV between the two phases at the given temperature and pressure.

engine A device used to convert the internal energy of a system into work.

Important Relations

In any quasi-static process,

$$dS = \frac{đQ}{T}. \tag{i}$$

For an isolated system in equilibrium,

$$S = \text{maximum}, \tag{ii}$$

$$P \propto e^{S/k}. \tag{iii}$$

For a system in equilibrium with a reservoir at constant temperature T' and pressure p'

$$G = \text{minimum}, \tag{iv}$$

$$P \propto e^{-G/kT'}. \tag{v}$$

For equilibrium between two phases

$$g_1 = g_2; \tag{vi}$$

along a phase-equilibrium curve,

$$\frac{dp}{dT} = \frac{\Delta S}{\Delta V} = \frac{L}{T\,\Delta V}. \tag{vii}$$

Suggestions for Supplementary Reading

F. Reif, *Fundamentals of Statistical and Thermal Physics* (McGraw-Hill Book Company, New York, 1965). Chapter 5 deals with applications of the thermodynamic laws; Chap. 8 deals with equilibrium between different phases and with chemical equilibrium between different kinds of molecules.

Completely macroscopic discussions of classical thermodynamics:

M. W. Zemansky, *Heat and Thermodynamics*, 4th ed. (McGraw-Hill Book Company, New York, 1957).

E. Fermi, *Thermodynamics* (Dover Publications, Inc., New York, 1957).

Applications:

J. F. Sandfort, *Heat Engines* (Anchor Books, Doubleday & Company, Inc., Garden City, N.Y., 1962). A discussion of heat engines through history up to modern times.

A. L. Lehninger, *Bioenergetics* (W. A. Benjamin, Inc., New York, 1965). Note especially chaps. 1–4. This interesting book, dealing with applications to biology, is suitable even for those with almost no background in biology.

Historical and biographical accounts:

S. Carnot, *Reflections on the Motive Power of Fire*, edited by E. Mendoza (Dover Publications, Inc., New York, 1960). A reprint and translation of Carnot's original writings. This book also contains a short historical and biographical introduction by the editor.

Problems

7.1 *Alternative derivation of the equation of state of an ideal gas*

According to the result of Prob. 3.8, the number $\Omega(E)$ of states accessible to N atoms of a monatomic ideal gas, enclosed in a volume V and having an energy between E and $E + \delta E$, is given by the proportionality

$$\Omega \propto V^N E^{(3/2)N}.$$

Use this relation to calculate the mean pressure $\bar{p}$ of this gas by the general relation (15). Show that you thus obtain the familiar equation of state of an ideal gas.

7.2 *Adiabatic compression of a gas*

Consider a monatomic ideal gas which is thermally insulated. Suppose that, starting at a temperature of 400°K and at a pressure of one atmosphere, this gas is slowly compressed to one-third of its original volume.

(*a*) What is the final pressure of the gas?

(*b*) What is the final temperature of the gas?

7.3 *Work done on an ideal gas in a quasi-static adiabatic process*

A thermally insulated ideal gas has a molar specific heat c_V (at constant volume) independent of temperature. Suppose that this gas is compressed quasi-statically from an initial macrostate, where its volume is V_i and its mean pressure is $\bar{p}_i$, to a final macrostate, where its volume is V_f and its mean pressure is $\bar{p}_f$.

(*a*) Calculate directly the work done on the gas in this process, expressing your answer in terms of the initial and final pressures and volumes.

(*b*) Express your answer to part (*a*) in terms of the initial and final absolute temperatures T_i and T_f of the gas. Show that this result would follow immediately from a consideration of the change of internal energy of the gas.

7.4 *Specific heat difference $c_p - c_V$ of an ideal gas*

Consider an ideal gas enclosed in a vertical cylinder closed by a piston. The piston is free to move and supports a weight; thus the gas is always kept at the same pressure (equal to the weight of the piston divided by its area) irrespective of its volume.

(*a*) If the gas is kept at a constant pressure, use (43) to calculate the heat $đQ$ absorbed by it if its temperature is increased by an amount dT. Use this result to show that its molar specific heat c_p, measured at constant pressure, is related to its molar specific heat c_V, at constant volume, by $c_p = c_V + R$.

(*b*) What is the value of c_p for a monatomic gas such as helium?

(*c*) Show that the ratio c_p/c_V is equal to the quantity γ defined in (57). What is the value of this ratio in the case of a monatomic ideal gas?

7.5 *A quasi-static process involving an ideal gas*

An ideal diatomic gas at an absolute temperature T has an internal energy per mole equal to $\bar{E} = \frac{5}{2}RT$. A mole of this gas is taken quasi-statically first from the macrostate a to the macrostate b, and then from macrostate b to the macrostate c along the straight-line paths of Fig. 7.15.

(*a*) What is the molar heat capacity at constant volume of this gas?

(*b*) What is the work done by the gas in the process $a \to b \to c$?

(*c*) What is the heat absorbed by the gas in this process?

(*d*) What is its change of entropy in this process?

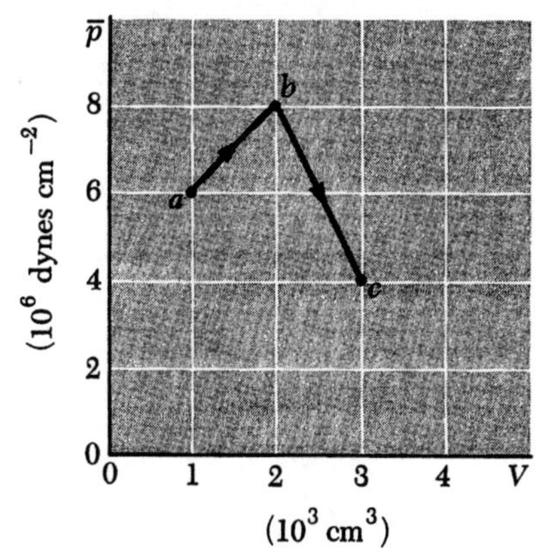

Fig. 7.15 A process indicated on a graph of mean pressure $\bar{p}$ versus volume V.

7.6 *Entropy change in an irreversible process*

Consider the gas of Prob. 5.8. Calculate the final entropy S of the gas in terms of its initial entropy S_0 before the piston was released. Show that the entropy change $\Delta S \equiv S - S_0$ is positive.

7.7 Equilibrium conditions for a system of fixed volume when it is in contact with a heat reservoir

Consider a system A whose only external parameter is its volume V which remains fixed. The system is in contact with a heat reservoir A' at the constant temperature T'.

(a) Use arguments similar to those of Sec. 7.5 to show that the equilibrium of A is characterized by the fact that the function

$$F \equiv \bar{E} - T'S$$

for this system must be a minimum. Here $\bar{E}$ is the mean energy and S the entropy of A. The function F is called its *Helmholtz free energy*.

(b) Show that the Gibbs free energy (76) of a system in contact with a reservoir at constant temperature T' *and* constant pressure p' can be expressed in terms of its Helmholtz free energy by the relation

$$G = F + p'V.$$

7.8 Triple point of ammonia

The vapor pressure $\bar{p}$ (in millimeters of mercury) of solid ammonia is given by $\ln \bar{p} = 23.03 - 3754/T$ while that of liquid ammonia is given by $\ln \bar{p} = 19.49 - 3063/T$. Use this information to answer the following questions:

(a) What is the temperature of the triple point of ammonia?

(b) What are the latent heats of sublimation and vaporization of ammonia at the triple point?

(c) What is the latent heat of melting of ammonia at the triple point?

7.9 Melting curve of helium near absolute zero

Helium remains a liquid down to absolute zero at atmospheric pressure, but becomes a solid at sufficiently high pressures. The density of the solid is, as usual, greater than that of the liquid. Consider the phase-equilibrium curve between the solid and liquid. In the limit as $T \to 0$, is the slope dp/dT of this curve positive, zero, or negative? (*Suggestion:* Use your knowledge of the general limiting behavior of the entropy as $T \to 0$.)

7.10 Intensity of an atomic beam produced by a vapor

An atomic beam of sodium (Na) atoms is produced by maintaining liquid sodium in an enclosure at some elevated temperature T. Atoms of Na from the vapor above the liquid escape by effusion through a narrow slit in the enclosure and thus give rise to an atomic beam of intensity I. (The intensity I is defined as the number of atoms in the beam which cross unit area per unit time.) The latent heat of vaporization per mole of liquid Na into a vapor of Na atoms is L. To estimate how sensitive the beam intensity is to fluctuations in the temperature of the enclosure, calculate the relative intensity change $I^{-1}(dI/dT)$ in terms of L and the absolute temperature T of the enclosure.

7.11 *Attainment of low temperatures by pumping on a liquid*

Liquid helium boils at a temperature T_0 (4.2°K) when its vapor pressure is equal to p_0, where $p_0 = 1$ atmosphere or 760 mm of mercury. The latent heat of vaporization per mole of the liquid is equal to L and approximately independent of temperature. ($L \approx 85$ joules/mole). The liquid is contained within a dewar which serves to insulate it thermally from the room-temperature surroundings. Since the insulation is not perfect, an amount of heat Q per second flows into the liquid and evaporates some of it. (This heat influx Q is essentially constant, independent of whether the temperature of the liquid is T_0 or less.) In order to reach low temperatures one can reduce the pressure of the He vapor over the liquid by pumping it away with a pump at room temperature T_r. (By the time it reaches the pump, the He vapor has warmed up to room temperature.) The pump has a maximum pumping speed such that it can remove a constant volume $\mathcal{V}$ of gas per second, irrespective of the pressure of the gas. (This is a characteristic feature of an ordinary rotary mechanical pump; it simply sweeps out a fixed volume of gas per revolution.)

(*a*) Calculate the minimum vapor pressure p_m which this pump can maintain over the surface of the liquid if the heat influx is Q.

(*b*) If the liquid is thus maintained in equilibrium with its vapor at this pressure p_m, calculate its approximate temperature T_m.

(*c*) To estimate how low a vapor pressure p_m, or how low a temperature T_m, one can achieve in practice, suppose that one has available a big pump with a pumping speed $\mathcal{V}$ of 70 liters/sec (1 liter $= 10^3$ cm^3). A typical heat influx is such that it evaporates about 50 cm^3 of liquid helium per hour (the density of the liquid being 0.145 gm/cm^3). Estimate the lowest temperature T_m which can be achieved in this experimental setup.

7.12 *Equilibrium between phases discussed in terms of chemical potential*

Consider a system consisting of two phases 1 and 2 maintained at a constant temperature T and pressure p by being in contact with a suitable reservoir. The total Gibbs free energy G of this system at the given temperature and pressure is then a function of the number N_1 of molecules in phase 1 and the number N_2 of molecules in phase 2; thus $G = G(N_1, N_2)$.

(*a*) Using very simple mathematics, show that the change ΔG in the free energy resulting from small changes ΔN_1 and ΔN_2 in the number of molecules in the two phases can be written in the form

$$\Delta G = \mu_1 \Delta N_1 + \mu_2 \Delta N_2 \qquad \text{(i)}$$

if one uses the convenient abbreviation

$$\mu_i \equiv \frac{\partial G}{\partial N_i}. \qquad \text{(ii)}$$

The quantity μ_i is called the *chemical potential* per molecule of the *i*th phase.

(*b*) Since G must be a minimum when the phases are in equilibrium, ΔG must then vanish if one molecule of phase 1 is transferred to phase 2. Show that the relation (i) thus yields the equilibrium condition

$$\mu_1 = \mu_2. \tag{iii}$$

(*c*) Using the relation (86), show that $\mu_i = g_i$, the Gibbs free energy per molecule of phase i. The result (iii) agrees thus with (87).

7.13 *Condition of chemical equilibrium*

Consider a chemical reaction such as

$$2\,CO_2 \rightleftarrows 2\,CO + O_2.$$

To avoid cumbersome writing, let us denote the CO_2 molecule by A_1, the CO molecule by A_2, and the O_2 molecule by A_3. Then the above chemical reaction becomes

$$2A_1 \rightleftarrows 2A_2 + A_3. \tag{i}$$

The system consisting of A_1, A_2, and A_3 molecules is supposed to be maintained at a constant temperature and pressure. If we denote by N_i the number of molecules of type i, the Gibbs free energy of this system is then a function of these numbers so that

$$G = G(N_1,N_2,N_3).$$

Since G must be minimum in equilibrium, ΔG must then vanish if, in accordance with the reaction (i), two A_1 molecules are transformed into two A_2 molecules and one A_3 molecule. By reasoning similar to that of the preceding problem, show that this equilibrium condition can be written in the form

$$2\mu_1 = 2\mu_2 + \mu_3 \tag{ii}$$

where

$$\mu_i \equiv \frac{\partial G}{\partial N_i}$$

is called the *chemical potential* per molecule of the type i.

7.14 *Refrigerators*

A refrigerator is a device which extracts heat from a system A and rejects it to some other system A' at a higher absolute temperature. Suppose that A is a heat reservoir at the temperature T and A' is another reservoir at the temperature T'.

(*a*) Show that, if $T' > T$, the transfer of heat q from A to A' involves a net decrease of entropy of the total system and is thus not realizable without auxiliary systems.

(*b*) If one wants to extract heat q from A and thus reduce its entropy, one must increase the entropy of A' by more than a compensating amount by rejecting to it an amount of heat q' greater than q. This can be accomplished by letting some system B do an amount of work w on the refrigerator mechanism M working in a cycle. One thus obtains the schematic diagram shown in Fig. 7.16 and understands why kitchen refrigerators need an external source of power to make them function. Use entropy considerations to show that

$$\frac{q}{q'} \leq \frac{T}{T'}.$$

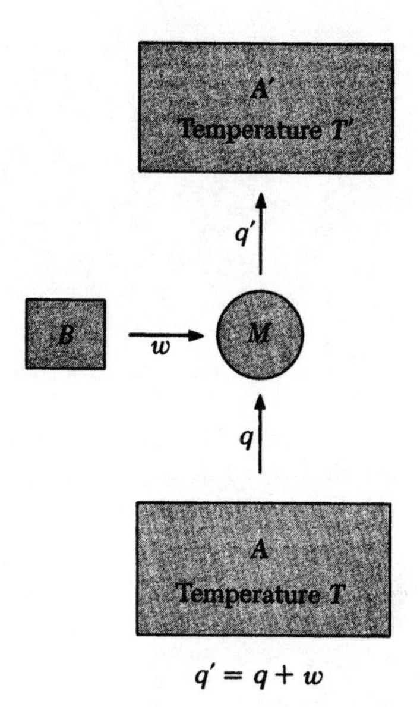

$q' = q + w$

Fig. 7.16 Schematic diagram of a refrigerator.

7.15 *Heat pumps*

Refrigeration cycles have been developed for heating buildings. The procedure is to design a device which absorbs heat from the surrounding earth or air outside the house, and then delivers heat at a higher temperature to the interior of the building. (Such a device is called a *heat pump.*)

(*a*) If a device is used in this way, operating between the outside absolute temperature T_0 and an interior absolute temperature T_i, what would be the maximum number of kilowatt-hours of heat that could be supplied to the building for every kilowatt-hour of electrical energy needed to operate the device? (*Suggestion:* Use entropy considerations.)

(*b*) Obtain a numerical answer for the case that the outside temperature is 0°C and the interior temperature is 25°C.

(*c*) Compare the cost of power necessary to operate this heat pump with the cost of power necessary to provide the same amount of heat to the interior of the house by means of an electrical resistance heater.

7.16 *Maximum work obtainable from two identical systems*

Consider two identical bodies A_1 and A_2, each characterized by a heat capacity C which is temperature-independent. The bodies are initially at temperatures T_1 and T_2, respectively, where $T_1 > T_2$. It is desired to operate an engine between A_1 and A_2 so as to convert some of their internal energy into work. As a result of the operation of the engine, the bodies ultimately will attain a common final temperature T_f.

(*a*) What is the total amount of work W done by the engine? Express your answer in terms of C, T_1, T_2, and T_f.

(*b*) Use arguments based upon entropy considerations to derive an inequality relating T_f to the initial temperatures T_1 and T_2.

(*c*) For given initial temperatures T_1 and T_2, what is the maximum amount of work obtainable from the engine?

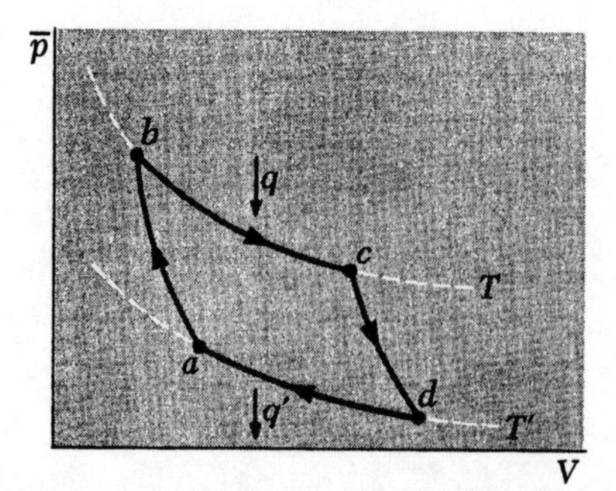

Fig. 7.17 Operation of a Carnot engine indicated on a diagram of mean pressure $\bar{p}$ versus volume V.

*7.17 *Ideal gas Carnot engine*

We wish to show explicitly that it is possible to design a highly idealized engine which can, in a cycle, extract heat q from some heat reservoir A at the absolute temperature T, reject heat q' to some heat reservoir A' at the lower absolute temperature T', and perform useful work $w = q - q'$ in the process. The simplest such engine is one (first considered by Sadi Carnot in 1824) which operates in a quasi-static manner. The cycle consists of four steps which take the engine from its initial macrostate a back to this state after passing through the intermediate macrostates b, c, d. The engine consists of ν moles of an ideal gas contained in a cylinder closed by a piston. The volume of the gas is denoted by V, its mean pressure by $\bar{p}$. The four steps of the cycle, shown in Fig. 7.17, are then as follows:

Step 1. $a \to b$: The engine, originally at the temperature T', is thermally insulated. Its volume is now decreased slowly from its initial value V_a until it reaches a value V_b where the temperature of the engine is T.

Step 2. $b \to c$: The engine is placed in thermal contact with the heat reservoir A at the temperature T. Its volume is now slowly changed from V_b to V_c, the engine remaining at the temperature T and absorbing some heat q from A.

Step 3. $c \to d$: The engine is again thermally insulated. Its volume is now increased slowly from V_c until it reaches a value V_d where the temperature of the engine is T'.

Step 4. $d \to a$: The engine is now placed in thermal contact with the heat reservoir A' at the temperature T'. Its volume is now slowly changed from V_d back to its original value V_a, the engine remaining at the temperature T' and rejecting some heat q' to A'.

Answer the following questions:

(*a*) What is the heat q absorbed in step 2? Express your answer in terms of V_b, V_c, and T.

(*b*) What is the heat q' rejected in step 4? Express your answer in terms of V_d, V_a, and T'.

(*c*) Calculate the ratio V_b/V_a in step 1 and the ratio V_d/V_c in step 3, and show that V_b/V_a is related to V_d/V_c.

(*d*) Use the preceding answer to calculate the ratio q/q' in terms of T and T'.

(*e*) Calculate the efficiency η of the engine and show that it agrees with the general result (109) valid for any quasi-static engine.

7.18 *Efficiency of a gasoline engine

In a gasoline engine a gas consisting of air, to which there is admixed a small amount of gasoline, is introduced in a cylinder closed by a movable piston. The gas is then subjected to a cyclic process which can be represented *approximately* by the steps shown in Fig. 7.18 where V denotes the volume and $\bar{p}$ the mean pressure of the gas. Here $a \to b$ represents the adiabatic compression of the air-gasoline mixture, $b \to c$ the rise in pressure due to the explosion of the mixture at constant volume (since the explosion is too rapid to allow the piston time to move), $c \to d$ the adiabatic expansion of the mixture during which it does useful work by moving the piston, and $d \to a$ the final cooling down of the gas at constant volume during the exhaust phase of the cycle.

To achieve an approximate analysis, assume that the previous cycle is carried out quasi-statically for a fixed amount of ideal gas having a constant molar specific heat c_V. Calculate thus for this engine its efficiency η (the ratio of work performed by the engine to its heat intake q_1). Express your answer in terms of V_1, V_2, and the quantity $\gamma \equiv 1 + R/c_V$.

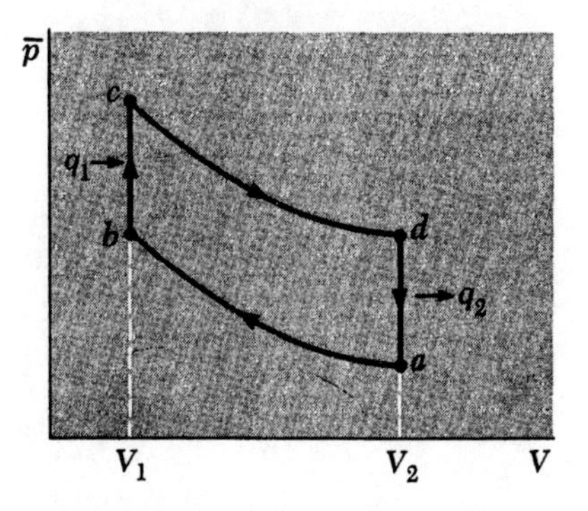

Fig. 7.18 Approximate representation of a gasoline-engine cycle on a diagram of mean pressure $\bar{p}$ versus volume V.

Chapter 8

Elementary Kinetic Theory of Transport Processes

Chapter 8 Elementary Kinetic Theory of Transport Processes

Up to now we have been concerned almost exclusively with systems in equilibrium. The postulate of equal a priori probabilities then provides a simple basis for the general quantitative discussion of such systems. In particular, it is not necessary to consider in detail the interactions which bring about the equilibrium situation; it is quite sufficient to know that such interactions exist. But although equilibrium situations are very important, they are rather special. Indeed, many problems of great physical interest deal with macroscopic systems which are *not* in equilibrium. Accordingly we shall devote this last chapter to a brief discussion of some of the considerations applicable to the simplest nonequilibrium situations.

In dealing with systems which are not in equilibrium, it is ordinarily necessary to investigate the specific interactions effective in bringing about the ultimate equilibrium of the system. As a result, the discussion of nonequilibrium processes tends to be considerably more difficult than that of equilibrium situations. The discussion, however, becomes relatively simple in the case of dilute gases. We shall, therefore, focus attention on dilute gases and shall treat these by the very simplest approximation methods. Although our calculations will therefore not be rigorously quantitative, they will yield many valuable insights and useful results by arguments of almost trivial simplicity. Such arguments are actually very useful in a wide variety of situations. First, they may apply equally well in other contexts, for example, in the discussion of nonequilibrium processes in solids. Second, they often yield fairly good numerical estimates and predict correctly the dependences on all significant parameters (such as temperature or pressure) in cases which are so complex as to make rigorous calculations very difficult.

Molecules in a gas interact with each other through mutual collisions. If such a gas is not in equilibrium initially, these collisions are responsible for bringing about the ultimate equilibrium situation where the Maxwell velocity distribution prevails. The discussion of a gas is particularly simple if the gas is sufficiently dilute so that the following conditions are satisfied:

(i) Each molecule spends a relatively large fraction of its time at distances far from other molecules so that it does not interact with them. In short, the time *between* collisions is much greater than the time involved *in* a collision.

(ii) The probability that three or more molecules approach each other sufficiently closely so as to interact with each other *simultaneously* is negligibly small compared to the probability that only two molecules approach each other sufficiently closely to interact. In short, triple collisions occur very rarely compared to two-particle collisions. Hence the analysis of collisions can be reduced to the relatively simple mechanical problem of only *two* interacting particles.

(iii) The mean separation between molecules is large compared to the typical de Broglie wavelength of a molecule. The behavior of a molecule between collisions can then be described adequately by the motion of a wave packet or classical particle trajectory, even though a quantum-mechanical calculation may be necessary to describe an actual collision between two molecules.

8.1 *Mean Free Path*

Let us begin by considering the collisions between molecules in a dilute gas. Our remarks are merely intended to review and refine some of the comments already made in Sec. 1.6. The collisions of a molecule with other molecules can be considered to occur at random. The probability that a molecule suffers a collision with some other molecule during any small time interval dt is thus assumed to be independent of its history of past collisions. Focus attention on a particular molecule at *any* instant of time. It has then some probability $P(t)$ of continuing to travel some time t before happening to collide with another molecule. The *mean* time τ which the molecule travels before suffering its next collision is called the *mean free time* of the molecule. (Since there is nothing special about the future compared to the past, τ is also the mean time which the molecule has traveled after suffering its preceding collision.) Similarly, the *mean* distance l which the molecule travels before suffering its next collision (or equivalently, the mean distance which it has traveled after its preceding collision) is called the *mean free path* of the molecule. Since all arguments of this chapter are intended to be approximate, we shall neglect details of the molecular velocity distribution. Accordingly we shall thus simply treat the molecules as if they all traveled in random directions with the same speed, equal to their mean speed $\bar{v}$. With these approximations, the mean free path l and mean free time τ are simply related by

$$l = \bar{v}\tau. \tag{1}$$

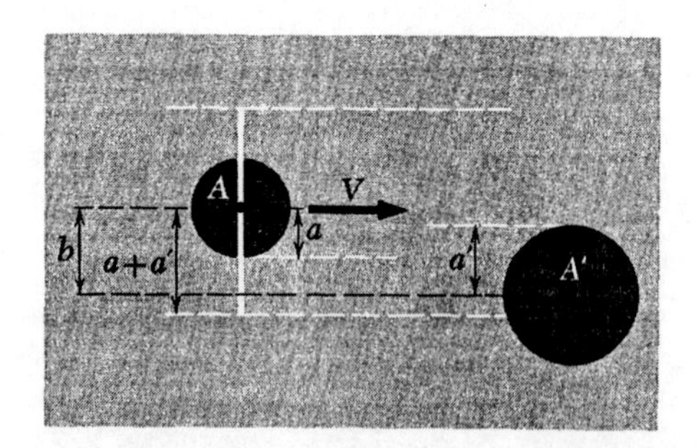

Fig. 8.1 Diagram illustrating a collision between two hard spheres having radii a and a'. The solid white line indicates an imaginary circular disk which is carried by the sphere of radius a and has a radius $(a + a')$.

The magnitude of the mean free path can be readily estimated as in Sec. 1.6 by examining the molecular collisions in greater detail. Consider thus a particular molecule A which approaches another molecule A' with a *relative* velocity **V** in such a way that the centers of the molecules would approach within a distance b of each other if they remained undeflected (see Fig. 8.1). If the forces between the two molecules are similar to those between two hard spheres of respective radii a and a', the molecules exert no force on each other as long as the distance R between their centers is such that $R > (a + a')$, but exert an extremely large force on each other if $R < (a + a')$. In this case it is clear from Fig. 8.1 that no forces between the molecules come into play as long as $b > (a + a')$, but that strong forces between them occur if $b < (a + a')$. In the latter case the velocities of the molecules are changed appreciably by their encounter and the molecules are said to have been *scattered* or to have *suffered a collision.* The condition necessary for the occurrence of a collision can be easily visualized by imagining that the molecule A carries with it a disk of radius $(a + a')$, this disk being centered about the center of the molecule A and oriented perpendicular to the relative velocity **V**. A collision between the two molecules will then occur only if the center of molecule A' lies within the volume swept out by the area σ of the imaginary circular disk carried by A. Here

$$\sigma = \pi(a + a')^2 \tag{2}$$

or, if the molecules are identical so that $a' = a$,

$$\sigma = \pi d^2 \tag{3}$$

where $d = 2a$ is the diameter of a molecule. The area σ is called the *total scattering cross section* characterizing the collision between the two molecules.

The forces between real molecules, while similar to those between hard spheres, are actually more complicated. The similarity to hard spheres is due to the very strong repulsion that occurs between two real molecules whenever they approach each other too closely; but, in addition, there also exists between two real molecules a weak attraction when they are somewhat further apart. The collision between two real molecules can still be described rigorously in terms of an effective area σ, the cross section for a collision, which can be calculated by the laws of quantum mechanics if the forces between the molecules are known. But simple relations of the form (2) or (3) no longer apply and the cross section σ is, in general, also a function of the relative speed V of the molecules. For purposes of making approximate estimates, however, the relations (2) or (3) are still useful (although the concept of a molecular radius is not well defined).

Let us now calculate approximately the mean free time τ of a molecule in a dilute gas consisting of n identical molecules per unit volume. We assume that the total scattering cross section σ is known. Focus attention on some particular molecule A at any instant. This molecule moves with a mean relative speed $\bar{V}$ with respect to any other typical molecule A' by which it can be scattered. The imaginary disk of area σ, carried by the molecule A traveling toward another molecule A', sweeps out in a time t a volume $\sigma(\bar{V}t)$. Hence this time t will be equal to the mean free time τ if the volume thus swept out contains, on the average, one other molecule; i.e., if

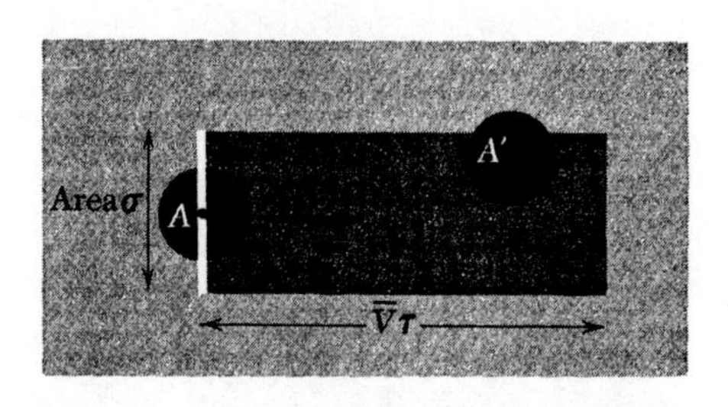

Fig. 8.2 Diagram illustrating the collisions suffered by a particular molecule A when it encounters another molecule whose center is located within the volume swept out by the area σ of the imaginary disk carried by A.

$$(\sigma\bar{V}\tau)n = 1.$$

Hence

$$\boxed{\tau = \frac{1}{n\sigma\bar{V}}}. \tag{4}$$

This result is eminently plausible. It merely asserts that the mean free time τ of a molecule is small (or equivalently, its collision rate τ^{-1} is large) if the number of molecules per unit volume is large so that there are more molecules with which a given molecule can collide; if the molecular diameter (or σ) is large so that any two molecules are more likely to scatter each other; and if the mean speed of the molecules relative to each other is large so that molecules are likely to encounter each other frequently.

According to (1), the mean free path l is thus given by

$$l = \bar{v}\tau = \frac{\bar{v}}{\bar{V}}\frac{1}{n\sigma}. \tag{5}$$

Since both of the colliding molecules move, their mean relative speed $\bar{V}$ is slightly different from the mean speed $\bar{v}$ of an individual molecule; hence $\bar{v}/\bar{V}$ differs somewhat from unity. To explore the difference, consider two molecules A and A' with respective velocities $\mathbf{v}$ and $\mathbf{v}'$. The velocity $\mathbf{V}$ of A relative to A' is then equal to

$$\mathbf{V} = \mathbf{v} - \mathbf{v}'.$$

Thus

$$\mathbf{V}^2 = \mathbf{v}^2 + \mathbf{v}'^2 - 2\mathbf{v}\cdot\mathbf{v}'. \tag{6}$$

If we take the average of both sides of this equation, $\overline{\mathbf{v} \cdot \mathbf{v}'} = 0$ since the cosine of the angle between $\mathbf{v}$ and $\mathbf{v}'$ is as likely to be positive as negative for molecules moving in random directions. Thus (6) becomes

$$\overline{V^2} = \overline{v^2} + \overline{v'^2}.$$

Neglecting the distinction between the average of a square and the square of an average (i.e., between a root-mean-square value and a mean value), this relation becomes approximately

$$\overline{V}^2 \approx \overline{v}^2 + \overline{v}'^2. \tag{7}$$

When all the molecules are identical, $\overline{v} = \overline{v}'$ and (7) reduces to

$$\overline{V} \approx \sqrt{2}\,\overline{v}. \tag{8}$$

Thus (5) becomes†

$$\boxed{l \approx \frac{1}{\sqrt{2}\,n\sigma}.} \tag{9}$$

The equation of state of an ideal gas allows us to express n in terms of the mean pressure $\overline{p}$ and absolute temperature T of the gas. Thus $\overline{p} = nkT$ and (9) becomes

$$l \approx \frac{kT}{\sqrt{2}\,\sigma\overline{p}}. \tag{10}$$

At a given temperature, the mean free path is thus inversely proportional to the pressure of the gas.

The magnitude of the mean free path in a gas at room temperature ($T \approx 300°$K) and atmospheric pressure ($\overline{p} \approx 10^6$ dynes/cm^2) can be readily estimated from (10). Using as a typical molecular radius $a \sim 10^{-8}$ cm, we obtain $\sigma \sim 12 \times 10^{-16}$ cm^2 and

$$l \sim 2 \times 10^{-5} \text{ cm}. \tag{11}$$

Since the mean speed $\overline{v}$ of a molecule is, by (6.33) or (1.30), of the order of 4×10^4 cm/sec, the mean free time of a molecule is

$$\tau = \frac{l}{\overline{v}} \sim 5 \times 10^{-10} \text{ sec}.$$

Thus a molecule collides approximately $\tau^{-1} \sim 10^9$ times per second with other molecules; this is a frequency corresponding to the micro-

† This relation is more accurate than the estimate (1.30) of Chap. 1 and is, indeed, an exact result for a gas of hard-sphere molecules moving with a Maxwell velocity distribution.

wave region of the electromagnetic spectrum. By virtue of (11) it also follows that

$$l \gg d \tag{12}$$

where $d \sim 10^{-8}$ cm is the molecular diameter. The relation (12) implies that gases under ordinary conditions are indeed sufficiently dilute so that a molecule travels a relatively long distance before encountering another molecule.

8.2 *Viscosity and Transport of Momentum*

Consider a macroscopic object which is immersed in a fluid (liquid or gas) at rest and which is not acted on by any external forces. If the object is in equilibrium, it is also at rest. On the other hand, if the object *is* moving through the fluid, it is not in equilibrium. The molecular interactions responsible for bringing about the equilibrium situation then manifest themselves by producing on a macroscopic scale a net frictional force acting on the moving object in such a way as to slow it down. This force is, to good approximation, proportional to the velocity of the object; it thus vanishes properly when the object is at rest. The actual magnitude of the force depends on a property of the fluid which is called its *viscosity*. Thus the force on the same object is much greater in molasses than in water; correspondingly, molasses is said to be much more viscous than water. We shall now define the concept of viscosity more precisely and try to illuminate its microscopic origin in the case of a dilute gas.

Definition of the coefficient of viscosity

Consider any fluid (liquid or gas). Imagine in this fluid some plane with its normal pointing along the z direction. Then the fluid below this plane (i.e., on the side of smaller z) is found to exert some mean force per unit area (or mean *stress*) $\mathbf{P}_z$ on the fluid above the plane. Conversely, it follows by Newton's third law that the fluid above the plane exerts a mean stress $-\mathbf{P}_z$ on the fluid below the plane. The mean stress normal to the plane, i.e., the z component of $\mathbf{P}_z$, measures just the mean pressure $\bar{p}$ in the fluid; to be precise, $P_{zz} = \bar{p}$. When the fluid is in equilibrium, so that it is at rest or moving with *uniform* velocity throughout, there is, by symmetry, no mean component of stress *parallel* to the plane. Thus $P_{zx} = 0$. Note that the quantity P_{zx} is labeled by two indices, the first of them designating the orientation

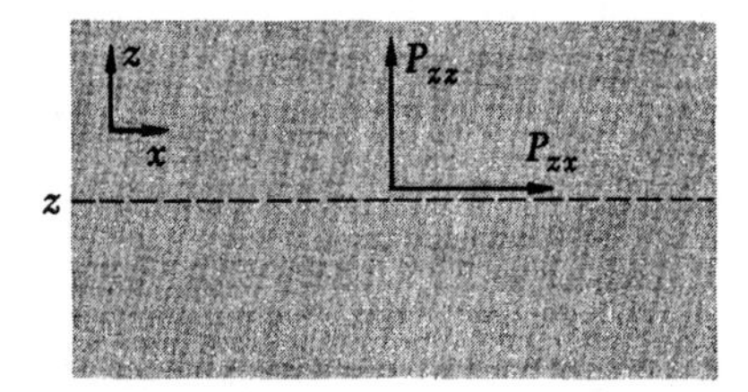

Fig. 8.3 A plane $z =$ constant in a fluid. The fluid below the plane exerts a force $\mathbf{P}_z$ on the fluid above it.

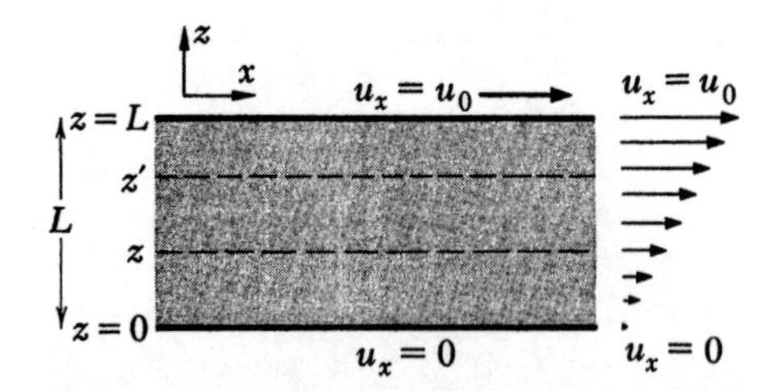

Fig. 8.4 A fluid contained between two plates. The lower plate is at rest while the upper one moves with velocity u_0 in the x direction; there exists then a velocity gradient $(\partial u_x/\partial z)$ in the fluid.

of the plane and the second the particular component of the force exerted across this plane.†

Consider now a simple nonequilibrium situation where the mean velocity $\mathbf{u}$ of the fluid (i.e., its macroscopic flow velocity) is not the same throughout the fluid. To be specific, consider the case where the fluid has a time-independent mean velocity u_x in the x direction, the magnitude of u_x depending on z so that $u_x = u_x(z)$. This kind of situation could be produced if the fluid is contained between two plates a distance L apart, the plate at $z = 0$ being stationary and the plate at $z = L$ moving in the x direction with constant velocity u_0. The layers of fluid immediately adjacent to the plates assume, to a good approximation, the respective velocities of the plates. Layers of fluid between the plates then have different mean velocities u_x, varying in magnitude between 0 and u_0. In this case the fluid exerts a tangential force on the moving plate, tending to slow it down so as to restore the equilibrium situation.

More generally, any layer of fluid below a plane $z =$ constant exerts a tangential stress P_{zx} on the fluid above it; i.e.,

$P_{zx} \equiv$ the mean force, in the x direction and per unit area of the plane, which the fluid below the plane exerts on the fluid above the plane. (13)

We have already seen that $P_{zx} = 0$ in the equilibrium situation where $u_x(z)$ does *not* depend on z. In the present nonequilibrium case where $\partial u_x/\partial z \neq 0$ one expects, therefore, that P_{zx} should be some function of derivatives of u_x with respect to z, this function being such that it vanishes when u_x is independent of z. But if $\partial u_x/\partial z$ is assumed to be relatively small, the leading term in a Taylor series expansion of P_{zx} should be an adequate approximation, i.e., one expects a linear relation of the form

$$\boxed{P_{zx} = -\eta \frac{\partial u_x}{\partial z}.} \tag{14}$$

Here the constant of proportionality η is called the *coefficient of viscosity* of the fluid. If u_x increases with increasing z, then the fluid below the plane tends to slow down the fluid above the plane and thus exerts on it a force in the $-x$ direction; i.e., if $\partial u_x/\partial z > 0$, then $P_{zx} < 0$. Hence the minus sign was introduced explicitly in (14) so as

† The quantity $P_{\alpha\gamma}$ (where α and γ can denote x, y, or z) is called the *pressure tensor*.

to make the coefficient η a positive quantity. By (14), the coefficient η is expressed in cgs units of gm cm^{-1} sec^{-1}.† The proportionality (14) between the stress P_{zx} and the velocity gradient $\partial u_x/\partial z$ is experimentally well satisfied by most liquids and gases if the velocity gradient is not too large.

Remark

Note the various forces acting in the x direction of the simple geometrical arrangement of Fig. 8.4. The fluid below the plane labeled by z exerts a force P_{zx} per unit area on the fluid above it. Since the fluid between this plane and any other plane labeled by z' moves without acceleration, the fluid above z' must exert a force $-P_{zx}$ per unit area on the fluid below z'. By Newton's third law, the fluid below z' thus also exerts a force P_{zx} per unit area on the fluid above z'. Thus the same force P_{zx} per unit area acts on the fluid above any plane and also on the top plate. Since P_{zx} is a constant independent of z, it also follows by (14) that $\partial u_x/\partial z =$ constant so that

$$\frac{\partial u_x}{\partial z} = \frac{u_0}{L}$$

and

$$P_{zx} = -\eta \frac{u_0}{L}.$$

Calculation of the coefficient of viscosity for a dilute gas

In the simple case of a dilute gas, the coefficient of viscosity can be calculated quite readily on the basis of microscopic considerations. Suppose that the gas has a mean velocity component u_x (which is assumed to be small compared to the mean thermal speed of the molecules) and that u_x is a function of z. Now consider any plane $z =$ constant. What is the microscopic origin of the stress P_{zx} acting across this plane? Qualitatively it can be understood by noting that molecules above the plane labeled by z in Fig. 8.4 have a somewhat larger x component of momentum than molecules below this plane. As molecules cross back and forth across this plane, they carry this x component of momentum with them. Hence the gas below the plane *gains* momentum in the x direction because molecules coming from above the plane carry a larger x component of momentum with them. Conversely, the gas above the plane *loses* momentum in the x direction because molecules coming from below the plane carry a smaller x component of momentum with them. But by Newton's second law, the force acting on a system is equal to the rate of change of momentum of this system. Hence {the force exerted on the gas above a plane by the gas below it}

† This combination of units is sometimes called the *poise* in honor of the physicist Poiseuille.

is merely equal to {the net momentum gain per unit time of the gas above the plane from the gas below it}. Hence the force P_{zx} of (13) is given by

$$P_{zx} = \text{the mean increase, per unit time and per unit area of the plane, of the } x \text{ component of momentum of the gas above the plane due to the net transport of momentum by molecules crossing this plane.} \tag{15}$$

Illustrative remark

An analogy may serve to illuminate this mechanism of viscosity by momentum transfer. Suppose two railroad trains move side by side along parallel tracks, the speed of one train being greater than that of the other. Imagine that workers on each train constantly pick up sandbags from their train and throw them onto the other train. There is then a transfer of momentum between the trains so that the slower train tends to be accelerated and the faster train to be decelerated.

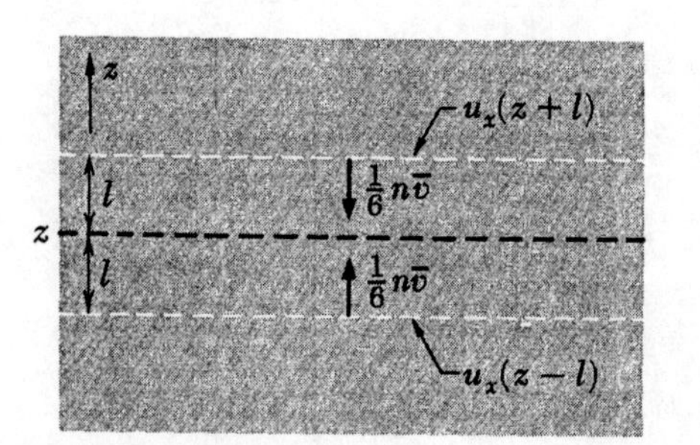

Fig. 8.5 Momentum transport by molecules crossing a plane.

To give an approximate simple calculation of the coefficient of viscosity, let us assume that all molecules move with the same speed equal to their mean speed $\bar{v}$. If there are n molecules per unit volume, one-third of them have velocities predominantly along the z direction. Half of these, or $\frac{1}{6}n$ molecules per unit volume, have a velocity $\bar{v}$ in the $+z$ direction, while the other half have a velocity $\bar{v}$ in the $-z$ direction. Consider now the plane labeled by z. There are then $\frac{1}{6}n\bar{v}$ molecules which in unit time cross a unit area of this plane from below; similarly, there are $\frac{1}{6}n\bar{v}$ molecules which in unit time cross a unit area of this plane from above. But the definition of the mean free path l implies that molecules which cross the plane from below have, on the average, experienced their preceding collision at a distance l below the plane. Since the mean velocity $u_x = u_x(z)$ is a function of z, the molecules at this position $(z - l)$ had, on the average, a mean x component of velocity $u_x(z - l)$. Thus each such molecule of mass m transports across the plane a mean x component of momentum $m\, u_x(z - l)$. Hence we conclude that†

$$\begin{bmatrix}\text{the mean } x \text{ component of momen-}\\ \text{tum transported per unit time per}\\ \text{unit area across the plane in the}\\ \textit{upward} \text{ direction}\end{bmatrix} = (\tfrac{1}{6}n\bar{v})[m\, u_x(z - l)]. \tag{16}$$

† To avoid any possible confusion, keep in mind that the symbol $u_x(z - l)$ denotes the value of the mean velocity u_x at the position $(z - l)$; it does *not* denote a product.

Similarly, in considering molecules coming from above the plane where they suffered their preceding collision at $(z + l)$, we obtain for

$$\begin{bmatrix}\text{the mean } x \text{ component of momen-}\\ \text{tum transported per unit time per}\\ \text{unit area across the plane in the}\\ \textit{down}\text{ward direction}\end{bmatrix} = (\tfrac{1}{6}n\bar{v})[m\, u_x(z + l)]. \quad (17)$$

By subtracting (17) from (16), we obtain the *net* molecular transport of mean x component of momentum per unit time per unit area from below to above the plane, i.e., the stress P_{zx} described in (15) or (13). Thus

$$P_{zx} = (\tfrac{1}{6}n\bar{v})[m\, u_x(z - l)] - (\tfrac{1}{6}n\bar{v})[m\, u_x(z + l)]$$

or

$$P_{zx} = \tfrac{1}{6}n\bar{v}m[u_x(z - l) - u_x(z + l)]. \quad (18)$$

But since the mean free path l is very small (compared to dimensions over which the velocity gradient $\partial u_x/\partial z$ varies appreciably), we can write to excellent approximation

$$u_x(z + l) = u_x(z) + \frac{\partial u_x}{\partial z} l$$

and

$$u_x(z - l) = u_x(z) - \frac{\partial u_x}{\partial z} l.$$

Hence

$$P_{zx} = \tfrac{1}{6}n\bar{v}m\left(-2\frac{\partial u_x}{\partial z} l\right) \equiv -\eta \frac{\partial u_x}{\partial z} \quad (19)$$

where

$$\boxed{\eta = \tfrac{1}{3}n\bar{v}ml.} \quad (20)$$

The relation (19) shows that P_{zx} is indeed proportional to the velocity gradient $\partial u_x/\partial z$, as expected by (14); furthermore, (20) provides an explicit approximate expression for the viscosity coefficient η in terms of the microscopic parameters characterizing the molecules of the gas.

Our calculation has been very simplified and has made no attempt to calculate carefully the pertinent averages of various quantities. Hence the factor $\frac{1}{3}$ in (20) is not to be trusted too much; the constant of proportionality given by a more careful calculation might be somewhat different. On the other hand, the essential dependence of η on the parameters n, $\bar{v}$, m, and l ought to be correct.

Discussion

The result (20) leads to some interesting predictions. By (9),

$$l = \frac{1}{\sqrt{2}n\sigma}. \tag{21}$$

Thus the factor n cancels in (20) and we are left with

$$\eta = \frac{1}{3\sqrt{2}}\frac{m}{\sigma}\bar{v}. \tag{22}$$

A sufficiently accurate result for the mean speed $\bar{v}$ can be obtained from the equipartition theorem which asserts that

$$\frac{1}{2}m\overline{v_x^2} = \frac{1}{2}kT \qquad \text{or} \qquad \overline{v_x^2} = \frac{kT}{m}.$$

Hence $$\overline{v^2} = \overline{v_x^2} + \overline{v_y^2} + \overline{v_z^2} = 3\overline{v_x^2} = \frac{3kT}{m}$$

since $\overline{v_x^2} = \overline{v_y^2} = \overline{v_z^2}$ by symmetry. In the approximate calculations of this chapter it is superfluous to make a distinction between the mean speed $\bar{v}$ and the root-mean-square speed $(\overline{v^2})^{1/2}$. Thus we can write with sufficient accuracy

$$\bar{v} \approx \sqrt{\frac{3kT}{m}}. \tag{23}$$

Irrespective of the precise value of the constant factor in (23), the mean speed of a molecule depends only on the temperature, but not on n, the number of molecules per unit volume. Hence the coefficient of viscosity (22) is seen to be independent of n. At a given temperature T, it is thus also independent of the gas pressure $\bar{p} = nkT$.

This is a remarkable result. It asserts that, in the situation illustrated in Fig. 8.4, the viscous retarding force exerted by the gas on the moving upper plate is the same whether the pressure of the gas between the two plates is, for example, equal to 1 mm of mercury or is increased to 1000 mm of mercury. At first sight such a conclusion seems strange, since a naïve intuition might lead one to expect that the tangential force transmitted by the gas should be proportional to the number of gas molecules present. The apparent paradox is, however, easily resolved by the following observation. If the number of gas molecules is doubled, there are indeed twice as many molecules available to transport momentum from one plate to the other; but the mean free path of each molecule is then also halved so that it can transport a given momentum over only half its previous distance. Thus the net rate of momentum transfer is left unchanged. The fact that the viscosity η of a gas at a given temperature is independent of its density was first derived by Maxwell in 1860 and was confirmed by him experimentally.

It is clear, however, that this result cannot hold over an arbitrarily large density range of the gas. Indeed, we made two assumptions in deriving the relation (20):

(i) We assumed that the gas is sufficiently dilute so that there is negligible probability that three or more molecules simultaneously come so close together as to interact appreciably. Correspondingly, we considered only two-particle collisions. This assumption is justified if the density n of the gas is sufficiently low so that

$$l \gg d \tag{24}$$

where $d \sim \sigma^{1/2}$ is a measure of the molecular diameter.

(ii) On the other hand, we assumed that the gas is dense enough so that the molecules collide predominantly with other molecules rather than with the walls of the container. This assumption implies that n is sufficiently large so that

$$l \ll L \tag{25}$$

where L is the smallest linear dimension of the container (e.g., L is the spacing between the plates in Fig. 8.4).

In the limiting case of a perfect vacuum when $n \to 0$, the tangential force on the moving plate of Fig. 8.4 must vanish since there is no gas to transmit any force. If n is made so small that the condition (25) is violated, the viscosity η must thus ultimately decrease and approach zero. Indeed, when the mean free path (21) due to collisions with other molecules becomes greater than the container dimension L, a molecule collides predominantly with the container walls rather than with other molecules. Its effective mean free path l then becomes approximately equal to L (so that it no longer depends on the number of other molecules present) and η in (22) becomes proportional to n.

Note, however, that the range of densities where both conditions (24) and (25) are simultaneously satisfied is quite large, because $L \gg d$ in usual macroscopic experiments. Thus the coefficient of viscosity η of a gas is independent of its pressure over a very considerable pressure range.

Now let us discuss the temperature dependence of η. If the scattering of molecules is similar to that of hard spheres, the cross section σ, as given by (2), is merely a number independent of T. Then it follows by (22) that the temperature dependence of η is the same as that of $\bar{v}$; i.e.,

$$\eta \propto T^{1/2}. \tag{26}$$

More generally, σ depends on the mean relative speed $\bar{V}$ of the molecules. Since $\bar{V} \propto T^{1/2}$, σ also becomes temperature-dependent. As a result, η tends to vary with temperature more rapidly than in (26), somewhat more like $T^{0.7}$. This can be qualitatively understood since there exists, in addition to a strong repulsive interaction between two molecules, a weak long-range attractive interaction. This last interaction tends to increase the scattering probability of a molecule, but becomes less effective at high temperatures where the molecule has a large velocity and is thus less readily deflected. Hence the scattering cross section σ tends to *decrease* with increasing temperature. As T increases, the viscosity $\eta \propto T^{1/2}/\sigma$ tends, therefore, to increase with temperature *more* rapidly than $T^{1/2}$.

Note that the viscosity of a gas *increases* as the temperature is raised. This behavior is quite different from that of the viscosity of a liquid, which generally *decreases* rapidly with increasing temperature. The reason for the difference is that molecules in a liquid are close together. Hence, momentum transfer across a plane in the liquid occurs by direct forces between molecules on adjacent sides of the plane, as well as by virtue of molecular motion across this plane.

Finally, we estimate the magnitude of η for a typical gas at room temperature. By (22), η is of the order of the mean momentum $m\bar{v}$ of a molecule divided by a typical molecular area. For nitrogen (N_2) gas, $m = 28/(6 \times 10^{23}) = 4.7 \times 10^{-23}$ gm so that the mean momentum of a molecule at $T=300°$K is $m\bar{v} \approx \sqrt{3mkT} = 2.4 \times 10^{-18}$ gm cm sec^{-1}. Assuming a molecular diameter of the order of $d \sim 2 \times 10^{-8}$ cm, $\sigma \sim \pi d^2 \approx 1.2 \times 10^{-15}$. Hence (22) leads to the estimate

$$\eta = \frac{1}{3\sqrt{2}} \frac{m\bar{v}}{\sigma} \sim 5 \times 10^{-4} \text{ gm cm}^{-1} \text{ sec}^{-1}.$$

For comparison, the measured value of the coefficient of viscosity η for N_2 gas at 300°K is 1.78×10^{-4} gm cm^{-1} sec^{-1}.

Combining (22) and (23), our approximate expression for the viscosity coefficient becomes

$$\eta \approx \frac{1}{\sqrt{6}} \frac{\sqrt{mkT}}{\sigma}. \tag{27}$$

8.3 *Thermal Conductivity and Transport of Energy*

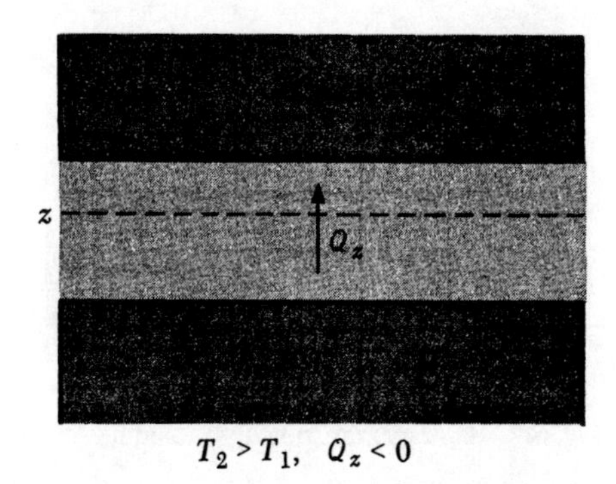

Fig. 8.6 A substance in thermal contact with two bodies at respective absolute temperatures T_1 and T_2. If $T_2 > T_1$, heat flows in the $-z$ direction from the region of higher to that of lower temperature; thus Q_z must be negative.

Definition of the coefficient of thermal conductivity

Consider a substance in which the temperature is *not* uniform throughout. In particular, imagine that the temperature T is a function of the z coordinate so that $T = T(z)$. Then the substance is certainly not in a state of equilibrium. The tendency to approach equilibrium manifests itself by a flow of heat from the region of higher to that of lower absolute temperature. Consider a plane $z =$ constant. Then we are interested in the quantity

$$Q_z \equiv \text{the heat-crossing unit area of the plane per unit time in the } +z \text{ direction.} \tag{28}$$

The quantity Q_z is called the *heat flux density* in the z direction. If the temperature is uniform, $Q_z = 0$. If it is not uniform, arguments similar to those we used in discussing viscosity lead us to expect that Q_z should be (to good approximation) proportional to the temperature gradient $\partial T/\partial z$ if the latter is not too large. Thus one can write

$$\boxed{Q_z = -\kappa \frac{\partial T}{\partial z}.} \tag{29}$$

The constant of proportionality κ is called the *coefficient of thermal conductivity* of the particular substance under consideration. Since heat flows from the region of higher to that of lower temperature, $Q_z < 0$ if $\partial T/\partial z > 0$. The minus sign was introduced explicitly in (29) so as to make κ a positive quantity. The relation (29) is found to be well obeyed in practically all gases, liquids, and isotropic solids.

Calculation of the coefficient of thermal conductivity for a dilute gas

In the simple case of a dilute gas, the coefficient of thermal conductivity can be readily calculated by simple microscopic arguments similar to those used in discussing the viscosity of a gas. Consider in the gas a plane $z =$ constant where $T = T(z)$. The mechanism of heat transport is due to the fact that molecules cross this plane from above and below. If $\partial T/\partial z > 0$, a molecule coming from above has a mean energy $\bar{\epsilon}(T)$ which is larger than that of a molecule coming from below; hence there results a net transport of energy from the region above the plane to that below it. More quantitatively, there are again roughly

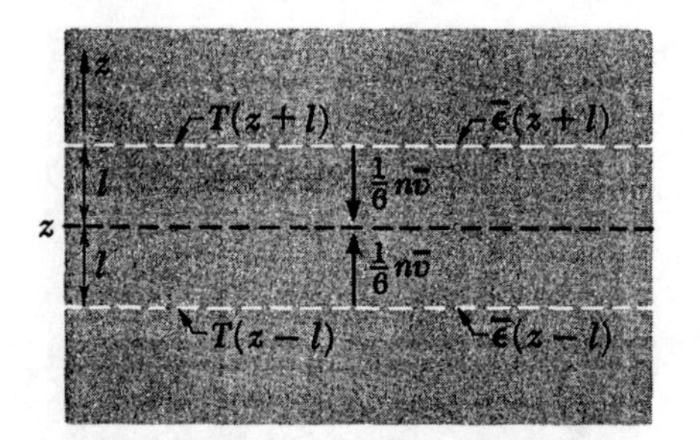

Fig. 8.7 Energy transport by molecules crossing a plane.

$\frac{1}{6}n\bar{v}$ molecules which in unit time cross unit area of this plane from below and an equal number of molecules which cross it from above.† Here n is the mean number of molecules per unit volume at the plane labeled by z, while $\bar{v}$ is their mean speed. Molecules which cross this plane from below have, on the average, experienced their last collision at a distance of one mean free path l below the plane. But since the temperature T is a function of z and since the mean energy $\bar{\epsilon}$ of a molecule depends on T, it follows that the mean energy $\bar{\epsilon}$ of a molecule depends on the position z of its last collision, i.e., $\bar{\epsilon} = \bar{\epsilon}(z)$. Thus molecules which cross the plane from below carry with them the mean energy $\bar{\epsilon}(z - l)$ assumed by them at their preceding collision at the position $(z - l)$. Thus it follows that

$$\begin{bmatrix}\text{the mean energy transported per unit} \\ \text{time per unit area across the plane from} \\ \text{below}\end{bmatrix} = \tfrac{1}{6}n\bar{v}\,\bar{\epsilon}(z - l). \tag{30}$$

Similarly, in considering molecules coming from above the plane where they suffered their preceding collision at the position $(z + l)$, it follows that

$$\begin{bmatrix}\text{the mean energy transported per unit} \\ \text{time per unit area across the plane from} \\ \text{above}\end{bmatrix} = \tfrac{1}{6}n\bar{v}\,\bar{\epsilon}(z + l). \tag{31}$$

By subtracting (31) from (30), we then obtain the *net* flux of energy Q_z crossing the plane from below in the $+z$ direction. Thus

$$Q_z = \tfrac{1}{6}n\bar{v}\,[\bar{\epsilon}(z - l) - \bar{\epsilon}(z + l)]$$

$$= \tfrac{1}{6}n\bar{v}\left[\left\{\bar{\epsilon}(z) - l\frac{\partial\bar{\epsilon}}{\partial z}\right\} - \left\{\bar{\epsilon}(z) + l\frac{\partial\bar{\epsilon}}{\partial z}\right\}\right]$$

or
$$Q_z = \tfrac{1}{6}n\bar{v}\left[-2l\frac{\partial\bar{\epsilon}}{\partial z}\right] = -\tfrac{1}{3}n\bar{v}l\frac{\partial\bar{\epsilon}}{\partial T}\frac{\partial T}{\partial z} \tag{32}$$

† Since the thermal conductivity of a gas is measured under steady-state conditions where there is no convective motion of the gas, the number of molecules crossing unit area of any plane per second from one side must always be equal to the number of molecules crossing this plane per second from the opposite direction. In our simple discussion we can thus assume that the product $n\bar{v}$ is constant and may disregard the fact that the temperature gradient causes n and $\bar{v}$ to have slightly different values above and below the plane.

since $\bar{\epsilon}$ depends on z through the temperature T. Let us introduce the abbreviation

$$c \equiv \frac{\partial \bar{\epsilon}}{\partial T} \tag{33}$$

which is the heat capacity (at constant volume) per *molecule*. Then (32) becomes

$$Q_z = -\kappa \frac{\partial T}{\partial z} \tag{34}$$

where

$$\boxed{\kappa = \tfrac{1}{3} n \bar{v} c l.} \tag{35}$$

The relation (34) shows that Q_z is indeed proportional to the temperature gradient, as expected by (29); furthermore, (35) provides an explicit expression for the thermal conductivity κ of the gas in terms of fundamental molecular quantities.

Discussion

The particular numerical factor $\frac{1}{3}$ obtained in (35) as a result of our simplified calculation is again not very trustworthy. The result (35) does, however, predict correctly the dependence of κ on all the significant parameters. Since $l \propto n^{-1}$, the density n again cancels. Using (21), the thermal conductivity (35) becomes thus

$$\kappa = \frac{1}{3\sqrt{2}} \frac{c}{\sigma} \bar{v}. \tag{36}$$

At a given temperature, the thermal conductivity κ is, therefore, *independent* of the pressure of the gas. This result is due to reasons identical to those mentioned in connection with the similar property of the viscosity coefficient η. It is again valid in a pressure range where the mean free path l satisfies the condition $d \ll l \ll L$ (where d denotes the molecular diameter and L the smallest dimension of the container).

In the case of a monatomic gas the equipartition theorem yields $\bar{\epsilon} = \frac{3}{2}kT$. The heat capacity c per molecule is then simply equal to $c = \frac{3}{2}k$.

Since $\bar{v} \propto T^{1/2}$ and since c is usually temperature-independent, Eq. (36) applied to molecules which interact like hard spheres gives the temperature dependence

$$\kappa \propto T^{1/2}. \tag{37}$$

More generally, σ also tends to vary with the temperature in the manner discussed in the last section in connection with the viscosity. As a result, κ increases again somewhat more rapidly with increasing temperature than is indicated by (37).

An estimate of the order of magnitude of κ for a gas at room temperature can readily be obtained by substituting typical numbers into (36). A representative value is the measured thermal conductivity of argon at 273°K, namely $\kappa = 1.65 \times 10^{-4}$ watts cm^{-1} deg^{-1}.

Using for $\bar{v}$ the result (23), the relation (35) yields for the thermal conductivity the approximate expression

$$\kappa \approx \frac{1}{\sqrt{6}} \frac{c}{\sigma} \sqrt{\frac{kT}{m}}. \tag{38}$$

Finally, comparison between the expressions (35) for the thermal conductivity κ and (20) for the viscosity η shows that these are quite similar in form. Indeed, one obtains for their ratio the relation

$$\frac{\kappa}{\eta} = \frac{c}{m}. \tag{39}$$

Equivalently, multiplying both numerator and denominator by Avogadro's number N_a, (39) becomes

$$\frac{\kappa}{\eta} = \frac{c_V}{\mu}$$

where $c_V = N_a c$ is the molar specific heat of the gas at constant volume and where $\mu = N_a m$ is its molecular weight. Thus there exists a very simple relation between the two transport coefficients κ and η, a relation which can readily be checked experimentally. One finds that the ratio $(\kappa/\eta)(c/m)^{-1}$ lies somewhere in the range between 1.3 and 2.5, instead of being unity as predicted by (39). In view of the very simplified nature of the arguments leading to our expressions for η and κ, there is greater justification for being pleased by the extent of agreement than there is cause for surprise at the discrepancy. Indeed, part of the latter is readily explained by the mere fact that our calculation did not take into account effects due to the distribution of molecular velocities. Thus faster molecules cross a given plane more frequently than slower ones. In the case of thermal conductivity these faster molecules also transport more kinetic energy; but in the case of viscosity they do not carry any greater mean x component of momentum. Thus the ratio κ/η should indeed be somewhat larger than that given by (39).

8.4 *Self-diffusion and Transport of Molecules*

Definition of the coefficient of self-diffusion

Consider a substance consisting of similar molecules, but assume that a certain number of these molecules are labeled in some way. For example, some of the molecules might be labeled by the fact that their nuclei are radioactive. Let n_1 be the mean number of labeled molecules per unit volume. In an equilibrium situation the labeled molecules would be distributed uniformly throughout the available volume, so that n_1 is independent of position. Suppose, however, that their distribution is not uniform, so that n_1 *does* depend on position; e.g., n_1 may depend on z so that $n_1 = n_1(z)$. (The *total* mean number n of molecules per unit volume is, however, assumed to remain constant so that there is no net motion of all the molecules of the substance.) This is not an equilibrium situation. The labeled molecules will, therefore, tend to move in such a way as to attain the ultimate equilibrium situation where they are uniformly distributed. Considering a plane $z =$ constant, we shall denote the *flux density* of the labeled particles by J_z, so that

$J_z \equiv$ the mean number of labeled molecules crossing unit area of the plane per unit time in the $+z$ direction. (40)

If n_1 is uniform, $J_z = 0$. If n_1 is not uniform, we expect that J_z should, to good approximation, be proportional to the concentration gradient $\partial n_1/\partial z$ of labeled molecules. Thus we can write

$$\boxed{J_z = -D\frac{\partial n_1}{\partial z}.} \tag{41}$$

The constant of proportionality D is called the *coefficient of self-diffusion* of the substance. If $\partial n_1/\partial z > 0$, the flow of labeled particles is in the $-z$ direction so as to equalize the concentration, i.e., $J_z < 0$. Hence the minus sign was introduced explicitly in (41) so as to make D a positive quantity. The relation (41) is found to describe quite adequately the self-diffusion of molecules in gases, liquids, or isotropic solids.†

† One speaks of *self*-diffusion if the diffusing molecules are, except for being labeled, identical to the remaining molecules of the substance. The more general and complicated situation would be that of *mutual* diffusion where the molecules are *unlike*, e.g., the diffusion of helium molecules in argon gas.

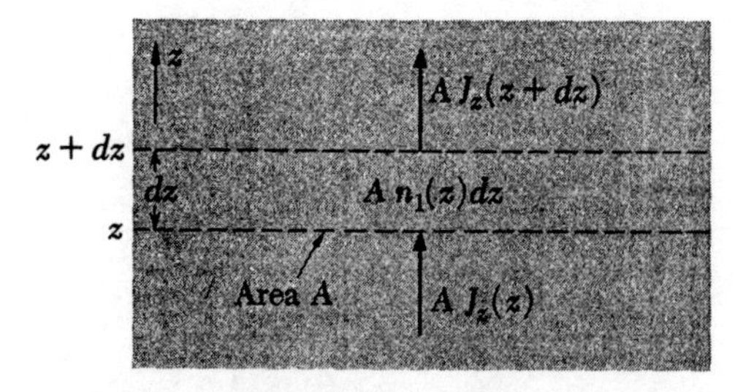

Fig. 8.8 Diagram illustrating the conservation of the number of labeled molecules during diffusion.

The diffusion equation

It is useful to point out that the quantity n_1 satisfies, by virtue of the relation (41), a simple differential equation. Consider a one-dimensional problem where $n_1(z,t)$ is the mean number of labeled molecules per unit volume located at time t near the position z. Focus attention on a slab of substance of thickness dz and of area A. Since the total number of labeled molecules is conserved, we can make the statement that {the increase per unit time in the number of labeled molecules contained within the slab} must be equal to {the number of labeled molecules entering the slab per unit time through its surface at z} minus {the number of labeled molecules leaving the slab per unit time through its surface at $(z + dz)$}. In symbols,

$$\frac{\partial}{\partial t}(n_1 A\, dz) = A\, J_z(z) - A\, J_z(z + dz).$$

Hence

$$\frac{\partial n_1}{\partial t} dz = J_z(z) - \left[J_z(z) + \frac{\partial J_z}{\partial z} dz\right]$$

or

$$\frac{\partial n_1}{\partial t} = -\frac{\partial J_z}{\partial z}. \tag{42}$$

This equation expresses merely the conservation of the number of labeled molecules. Using the relation (41), this becomes

$$\boxed{\frac{\partial n_1}{\partial t} = D\frac{\partial^2 n_1}{\partial z^2}.} \tag{43}$$

This is the desired partial differential equation, the *diffusion equation*, satisfied by $n_1(z,t)$.

Calculation of the coefficient of self-diffusion for a dilute gas

In the simple case of a dilute gas, the coefficient of self-diffusion can readily be calculated by mean-free-path arguments similar to those used in the last two sections. Consider a plane $z =$ constant in the gas. Since $n_1 = n_1(z)$, the mean number of labeled molecules which in unit time cross a unit area of this plane from below is equal to $\frac{1}{6}\bar{v}\, n_1(z - l)$; the mean number of labeled molecules which in unit time cross a unit area of this plane from above is $\frac{1}{6}\bar{v}\, n_1(z + l)$. Hence one obtains for the net flux of labeled molecules crossing unit area of the plane from below in the $+z$ direction

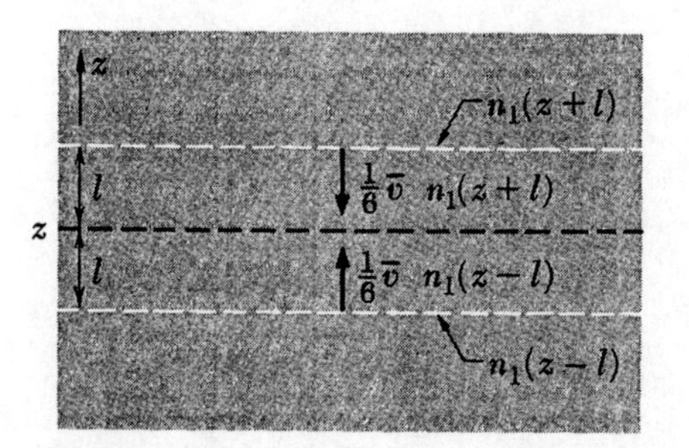

Fig. 8.9 Transport of labeled molecules across a plane.

$$J_z = \tfrac{1}{6}\bar{v}\, n_1(z - l) - \tfrac{1}{6}\bar{v}\, n_1(z + l)$$

$$= \tfrac{1}{6}\bar{v}[n_1(z - l) - n_1(z + l)] = \tfrac{1}{6}\bar{v}\left[-2\frac{\partial n_1}{\partial z} l\right]$$

or

$$J_z = -D\frac{\partial n_1}{\partial z} \tag{44}$$

where

$$\boxed{D = \tfrac{1}{3}\bar{v}l.} \tag{45}$$

Thus (44) shows explicitly that J_z is proportional to the concentration gradient, in accordance with the general relation (41); furthermore, (45) provides an approximate expression for the coefficient of self-diffusion in terms of fundamental molecular quantities.

To express D in more explicit form one need only use the relations (10) and (23). Thus

$$l = \frac{1}{\sqrt{2}n\sigma} = \frac{1}{\sqrt{2}\,\sigma}\frac{kT}{\bar{p}}$$

and

$$\bar{v} \approx \sqrt{\frac{3kT}{m}}.$$

Hence

$$D \approx \frac{1}{\sqrt{6}}\frac{1}{\bar{p}\sigma}\sqrt{\frac{(kT)^3}{m}}. \tag{46}$$

The coefficient of self-diffusion D *does*, therefore, depend on the pressure of the gas. At a fixed temperature T,

$$D \propto \frac{1}{n} \propto \frac{1}{\bar{p}}. \tag{47}$$

Also at a fixed pressure,

$$D \propto T^{3/2} \tag{48}$$

if the molecules scatter like hard spheres so that σ is a constant independent of T.

By virtue of (45), the order of magnitude of D at room temperature and atmospheric pressure is $\frac{1}{3}\bar{v}l \sim \frac{1}{3}(5\times10^4)(3\times10^{-5}) \sim 0.5\ \text{cm}^2\ \text{sec}^{-1}$. The experimentally measured value for N_2 gas at 273°K and 1 atmosphere pressure is $0.185\ \text{cm}^2\ \text{sec}^{-1}$.

Comparison between (45) and the coefficient of viscosity η in (20) yields the relation

$$\frac{D}{\eta} = \frac{1}{nm} = \frac{1}{\rho} \tag{49}$$

where ρ is the mass density of the gas. Experimentally, one finds that the ratio $(D\rho/\eta)$ lies in the range between 1.3 and 1.5 instead of being unity as predicted by (49). In view of the approximate nature of our simple calculations, this extent of agreement between theory and experiment can be regarded as quite satisfactory.

Diffusion regarded as a random walk problem

Suppose that N_1 labeled molecules are introduced at some initial time $t = 0$ near the plane $z = 0$. As time goes on, these molecules proceed to diffuse and thus to spread out in space as shown in Fig. 8.10. The number $n_1(z,t)$ of molecules per unit volume at any position z and at any time t can then be predicted by solving the diffusion equation (43).

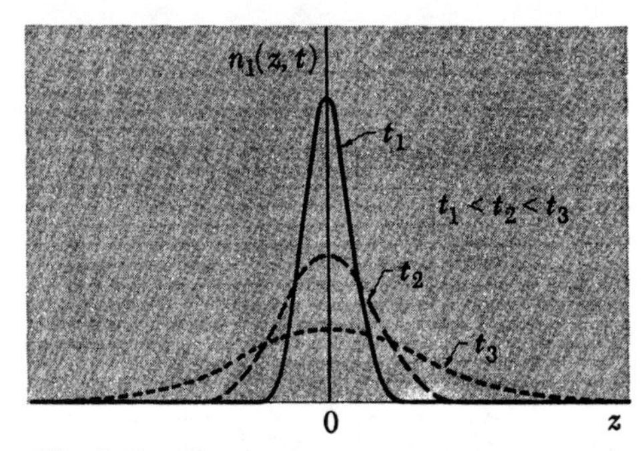

Fig. 8.10 The number $n_1(z,t)$ of labeled molecules per unit volume as a function of z at various times t after molecules are introduced at the time $t = 0$ near the plane $z = 0$. The areas under all the curves are the same and equal to the total number N_1 of labeled molecules.

Alternatively, one can simply look upon the diffusion process as a random walk of the labeled molecules. It is then possible to apply our discussion of Chap. 2 to gain immediate insight into the main features of this diffusion process. We assume that successive displacements suffered by a labeled molecule between collisions are statistically independent and shall denote by s_i the z component of the ith displacement of the molecule. If the molecule starts at $z = 0$, the z component of its position vector after N displacements is then given by

$$z = \sum_{i=1}^{N} s_i. \tag{50}$$

By virtue of the random direction of each displacement, the mean value of each displacement vanishes, i.e., $\bar{s}_i = 0$. The mean value of the sum then also vanishes so that $\bar{z} = 0$. Accordingly the curves in Fig. 8.10 are symmetrical about the value $z = 0$. Analogously to (2.49), the relation (50) yields for the dispersion of z the result

$$\overline{z^2} = \sum_i \overline{s_i^2} + \sum_i \sum_{\substack{j \\ i \neq j}} \overline{s_i s_j}. \tag{51}$$

But by virtue of the statistical independence of the displacements, $\overline{s_i s_j} = \bar{s}_i\, \bar{s}_j = 0$. Hence (51) reduces to

$$\overline{z^2} = N\,\overline{s^2}. \tag{52}$$

If a molecule has a velocity **v**, the z component of its displacement in a time t' is $s = v_z t'$. The mean square displacement during the mean free time τ available for an individual displacement is thus approximately

$$\overline{s^2} \approx \overline{v_z^2}\tau^2 = \tfrac{1}{3}\overline{v^2}\tau^2 \tag{53}$$

where we have put $\overline{v^2} = \overline{v_x^2} + \overline{v_y^2} + \overline{v_z^2} = 3\overline{v_z^2}$ since $\overline{v_x^2} = \overline{v_y^2} = \overline{v_z^2}$ by symmetry. Furthermore, the total number N of molecular displacements occurring in a total time t must be approximately equal to t/τ. Hence (52) yields for the mean square z component of displacement of a labeled molecule during a time t the approximate result

$$\overline{z^2} \approx \frac{t}{\tau}\left(\tfrac{1}{3}\overline{v^2}\tau^2\right) = \left(\tfrac{1}{3}\overline{v^2}\tau\right)t. \tag{54}$$

The width of the curves of Fig. 8.10 is measured by the square root of $\overline{z^2}$, i.e., by the standard deviation

$$\Delta z \equiv (\overline{z^2})^{1/2} \propto t^{1/2}.$$

The extent of the spreading of the labeled molecules, represented by the width of the curves of Fig. 8.10, thus grows in time proportionally to $N^{1/2}$ or to $t^{1/2}$. This result reflects merely the statistical character of the diffusion process. The relation (54) can be shown to be consistent with the predictions of the diffusion equation (43) and the magnitude of the diffusion constant given by (45).

8.5 *Electrical Conductivity and Transport of Charge*

Consider a system (liquid, solid, or gas) containing charged particles which are free to move. If a small uniform electric field $\mathcal{E}$ is applied in the z direction, a nonequilibrium situation results in which an electric current density j_z is set up in this direction. Considering any plane $z =$ constant, the current density is defined as

$j_z \equiv$ the mean electric charge crossing unit area of this plane per unit time in the $+z$ direction. (55)

The current density vanishes, of course, in equilibrium when $\mathcal{E} = 0$ so that no applied forces act on the charged particles. If the electric field $\mathcal{E}$ is sufficiently small, we then expect a linear relation of the form

$$\boxed{j_z = \sigma_e \mathcal{E}} \tag{56}$$

where the constant of proportionality σ_e is called the *electrical conductivity* of the system. The relation (56) is called *Ohm's law.*†

Consider now a dilute gas of particles having mass m and charge q and interacting with some other system of particles by which they can be scattered. A particularly simple case would be that of a relatively small number of ions (or electrons) in a gas where these ions are scattered predominantly by collisions with the neutral gas molecules. Another case is that of the electrons in a metal where these electrons are scattered by the vibrating atoms of the solid or by impurity atoms in the solid.‡ When an electric field $\mathcal{E}$ is applied in the z direction, it gives rise to a mean z component of velocity $\bar{v}_z$ of the charged particles. The mean number of such particles crossing a unit area (perpendicular

† Do not confuse the symbol σ_e for electrical conductivity with the symbol σ denoting the scattering cross section.

‡ The case of electrons in metals involves, however, some subtleties since these electrons (as we showed at the end of Sec. 6.3) do not obey the classical Maxwell distribution of velocities. Instead, they are described by the so-called *Fermi-Dirac distribution* which is a consequence of the rigorous quantum-mechanical treatment of the electron gas.

to the z direction) per unit time is then given by $n\bar{v}_z$ if n is the mean number of charged particles per unit volume. Since each particle carries a charge q, one thus obtains

$$j_z = nq\bar{v}_z. \tag{57}$$

It only remains to calculate $\bar{v}_z$. Let us measure time from the instant $t = 0$ immediately after the particle's last collision. The equation of motion of the particle between this collision and the next one is

$$m\frac{dv_z}{dt} = q\mathcal{E}.$$

Hence

$$v_z = \frac{q\mathcal{E}}{m}t + v_z(0). \tag{58}$$

In order to calculate the mean value $\bar{v}_z$, we must first average (58) over all of the possible velocities $v_z(0)$ of the particle immediately after a collision, and must then average (58) over all possible times t which the particle travels before its next collision. We shall assume that each collision is sufficiently effective to restore the particle to thermal equilibrium immediately afterward; the velocity $\mathbf{v}(0)$ immediately after a collision then has random direction so that $\bar{v}_z(0) = 0$ irrespective of the particle's past history before that collision.† Since the mean value of the time elapsed before the next collision of the particle is, by definition, the mean free time τ, the average of (58) then yields simply

$$\bar{v}_z = \frac{q\mathcal{E}}{m}\tau. \tag{59}$$

The expression (57) for the current density becomes thus

$$j_z = \sigma_e\mathcal{E} \tag{60}$$

where

$$\boxed{\sigma_e \equiv \frac{nq^2}{m}\tau.} \tag{61}$$

Thus j_z is indeed proportional to $\mathcal{E}$, as expected by (56); furthermore, (61) provides an explicit expression for the electrical conductivity σ_e in terms of the microscopic parameters characterizing the gas. The relation (61) is generally valid, even for electrons in metals.

† One can expect this to be a very good approximation if the charged particle suffers collisions with particles of much larger mass. Otherwise the charged particle retains after each collision some memory of the z component of velocity which it had before that collision. We shall neglect any corrections due to such "persistence-of-velocity" effects.

In the case of conductivity due to a small number of ions in a gas, the collisions limiting the mean path of an ion are predominantly those with neutral gas molecules.† Denote by σ the total scattering cross section of an ion by a molecule and assume that there are, per unit volume, n_1 molecules of mass $m_1 \gg m$. The thermal speed of the ions is then much greater than that of the molecules, and the mean relative speed of an ion-molecule encounter is simply the mean ion speed $\bar{v}$. Thus the mean free time τ of an ion is, by (4), simply equal to

$$\tau = \frac{1}{n_1 \sigma \bar{v}}.$$

Using the expression (23) for $\bar{v}$, Eq. (61) becomes approximately

$$\sigma_e = \frac{nq^2}{n_1 m \sigma \bar{v}} = \frac{1}{\sqrt{3}} \frac{nq^2}{n_1 \sigma \sqrt{mkT}}. \tag{62}$$

† Collisions between two like ions would actually not affect the electrical conductivity even if such collisions occurred frequently. The reason is that in any such collision the total momentum of the colliding ions is conserved. If the ions are identical so that they have the same mass, the vector sum of their velocities therefore remains unchanged in a collision. Thus, since both ions carry the same charge, they simply exchange roles in carrying the electric current.

Summary of definitions

mean free time The mean time which a molecule travels before suffering its next collision.

mean free path The mean distance which a molecule travels before suffering its next collision.

total scattering cross section The effective area determining the probability that a molecule incident on another molecule will be scattered by it.

stress Force per unit area.

viscosity The coefficient of viscosity η is defined by the equation

$$P_{zx} = -\eta \frac{\partial u_x}{\partial z}$$

which relates the stress P_{zx} across a surface in a fluid to the gradient of the mean fluid velocity u_x.

thermal conductivity The coefficient of thermal conductivity κ is defined by the equation

$$Q_z = -\kappa \frac{\partial T}{\partial z}$$

which relates the heat flux density Q_z to the gradient of the temperature T.

self-diffusion The coefficient of self-diffusion D is defined by the equation

$$J_z = -D \frac{\partial n_1}{\partial z}$$

which relates the flux density J_z of labeled particles to the gradient of their number n_1 per unit volume.

electrical conductivity The coefficient of electrical conductivity σ_e is defined by the equation

$$j_z = \sigma_e \mathcal{E}$$

which relates the current density j_z to the electric field $\mathcal{E}$ (i.e., to the gradient of the electrical potential).

Important Relations

Mean free path:

$$l \approx \frac{1}{\sqrt{2} n \sigma}.$$

Suggestions for Supplementary Reading

The discussion of transport processes in this chapter has been only a very brief introduction to a vast and important field. The theory can be refined by developing methods designed to obtain results of greater quantitative accuracy. Furthermore, the realm of important applications is very large, particularly when one is dealing with charged particles such as electrons in metals or in plasmas. (Typical problems of interest might be the electrical conductivity as a function of temperature, the dielectric constant and dielectric loss factor as a function of frequency, "thermoelectric effects" resulting in the appearance of electrical fields as a result of temperature differences, etc.) The following books carry the theory of transport processes appreciably further than we have done in this chapter and give references to further works on the subject.

F. Reif, *Fundamentals of Statistical and Thermal Physics* (McGraw-Hill Book Company, New York, 1965). Chapter 12 discusses the most elementary theory; chaps. 13 and 14 present more sophisticated treatments.

R. D. Present, *Kinetic Theory of Gases* (McGraw-Hill Book Company, New York, 1958). Chapter 3 describes the most elementary theory; chaps. 8 and 11 give the more sophisticated treatments.

Problems

8.1 *Tossing of a coin*

Consider the tossing of a coin which has probability $\frac{1}{2}$ of landing on either one of its two faces. Focus attention on this coin after any toss.

(*a*) What is the mean number of tosses of the coin before the next appearance of a "head"?

(*b*) What is the mean number of tosses of the coin since the preceding appearance of a "head"?

(*c*) Suppose that a "head" has just been obtained in the preceding toss. How would the answer to part (*a*) be affected by this information?

8.2 *Analogy between mean free time arguments and the preceding problem*

Consider a gas of such density that the mean free time of a molecule is τ. Focus attention on a particular molecule at any instant of time.

(*a*) What is the mean time traveled by this molecule before it suffers its next collision?

(*b*) What is the mean time traveled by this molecule since it suffered its preceding collision?

(*c*) Suppose that the molecule has just experienced a collision. How would the answer to part (*a*) be affected by this information?

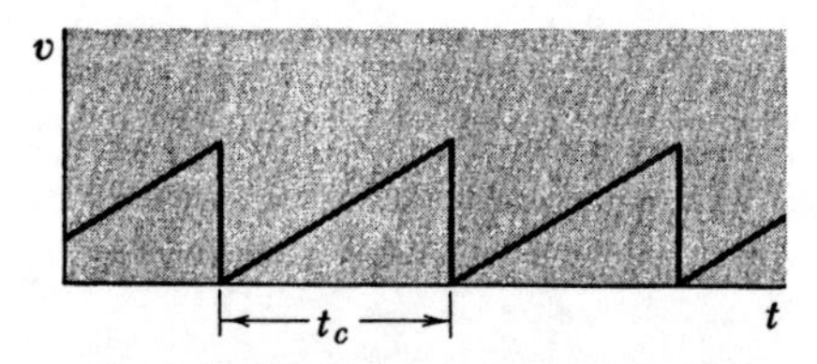

Fig. 8.11 Diagram of velocity v as a function of time t for a simple model of an ion in a gas.

8.3 *Mean free time and time between collisions*

Consider an ion of charge q and mass m immersed in a gas and subject to an electric field $\mathcal{E}$ in the z direction. For purposes of simplicity, assume the following model: An ion travels in the z direction, starting always from rest after each collision and moving with an acceleration $a = q\mathcal{E}/m$ for a fixed time t_c; it is then brought to rest by the next collision and the whole process starts over again. A plot of the velocity v of an ion as a function of time would then have the appearance shown in Fig. 8.11.

(*a*) Consider an ensemble of such ions at any arbitrary instant of time. What is the mean time τ which elapses before an ion suffers its next collision? Express your answer in terms of the time t_c between successive collisions.

(*b*) What is the mean time which has elapsed since an ion suffered its preceding collision? Express your answer in terms of t_c.

(*c*) What is the maximum velocity ever attained by an ion? What is its mean velocity $\bar{v}$? Express your result in terms of t_c; express it also in terms of the mean time τ of part (*a*). Compare your answer with (59).

(*d*) What is the distance s traversed by an ion in a time t_c, starting from rest? If we define the mean velocity $\bar{v}$ of the ion by $\bar{v} = s/t_c$, what is the value of $\bar{v}$ thus calculated? Express your result in terms of t_c; express it also in terms of τ. Compare it with the result of part (*c*).

8.4 *Millikan oil-drop experiment*

The Millikan oil-drop experiment, which was the first experiment to measure the charge of the electron, compared the electrical force on a small charged oil drop with the force exerted on this drop by gravity. The experiment thus required a knowledge of the weight of the drop. This weight can be obtained by observing the drop falling with a constant terminal velocity when the gravitational force on the drop is balanced by the frictional force due to the viscosity of the surrounding air. (The air is at atmospheric pressure so that the mean free path of the air molecules is much smaller than the diameter of the drop.)

The terminal velocity with which the oil drop falls is inversely proportional to the viscosity of the air. If the temperature of the air increases, does the terminal velocity of the drop increase, decrease, or remain the same? What happens when the atmospheric pressure increases?

8.5 *Rotating-cylinder viscometer*

It is desired to measure the coefficient of viscosity η of air at room temperature since this parameter is essential for determining the electronic charge by Millikan's oil-drop experiment. It is proposed to perform the measurement in a viscometer consisting of a stationary inner cylinder (of radius R and length L) supported by a torsion fiber, and of an outer cylinder [of slightly larger inner radius $(R + \delta)$] rotating slowly with angular velocity ω. The narrow annular region of thickness δ (where $\delta \ll R$) is filled with air and one measures the torque G on the inner cylinder. (See Fig. 8.12.)

(*a*) Find the torque G in terms of η and the parameters of this experimental apparatus.

(*b*) To determine what quartz fiber is needed, estimate the magnitude of the viscosity of air from first principles and use this result to estimate the magnitude of the torque that has to be measured in an apparatus of this kind. Take as dimensions $R = 2$ cm, $\delta = 0.1$ cm, $L = 15$ cm, and $\omega = 2\pi$ radians/second.

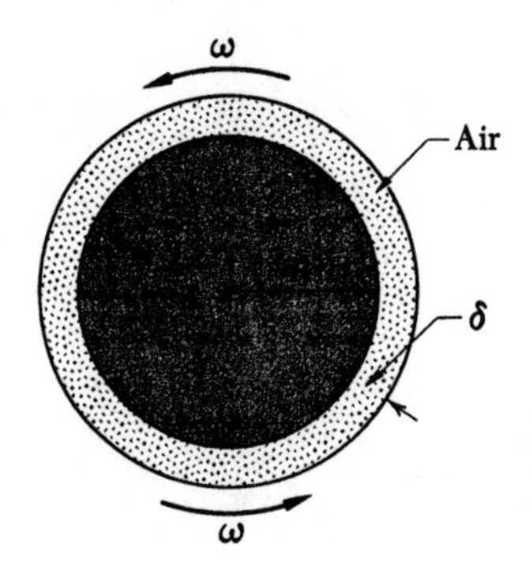

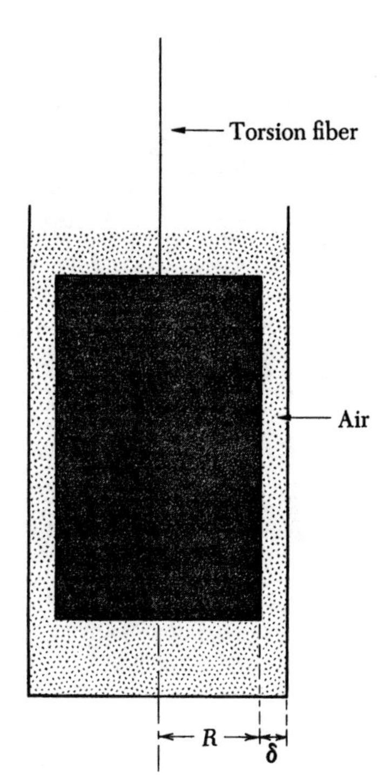

Fig. 8.12 Top view and side view of a rotating-cylinder viscometer.

8.6 *Estimate of the viscosity coefficient of Ar gas*

Estimate the magnitude of the coefficient of viscosity η of argon (Ar) gas at 25°C and 1 atmosphere pressure. To estimate the size of an argon atom, consider the atoms as hard spheres which touch each other in solid argon at low temperatures. Studies of x-ray diffraction show that the crystal structure of solid argon is face-centered cubic, i.e., the Ar atoms occupy the corners and midpoints of the faces of a regular array of cubes. The density of solid argon is 1.65 gm/cm^3 and the atomic weight of Ar is 39.9. Compare your estimate with the experimentally observed value of $\eta = 2.27 \times 10^{-4}$ gm cm^{-1} sec^{-1}.

8.7 *Effect of a velocity-dependent scattering cross section*

Suppose that the molecules of a gas interact with each other through a radial force F which depends on the intermolecular separation R according to $F = CR^{-s}$ where s is some positive integer and C is a constant.

(*a*) Use arguments of dimensional analysis to show how the total scattering cross section σ of the molecules depends on their relative speed V. Assume a classical calculation so that σ can only depend on V, the molecular mass m, and the force constant C.

(*b*) How does the coefficient of viscosity η of this gas depend on the absolute temperature T?

8.8 *Degree of vacuum necessary to achieve thermal insulation*

Consider a cylindrical dewar of the usual double-walled construction shown in Fig. 5.4. The outer diameter of the inner wall is 10 cm, the inner diameter of the outer wall is 10.6 cm. The dewar contains a mixture of ice and water; the outside of the dewar is at room temperature, i.e., about 25°C.

(*a*) If the space between the two walls of the dewar contains helium (He) gas at atmospheric pressure, calculate approximately the heat influx (in watts per cm height of the dewar) due to heat conduction by the gas. (A reasonable estimate for the radius of a He atom is about 10^{-8} cm.)

(*b*) Estimate roughly to what value (in mm of Hg) the pressure of the gas between the walls must be reduced before the heat influx due to conduction is reduced below the value calculated in part (*a*) by a factor of 10.

8.9 *Comparison between transport coefficients*

The coefficient of viscosity of helium (He) gas at $T = 273$°K and 1 atmosphere is η_1, that of argon (Ar) gas is η_2. The atomic weights of these monatomic gases are μ_1 and μ_2, respectively.

(*a*) What is the ratio σ_2/σ_1 of the Ar-Ar atom total scattering cross section σ_2 as compared to that of the He-He atom total scattering cross secion σ_1?

(*b*) What is the ratio κ_2/κ_1 of the thermal conductivity κ_2 of Ar gas compared to the thermal conductivity κ_1 of He gas when $T = 273$°K?

(*c*) What is the ratio D_2/D_1 of the diffusion coefficients of these gases when $T = 273$°K?

(*d*) The atomic weights of He and Ar are respectively $\mu_1 = 4$ and $\mu_2 = 40$. The measured viscosities at 273°K are respectively $\eta_1 = 1.87 \times 10^{-4}$ and $\eta_2 = 2.105 \times 10^{-4}$ gm cm^{-1} sec^{-1}. Use this information to calculate approximate values of the cross sections σ_1 and σ_2.

(*e*) If the atoms are assumed to scatter like hard spheres, estimate the diameter d_1 of a He atom and the diameter d_2 of an Ar atom.

8.10 *Mixing of isotopes by diffusion*

It is desired to do an experiment on an isotopic mixture of N_2 gas. For this purpose one takes a spherical storage vessel, 1 meter in diameter, containing $^{14}N_2$ gas at room temperature and atmospheric pressure, and introduces through a valve at one side of the container a small amount of $^{15}N_2$ gas. In the absence of any convection in the gas, make a *rough* estimate of how long one has to wait before one can be reasonably sure that the $^{14}N_2$ and $^{15}N_2$ molecules are uniformly mixed throughout the container.

8.11 *Effect of interplanetary gas on a spacecraft*

A spacecraft, in the shape of a cube of edge length L, moves through outer space with a velocity v parallel to one of its edges. The surrounding gas consists of molecules of mass m at a temperature T, the number n of molecules per unit volume being very small so that the mean free path of the molecules is much larger than L. Assuming that collisions of the molecules with the spacecraft are elastic, estimate the mean retarding force exerted on the spacecraft by its collisions with the interplanetary gas. Assume that v is small compared to the mean speed of the gas molecules and ignore the distribution of velocities of these molecules. If the mass of the spacecraft is M and it is not subject to external forces, after approximately how long a time will the velocity of the spacecraft be reduced to half its original value?

8.12 *Probability that a molecule survives a time t without collision*

Focus attention on a particular molecule in a gas at any instant of time. Let

$w\,dt \equiv$ the probability that the molecule suffers a collision during a time interval dt.

Consider now the survival probability

$P(t) \equiv$ the probability that the molecule survives a time t without suffering a collision.

Clearly $P(0) = 1$ since it is certain that the molecule survives for a vanishingly short time; it is equally clear that $P(t) \to 0$ as $t \to \infty$ since the molecule must sooner or later suffer a collision.

The survival probability $P(t)$ must be related to the collision probability $w\,dt$. Indeed, the probability $P(t + dt)$ that the molecule survives a time $(t + dt)$ without a collision must be equal to the probability $P(t)$ that it survives a time t without a collision multiplied by the probability $(1 - w\,dt)$ that it does not suffer a collision in the subsequent time interval between t and $t + dt$. By writing down this relation, obtain a differential equation for $P(t)$. Solving this equation, and using the fact that $P(0) = 1$, show that $P(t) = e^{-wt}$.

8.13 *Calculation of the mean free time τ*

The probability $\mathcal{P}(t)\,dt$ that a molecule, having survived without a collision for a time t, suffers a collision in the time between t and $t + dt$ is simply $P(t)\,w\,dt$.

(*a*) Show that this probability is properly normalized in the sense that

$$\int_0^\infty \mathcal{P}(t)\,dt = 1.$$

This merely asserts that the probability is unity that the molecule suffers a collision at *some* time.

(*b*) Use the probability $\mathcal{P}(t)\,dt$ to show that the mean time $\bar{t} \equiv \tau$ during which a molecule survives before suffering a collision is given by $\tau = 1/w$.

(*c*) Express the mean square time $\overline{t^2}$ in terms of τ.

8.14 *Differential equation for heat conduction*

Consider a general situation where the temperature T of a substance is a function of the time t and of the spatial coordinate z. The density of the substance is ρ, its specific heat *per unit mass* is c, and its thermal conductivity is κ. By macroscopic reasoning similar to that used in deriving the diffusion equation (43), obtain the general partial differential equation which must be satisfied by the temperature $T(z,t)$.

8.15 *Apparatus for measuring the thermal conductivity of a gas

A long cylindrical wire of radius a and electrical resistance R per unit length is stretched along the axis of a long cylindrical container of radius b. This container is maintained at a fixed temperature T_0 and is filled with a gas having a thermal conductivity κ. Calculate the temperature difference ΔT between the wire and the container walls when a small constant electrical current I is passed through the wire. Show thus that a measurement of ΔT provides a means for determining the thermal conductivity of the gas. Assume that a steady state condition has been reached so that the temperature T at any point is independent of time. (*Suggestion:* Consider the condition which must be satisfied by any cylindrical shell of the gas contained between radius r and radius $r + dr$.)

8.16 *Viscous flow through tubes

A fluid of viscosity η flows through a tube of length L and radius a as a result of a pressure difference, the pressure being p_1 at one end and p_2 at the other end of the tube. Write down the conditions which must prevail so that a cylinder of fluid of radius r moves without acceleration under the influence of the pressure difference and the shearing force due to the viscosity of the liquid. Hence derive an expression for the mass $\mathcal{M}$ of fluid flowing per second through the tube in each of the following two cases:

(*a*) The fluid is an incompressible liquid of density ρ.

(*b*) The fluid is an ideal gas of molecular weight μ at an absolute temperature T. (These results are known as Poiseuille's formulas.) Assume that the layer of fluid in contact with the walls of the tube is at rest. Note also that the same mass of fluid must cross any cross-sectional area of the tube per unit time.

Appendix

Appendix

A.1 Gaussian Distribution

Consider the binomial distribution derived in (2.14),

$$P(n) = \frac{N!}{n!(N-n)!} p^n q^{N-n} \tag{1}$$

where $q = 1 - p$. When N is large, the calculation of the probability $P(n)$ appears difficult since it requires the computation of the factorials of large numbers. It then becomes possible, however, to use approximations which allow us to transform the expression (1) into a particularly simple form.

The simplifying feature, already pointed out in Sec. 2.3, is that the probability $P(n)$ tends to exhibit a maximum which is quite pronounced when N is large. The probability $P(n)$ thus becomes negligibly small whenever n differs appreciably from the particular value $n = \tilde{n}$ where P is maximum. Hence the region ordinarily of interest, that where the probability $P(n)$ is *not* negligible, consists only of those values of n which do not differ very much from $\tilde{n}$. But in this relatively small region an approximate expression for $P(n)$ can readily be found. This expression can then be used for all values of n where the probability P is not negligibly small, i.e., in the entire domain where a knowledge of P is ordinarily of interest.

It is thus sufficient to investigate the behavior of $P(n)$ near the position $\tilde{n}$ of its maximum. Note first that, unless $p \approx 0$ or $q \approx 0$, this value $\tilde{n}$ is neither very close to 0 nor to N; thus $\tilde{n}$ itself is also a large number when N is large. The numbers n in the region of interest near $\tilde{n}$ are thus also large. But when n is large, $P(n)$ changes relatively little when n changes by unity; i.e.,

$$|P(n+1) - P(n)| \ll P(n)$$

so that P is a slowly varying function of n. It is, therefore, a good approximation to regard P as a smooth function of a continuous variable n, although only integral values of n are of physical relevance. A second useful observation concerns the fact that the logarithm of P is a much more slowly varying function of n than P itself. Instead of dealing directly with P, it is thus easier to investigate the behavior of $\ln P$ and to find for $\ln P$ a good approximation valid in a large domain of the variable n.

Taking the logarithm of (1), we thus obtain

$$\ln P = \ln N! - \ln n! - \ln (N-n)! + n \ln p + (N-n) \ln q. \qquad (2)$$

The particular value $n = \tilde{n}$ where P has its maximum is then determined by the condition

$$\frac{dP}{dn} = 0$$

or, equivalently, by the condition that $\ln P$ is maximum,

$$\frac{d \ln P}{dn} = \frac{1}{P}\frac{dP}{dn} = 0. \qquad (3)$$

To differentiate the expression (2), we note that all the numbers occurring as factorials are large compared to unity. We can, therefore, apply to each of these the approximation (M.7) which asserts that for any number m, large enough so that $m \gg 1$,

$$\frac{d \ln m!}{dm} \approx \ln m. \qquad (4)$$

Differentiation of (2) with respect to n then yields to good approximation

$$\frac{d \ln P}{dn} = -\ln n + \ln (N-n) + \ln p - \ln q. \qquad (5)$$

To find the maximum of P we equate the expression (5) to zero in accordance with (3). Thus

$$\ln \left[\frac{(N-n)}{n}\frac{p}{q}\right] = 0$$

or

$$\frac{(N-n)}{n}\frac{p}{q} = 1.$$

Hence

$$(N-n)p = nq$$

or

$$Np = n(p+q).$$

Since $p + q = 1$, the value $n = \tilde{n}$ where P has its maximum is then given by

$$\boxed{\tilde{n} = Np.} \qquad (6)$$

To investigate the behavior of $\ln P$ near its maximum, we need only expand it in a Taylor series about the value $\tilde{n}$. Thus we can write

$$\ln P(n) = \ln P(\tilde{n}) + \left[\frac{d \ln P}{dn}\right] y + \frac{1}{2!}\left[\frac{d^2 \ln P}{dn^2}\right] y^2 + \frac{1}{3!}\left[\frac{d^3 \ln P}{dn^3}\right] y^3 + \cdots \tag{7}$$

where

$$y \equiv n - \tilde{n} \tag{8}$$

and where the square brackets indicate that the derivatives are to be evaluated at $n = \tilde{n}$. The first derivative vanishes since we are expanding about a maximum where (3) is satisfied. The other derivatives can be found by successive differentiation of (5). In particular,

$$\frac{d^2 \ln P}{dn^2} = -\frac{1}{n} - \frac{1}{N-n} = -\frac{N}{n(N-n)}.$$

Evaluating this derivative where $n = \tilde{n}$, i.e., where $n = Np$ and $N - n = N(1 - p) = Nq$, we find

$$\left[\frac{d^2 \ln P}{dn^2}\right] = -\frac{1}{Npq}.$$

Hence (7) becomes

$$\ln P(n) = \ln P(\tilde{n}) - \frac{y^2}{2Npq} + \cdots$$

or

$$P(n) = \tilde{P} e^{-y^2/2Npq} \cdots = \tilde{P} e^{-(n-\tilde{n})^2/2Npq} \cdots \tag{9}$$

where we have written $\tilde{P} \equiv P(\tilde{n})$.

Note that the probability $P(n)$ in (9) becomes negligible compared to its maximum value $\tilde{P}$ when y becomes so large that $y^2/(Npq) \gg 1$ or $|y| \gg (Npq)^{1/2}$, since the exponential factor is then very much smaller than unity. The probability $P(n)$ is thus only appreciable in the domain where $|y| \lesssim (Npq)^{1/2}$. But there y is ordinarily sufficiently small so that the terms in (7) which involve y^3 and higher powers of y are negligible compared to the leading term, involving y^2, which we retained.† Hence we can conclude that (9) is indeed a good approximation to the probability $P(n)$ in the entire region where this probability has appreciable magnitude.

† This statement is true to the extent that $(Npq)^{1/2} \gg 1$. See Prob. P. 3.

The value of the constant $\tilde{P}$ in (9) can be expressed directly in terms of p and q by using the normalization condition

$$\sum_n P(n) = 1 \tag{10}$$

where the summation is over all possible values of n. Since $P(n)$ changes only slightly between the successive integral values of n, this sum can be replaced by an integration. A range of n of magnitude dn (much larger than unity) contains dn possible values of $P(n)$. Hence the condition (10) becomes

$$\int P(n)\,dn = \int_{-\infty}^{\infty} \tilde{P} e^{-(n-\tilde{n})^2/2Npq}\,dn$$

$$= \tilde{P} \int_{-\infty}^{\infty} e^{-y^2/2Npq}\,dy = 1. \tag{11}$$

Here we have made the simplification of extending the range of integration from $-\infty$ to $+\infty$. This is an excellent approximation since $P(n)$ is negligibly small anyhow whenever $|n - \tilde{n}|$ becomes sufficiently large. Using the relation (M.23), the integral in (11) yields simply

$$\tilde{P}\sqrt{2\pi Npq} = 1.$$

Thus

$$\tilde{P} = \frac{1}{\sqrt{2\pi Npq}}. \tag{12}$$

Using this result and the value $\tilde{n} = Np$ given by (6), the expression (9) for the probability $P(n)$ then becomes

$$\boxed{P(n) = \frac{1}{\sqrt{2\pi Npq}} e^{-(n-Np)^2/2Npq}.} \tag{13}$$

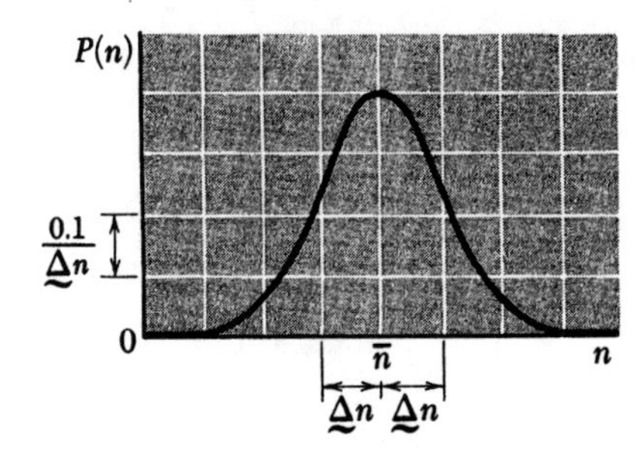

Fig. A.1 Gaussian distribution showing as a smooth curve the values of the probability $P(n)$ as a function of n. The probability $W(x)$ that n assumes a value between $\tilde{n} - x$ and $\tilde{n} + x$ is given by the area lying in this range below the curve. If Δn denotes the standard deviation of n, computation shows that

$$W(\Delta n) = 0.683, \quad W(2\Delta n) = 0.954,$$

and

$$W(3\Delta n) = 0.997.$$

Note that this expression is much easier to evaluate than (1) since it does not require the calculation of any factorials.

A probability of the functional form given by the right side of (9) or (13) is known as a *Gaussian distribution*. The essential argument leading to this distribution, i.e., a power series expansion of a logarithm, is widely applicable. It is therefore not surprising that Gaussian distributions occur very frequently in statistical arguments whenever the numbers under consideration are large.

The expression (13) for $P(n)$ can be readily used to calculate various mean values of n. The calculation of sums can be reduced to an evaluation of equivalent integrals in the same manner as that used in calculating the normalization condition (10) by means of (11). Thus one finds

$$\begin{aligned}\bar{n} &\equiv \sum_n P(n)n \\ &= (2\pi Npq)^{-1/2}\int_{-\infty}^{\infty} e^{-(n-Np)^2/2Npq}\, n\, dn = (2\pi Npq)^{-1/2}\int_{-\infty}^{\infty} e^{-y^2/2Npq}(\tilde{n}+y)\, dy \\ &= \tilde{n}(2\pi Npq)^{-1/2}\int_{-\infty}^{\infty} e^{-y^2/2Npq}\, dy + (2\pi Npq)^{-1/2}\int_{-\infty}^{\infty} e^{-y^2/2Npq}\, y\, dy.\end{aligned}$$

Here the first integral is the same as that occurring in (11) and has the value $(2\pi Npq)^{1/2}$. The second integral vanishes by symmetry since its integrand is odd (i.e., is of opposite sign for $+y$ and $-y$) so that contributions to the integral near $+y$ and $-y$ cancel each other. Thus we are left simply with

$$\bar{n} = \tilde{n} = Np. \tag{14}$$

This shows that the mean value of n is equal to the value $\tilde{n} = Np$ where the probability P is maximum.

Similarly one finds for the dispersion of n the result

$$\begin{aligned}\overline{(\Delta n)^2} &= \overline{(n-\bar{n})^2} = \sum_n P(n)(n-Np)^2 = (2\pi Npq)^{-1/2}\int_{-\infty}^{\infty} e^{-(n-Np)^2/2Npq}(n-Np)^2\, dn \\ &= (2\pi Npq)^{-1/2}\int_{-\infty}^{\infty} e^{-y^2/2Npq} y^2\, dy.\end{aligned}$$

Using (M.26), this integral yields

$$\overline{(\Delta n)^2} = Npq. \tag{15}$$

The standard deviation of n is then†

$$\underline{\Delta} n = \sqrt{Npq}. \tag{16}$$

The Gaussian distribution (13) can thus be expressed solely in terms of the two parameters $\bar{n}$ and $\underline{\Delta} n$ of (14) and (16)

$$P(n) = \frac{1}{\sqrt{2\pi}\,\underline{\Delta} n}\exp\left[-\frac{1}{2}\left(\frac{n-\bar{n}}{\underline{\Delta} n}\right)^2\right]. \tag{17}$$

Introducing the variable

$$z \equiv \frac{n-\bar{n}}{\underline{\Delta} n} \qquad \text{so that} \qquad n = \bar{n} + (\underline{\Delta} n)\, z,$$

Eq. (17) can also be expressed more compactly as

$$P(n)\,\underline{\Delta} n = \frac{1}{\sqrt{2\pi}} e^{-(1/2)z^2}.$$

Note that the Gaussian distribution is symmetric about its mean value, i.e., $P(n)$ has the same value for z and $-z$.

† Note that (14) and (15) agree properly with the results (2.66) and (2.67) derived in the text under the most general conditions of arbitrary magnitude of N.

A.2 *Poisson Distribution*

Consider again the binomial distribution derived in (2.14),

$$P(n) = \frac{N!}{n!(N-n)!} p^n (1-p)^{N-n}. \tag{18}$$

In the preceding section we showed that, when $N \gg 1$, the expression (18) can be approximated by a Gaussian distribution valid in the entire domain where the probability $P(n)$ is of appreciable magnitude (i.e., where it is not too far removed from the region of its maximum). We shall now examine an approximation of (18) valid in a different domain. This approximation becomes relevant when the probability p is sufficiently small so that

$$p \ll 1 \tag{19}$$

and when the number n of interest is sufficiently small so that

$$n \ll N. \tag{20}$$

In contrast to the situation envisaged in discussing the Gaussian approximation, the number n may here be arbitrarily small.

Let us now examine the approximations made possible by the conditions (19) and (20). We note first that

$$\frac{N!}{(N-n)!} = N(N-1)(N-2)\cdots(N-n+1).$$

Since $n \ll N$, each of the n factors on the right is essentially equal to N. Hence we obtain the approximate result

$$\frac{N!}{(N-n)!} \approx N^n. \tag{21}$$

We examine next the factor

$$y \equiv (1-p)^{N-n}$$

or, equivalently, its logarithm

$$\ln y = (N-n) \ln (1-p).$$

Since $n \ll N$, we can put $N - n \approx N$. Furthermore, the condition $p \ll 1$ allows us to approximate the logarithm by the first term in a Taylor expansion, i.e., to put $\ln (1-p) \approx -p$.

Hence
$$\ln y \approx -Np$$

or
$$y \equiv (1-p)^{N-n} \approx e^{-Np}. \tag{22}$$

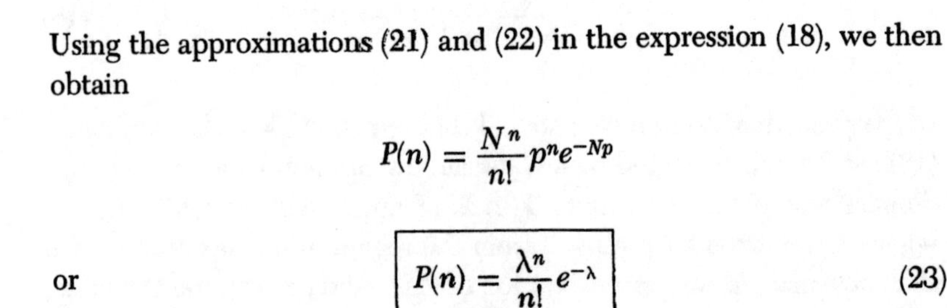

Using the approximations (21) and (22) in the expression (18), we then obtain

$$P(n) = \frac{N^n}{n!} p^n e^{-Np}$$

or
$$\boxed{P(n) = \frac{\lambda^n}{n!} e^{-\lambda}} \tag{23}$$

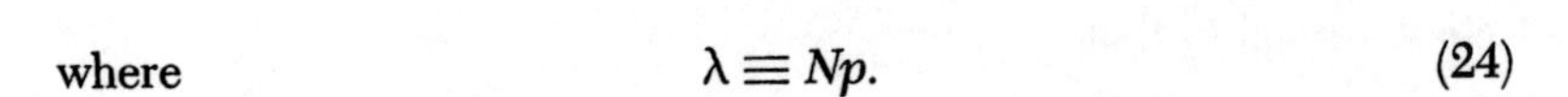

where
$$\lambda \equiv Np. \tag{24}$$

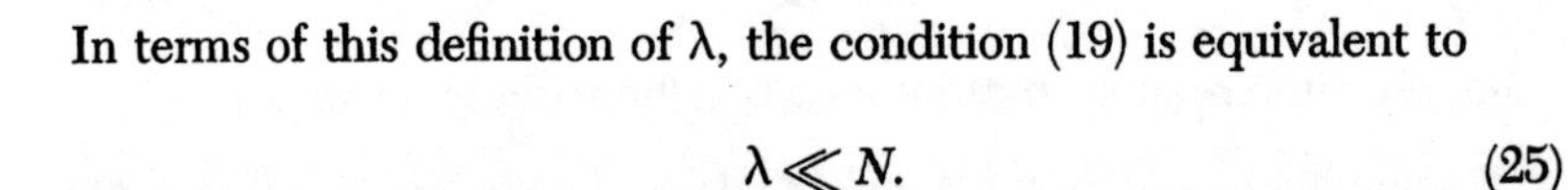

In terms of this definition of λ, the condition (19) is equivalent to

$$\lambda \ll N. \tag{25}$$

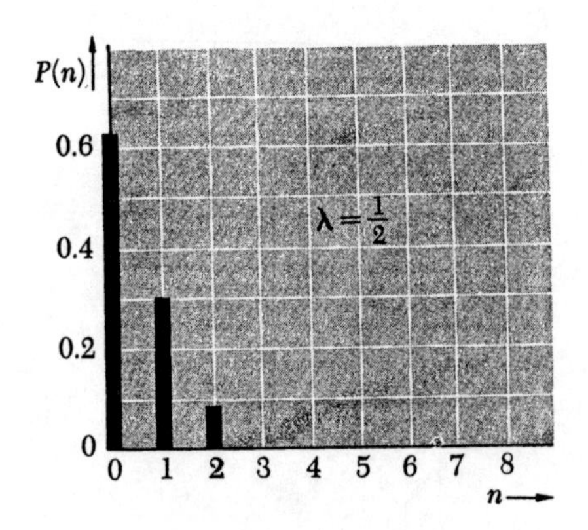

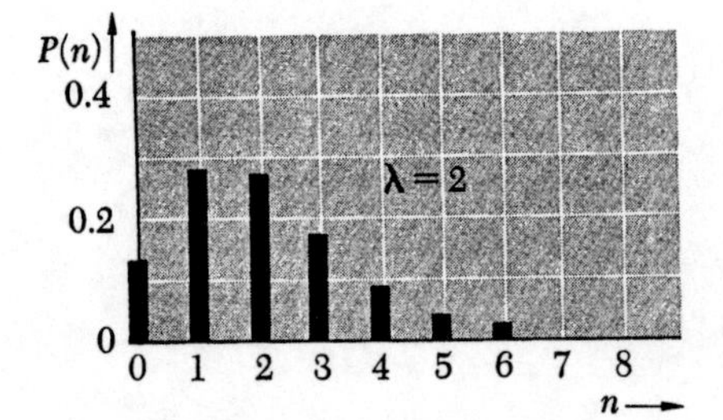

Fig. A.2 The Poisson distribution $P(n)$ of (23) as a function of n. The two cases shown correspond to a mean value $\bar{n} = \lambda$ of $\lambda = \frac{1}{2}$ and $\lambda = 2$.

The result (23) is called the *Poisson distribution.* Note that the factor $n!$ in the denominator causes $P(n)$ to decrease very rapidly when n becomes sufficiently large. Indeed, when $\lambda < 1$, λ^n itself is a decreasing function of n so that $P(n)$ decreases monotonically as a function of n. When $\lambda > 1$, λ^n is an increasing function of n so that the factor $\lambda^n/n!$, and hence also $P(n)$, tends to exhibit a maximum near $n \approx \lambda$ before falling off for larger values of n.† In any event, when $n \gg \lambda$, the probability $P(n)$ becomes negligibly small. In the entire region where $n \lesssim \lambda$ so that $P(n)$ is not negligible, the condition (25) then implies that $n \lesssim \lambda \ll N$. The requirement (20) used in deriving the Poisson distribution is thus automatically satisfied wherever the probability $P(n)$ is of appreciable magnitude.

The parameter λ defined in (24) is, by virtue of (2.66), equal to the mean value $\bar{n}$ of n. Thus

$$\lambda = \bar{n}. \tag{26}$$

Note incidentally that, for a *given* value of λ or $\bar{n}$, the condition (25) or (20), requiring that $p \ll 1$, becomes increasingly well satisfied as $N \to \infty$. The Poisson distribution thus becomes always applicable in this limit.

† When N is large and $\lambda \gg 1$, the Poisson distribution (23) reduces properly to a Gaussian distribution for values of n not too far removed from λ.

Explicit verification that $\lambda = \bar{n}$

The result (26) follows also directly from the Poisson distribution (23). Using the definition of the mean value we have

$$\bar{n} \equiv \sum_{n=0}^{N} P(n)n = e^{-\lambda} \sum_{n=0}^{N} \frac{\lambda^n}{n!} n.$$

No significant error results if this sum is extended to infinity since $P(n)$ becomes negligibly small when n is large. Noting that the term with $n = 0$ vanishes, we then obtain, putting $k = n - 1$,

$$\bar{n} = e^{-\lambda} \sum_{n=1}^{\infty} \frac{\lambda^n}{(n-1)!} = e^{-\lambda} \sum_{k=0}^{\infty} \frac{\lambda^{k+1}}{k!}$$

$$= e^{-\lambda} \lambda \sum_{k=0}^{\infty} \frac{\lambda^k}{k!} = e^{-\lambda} \lambda e^{\lambda}$$

since the last sum is simply the series expansion of the exponential function. Thus

$$\bar{n} = \lambda. \tag{27}$$

A.3 *Magnitude of Energy Fluctuations*

Consider two macroscopic systems A and A' in thermal interaction with each other. We shall use the notation of Sec. 4.1 and examine more closely the probability $P(E)$ that A has an energy between E and $E + \delta E$. In particular, we should like to investigate the behavior of $P(E)$ near the energy $E = \tilde{E}$ where it is maximum.

For this purpose we examine the slowly varying logarithm of $P(E)$ given by (4.6),

$$\ln P(E) = \ln C + \ln \Omega(E) + \ln \Omega'(E'), \tag{28}$$

and expand it in a Taylor series about the value $\tilde{E}$. Introducing the energy difference

$$\epsilon \equiv E - \tilde{E}, \tag{29}$$

the Taylor series for $\ln \Omega(E)$ becomes

$$\ln \Omega(E) = \ln \Omega(\tilde{E}) + \left[\frac{\partial \ln \Omega}{\partial E}\right]\epsilon + \frac{1}{2}\left[\frac{\partial^2 \ln \Omega}{\partial E^2}\right]\epsilon^2. \tag{30}$$

Here the square brackets indicate that the derivatives are all evaluated for $E = \tilde{E}$. Terms involving powers of ϵ greater than ϵ^2 have been neglected. By introducing the abbreviations

$$\beta \equiv \left[\frac{\partial \ln \Omega}{\partial E}\right] \tag{31}$$

and

$$\gamma \equiv -\left[\frac{\partial^2 \ln \Omega}{\partial E^2}\right] = -\left[\frac{\partial \beta}{\partial E}\right], \tag{32}$$

Eq. (30) can be written in the simple form

$$\ln \Omega(E) = \ln \Omega(\tilde{E}) + \beta\epsilon - \tfrac{1}{2}\gamma\epsilon^2. \tag{33}$$

The minus sign has been introduced in the definition (32) for the sake of convenience so that the parameter γ will turn out to be positive [in accordance with (4.32)].

A similar Taylor series can readily be written down for $\ln \Omega'(E')$, where $E' = E^* - E$. Expanding about the value $\tilde{E}' \equiv E^* - \tilde{E}$, we have

$$E' - \tilde{E}' = -(E - \tilde{E}) = -\epsilon.$$

Hence we obtain, analogously to (30),

$$\ln \Omega'(E') = \ln \Omega'(\tilde{E}') + \beta'(-\epsilon) - \tfrac{1}{2}\gamma'(-\epsilon)^2 \tag{34}$$

where $$\beta' \equiv \left[\frac{\partial \ln \Omega'}{\partial E'}\right]$$

and $$\gamma' \equiv -\left[\frac{\partial^2 \ln \Omega'}{\partial E'^2}\right] = -\left[\frac{\partial \beta'}{\partial E'}\right]$$

are defined analogously to (31) and (32) in terms of the derivatives evaluated at $E' = \tilde{E}'$. Adding (33) and (34), we then obtain

$$\ln \{\Omega(E)\Omega'(E')\} = \ln \{\Omega(\tilde{E})\Omega'(\tilde{E}')\} + (\beta - \beta')\epsilon - \tfrac{1}{2}(\gamma + \gamma')\epsilon^2. \tag{35}$$

At the value $E = \tilde{E}$ where $P(E) = C\Omega(E)\Omega'(E')$ is maximum, it follows by (4.8) that $\beta = \beta'$; hence the term linear in ϵ vanishes, as it should. Thus (28) can be written in the form

$$\ln P(E) = \ln P(\tilde{E}) - \tfrac{1}{2}\gamma_0\epsilon^2$$

or

$$\boxed{P(E) = P(\tilde{E})e^{-(1/2)\gamma_0(E-\tilde{E})^2}} \tag{36}$$

where† $$\gamma_0 \equiv \gamma + \gamma'. \tag{37}$$

The result (36) shows that the value of γ_0 must be positive to guarantee that the probability $P(E)$ has a maximum (rather than a minimum) at $E = \tilde{E}$. Indeed, it shows explicitly that $P(E)$ becomes negligibly small compared to its maximum value when $\tfrac{1}{2}\gamma_0(E - \tilde{E})^2 \gg 1$, i.e., when $|E - \tilde{E}| \gg \gamma_0^{-1/2}$. In other words, it is very improbable that

† Note that our whole argument has been similar to that used in Appendix A.1 and that (36) is indeed a Gaussian distribution.

the energy of A lies far outside the range $\tilde{E} \pm \Delta E$, where†

$$\Delta E = \gamma_0^{-1/2}. \tag{38}$$

The order of magnitude of ΔE can readily be estimated by using the definition (32) of γ and the approximate expression (3.38) for $\Omega(E)$. Thus we can write for any ordinary system with a ground state energy E_0

$$\ln \Omega \sim f(E - E_0) + \text{constant}.$$

Hence we obtain by the definitions (31) and (32), evaluated for $E = \tilde{E} = \bar{E}$,

$$\beta = \left[\frac{\partial \ln \Omega}{\partial E}\right] \sim \frac{f}{\bar{E} - E_0}$$

and

$$\gamma = -\left[\frac{\partial \beta}{\partial E}\right] \sim \frac{f}{(\bar{E} - E_0)^2} \sim \frac{\beta^2}{f}. \tag{39}$$

This last relation shows explicitly that γ is positive. It also shows that, for the given value of β at which the systems are in equilibrium with each other, the smaller system (i.e., the one with the smaller number of degrees of freedom) has the larger value of γ. In (37) the magnitude of γ_0 is thus predominantly determined by the *smaller* of the two systems. Suppose, for example, that A is much smaller than A' so that $\gamma \gg \gamma'$ and $\gamma_0 \approx \gamma$. Then it follows from (38) and (39) that

$$\Delta E \sim \frac{\bar{E} - E_0}{\sqrt{f}}. \tag{40}$$

Since f is a very large number in the case of a macroscopic system, Eq. (40) shows that the *relative* magnitude $\Delta E/(\bar{E} - E_0)$ of energy fluctuations is very small. This result is discussed in greater detail in Sec. 4.1 where Eq. (4.10) is based on (40).

† Since (36) depends only on the absolute value $|E - \tilde{E}|$ and is thus symmetric about the value $\tilde{E}$, the mean value of the energy must be equal to $\tilde{E}$, i.e., $\bar{E} = \tilde{E}$. This result is simply that already obtained in Eq. (A.17) for any Gaussian distribution. Similarly, it follows from (A.17) that ΔE in (38) is equal to the standard deviation of the energy E.

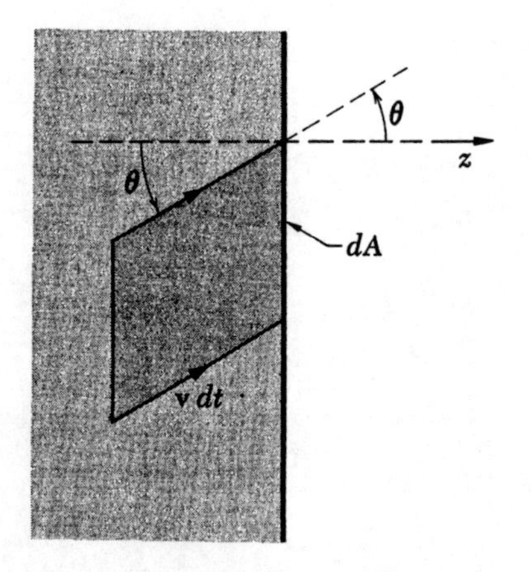

Fig. A.3 Molecules, with velocity between v and v + dv, colliding with an element of area dA of a wall. (Note that the height of the cylinder shrinks to zero as $dt \to 0$.)

A.4 *Molecular Impacts and Pressure in a Gas*

Consider a dilute gas in equilibrium. The number of molecules striking a small area dA of the container wall can then be readily calculated exactly. Choose the z axis so that it points along the outward normal of this area dA, as shown in Fig. A.3. Focus attention first on those molecules, located in the immediate vicinity of the wall, which have a velocity between $\mathbf{v}$ and $\mathbf{v} + d\mathbf{v}$. Such molecules travel a distance $\mathbf{v}\, dt$ during an infinitesimal time dt. Hence all such molecules which lie within the infinitesimal cylinder of cross-sectional area dA and of length $\mathbf{v}\, dt$ strike the wall within the time interval dt; the molecules lying outside this cylinder do not.† If θ denotes the angle between $\mathbf{v}$ and the z direction, the volume of this cylinder is merely

$$dA\, v\, dt \cos\theta = dA\, v_z\, dt$$

where $v_z \equiv v\cos\theta$ is the z component of the velocity $\mathbf{v}$. The mean number of molecules having a velocity between $\mathbf{v}$ and $\mathbf{v} + d\mathbf{v}$ and contained in this cylinder is thus

$$[f(\mathbf{v})\, d^3\mathbf{v}]\,[dA\, v_z\, dt]. \tag{41}$$

where $f(\mathbf{v})\, d^3\mathbf{v}$ is the mean number of molecules, per unit volume, with a velocity between $\mathbf{v}$ and $\mathbf{v} + d\mathbf{v}$. Since (41) gives the number of molecules which strike the area dA in the time dt,

$\mathcal{F}(\mathbf{v})\, d^3\mathbf{v} \equiv$ the mean number of molecules, with velocity between $\mathbf{v}$ and $\mathbf{v} + d\mathbf{v}$, which strike a *unit* area of the wall per *unit* time (42)

is simply obtained by dividing (41) by the area dA and the time dt.

Thus
$$\boxed{\mathcal{F}(\mathbf{v})\, d^3\mathbf{v} = f(\mathbf{v}) v_z\, d^3\mathbf{v}.} \tag{43}$$

Here $f(\mathbf{v})$ is given by the Maxwell distribution (6.21).

The *total* mean number $\mathcal{F}_0$ of molecules striking unit area of the wall per unit time can then be obtained by summing (i.e., integrating) (42) over all possible velocities of those molecules which strike the wall,

† Since the length $v\, dt$ of the cylinder can be considered arbitrarily small, only molecules located in the immediate vicinity of the wall are involved in this argument. Thus $v\, dt$ can be made much smaller than the molecular mean free path so that collisions between molecules need not be considered; i.e., any molecule located in the cylinder and traveling toward the wall will indeed strike the wall without being deflected by a collision before it gets there.

i.e., over all velocities for which v_z is positive so that the molecules move *toward* the wall and therefore collide with it. Thus†

$$\mathcal{F}_0 = \int_{v_z>0} f(\mathbf{v}) v_z \, d^3\mathbf{v}. \tag{44}$$

The result (43) also allows us to compute readily the mean force per unit area (or pressure) exerted by the molecules of the gas. The argument is merely an exact version of that given in Sec. 1.6. A molecule of velocity $\mathbf{v}$ has a z component of momentum equal to mv_z. Hence, the mean z component of momentum carried per unit time to unit area of the wall by all the molecules traveling *toward* the wall is given by multiplying the mean number $\mathcal{F}(\mathbf{v}) \, d^3\mathbf{v}$ of molecules of (42) by mv_z and summing over all molecules traveling toward the wall; i.e., this mean momentum is given by

$$\int_{v_z>0} \mathcal{F}(\mathbf{v}) \, d^3\mathbf{v} \, (mv_z) = m \int_{v_z>0} f(\mathbf{v}) v_z^2 \, d^3\mathbf{v}. \tag{45}$$

Since there is no preferred direction in the gas in equilibrium, the mean z component of momentum of molecules reflected from the wall must be equal and opposite to the mean z component of momentum (45) of molecules incident on the wall. The *net* mean z component of momentum transferred to unit area of the wall per unit time is thus merely twice the result (45); i.e., in accordance with Newton's second law, the mean force per unit area (or pressure) on the wall is given by

$$\bar{p} = 2m \int_{v_z>0} f(\mathbf{v}) v_z^2 \, d^3\mathbf{v}. \tag{46}$$

But $f(\mathbf{v})$ depends only on $|\mathbf{v}|$ so that the integrand has the same value for $+v_z$ as for $-v_z$. Hence the integral is just one-half as large as would be the case if it were extended over *all* values of $\mathbf{v}$ without restriction. Thus we can write

$$\bar{p} = m \int f(\mathbf{v}) v_z^2 \, d^3\mathbf{v} = mn \, \overline{v_z^2} \tag{47}$$

where
$$\overline{v_z^2} = \frac{1}{n} \int f(v) v_z^2 \, d^3\mathbf{v}$$

is, by definition, the mean value of $\overline{v_z^2}$. But, by symmetry,

$$\overline{v_x^2} = \overline{v_y^2} = \overline{v_z^2}$$

so that
$$\overline{v^2} = \overline{v_x^2} + \overline{v_y^2} + \overline{v_z^2} = 3\overline{v_z^2}.$$

† By integrating over all angles, (44) can be written in the form $\mathcal{F}_0 = \frac{1}{4} n \bar{v}$ where n is the mean number of molecules per unit volume and $\bar{v}$ is their mean speed.

Hence (47) becomes

$$\boxed{\bar{p} = \tfrac{1}{3}nm\overline{v^2} = \tfrac{2}{3}n\overline{\epsilon^{(k)}}} \tag{48}$$

where $\overline{\epsilon^{(k)}} = \frac{1}{2}m\overline{v^2}$ is the mean kinetic energy of a molecule. The relation (48) differs from the result (1.19) of our former crude calculation merely by containing $\overline{v^2}$ instead of $\bar{v}^2$. Since $\overline{\epsilon^{(k)}} = \frac{3}{2}kT$ by the equipartition theorem, (48) yields

$$\bar{p} = nkT, \tag{49}$$

the familiar equation of state for an ideal gas.

Mathematical Notes

Mathematical Notes

M.1 The Summation Notation

Suppose that x denotes a variable which can assume the discrete values $x_1, x_2, \ldots, x_m$. Then the sum

$$x_1 + x_2 + \cdots + x_m \equiv \sum_{i=1}^{m} x_m \tag{1}$$

is conveniently abbreviated by the compact notation on the right of the identity sign. Note that the symbol i used as a labeling index is quite arbitrary. One might equally well use any different symbol, say k, and write the definition (1) as

$$\sum_{i=1}^{m} x_i = \sum_{k=1}^{m} x_k.$$

Double sums are easily handled by this notation. For example, suppose that y is a variable which assumes the discrete values $y_1, y_2, \ldots, y_n$. Then the sum of the product $x_i y_j$ over all possible values of x and y is given by

$$\begin{aligned}\sum_{i=1}^{m}\sum_{j=1}^{n} x_i y_j &= x_1(y_1 + y_2 + \cdots + y_n)\\ &\quad + x_2(y_1 + y_2 + \cdots + y_n)\\ &\quad + \cdots\\ &\quad + x_m(y_1 + y_2 + \cdots + y_n)\\ &= (x_1 + x_2 + \cdots + x_m)(y_1 + y_2 + \cdots + y_n)\end{aligned}$$

or

$$\boxed{\sum_{i=1}^{m}\sum_{j=1}^{n} x_i y_j = \left(\sum_{i=1}^{m} x_i\right)\left(\sum_{j=1}^{n} y_j\right).} \tag{2}$$

M.2 Sum of a Geometric Series

Consider the sum

$$S_n \equiv a + af + af^2 + \cdots + af^n. \tag{3}$$

Here the right side is a *geometric series* where each term is obtained from the preceding one as a result of multiplication by f. This factor f

may be real or complex. To evaluate the sum (3), we multiply both sides by f to obtain

$$fS_n = af + af^2 + \cdots + af^n + af^{n+1}. \tag{4}$$

Subtraction of (4) from (3) then yields

$$(1 - f)S_n = a - af^{n+1}$$

or

$$\boxed{S_n = a\frac{1 - f^{n+1}}{1 - f}.} \tag{5}$$

If $|f| < 1$ and the geometric series (3) is infinite so that $n \to \infty$, the series converges. Indeed, in this case $f^{n+1} \to 0$ so that (3) becomes, for $n \to \infty$,

$$\boxed{S_\infty = \frac{a}{1 - f}.} \tag{6}$$

M.3 *Derivative of* ln $n!$ *for large* n

Consider $\ln n!$ when n is any large integer. Since $\ln n!$ changes then only by a small fraction of itself if n is changed by a small integer, it can be regarded as an almost continuous function of n. Increasing n by an increment of unity, we obtain

$$\frac{d\ln n!}{dn} \approx \frac{\ln (n+1)! - \ln n!}{1} = \ln\left[\frac{(n+1)!}{n!}\right] = \ln (n+1).$$

Since $n \gg 1$, $n + 1 \approx n$. Hence we obtain the general result that

if $n \gg 1$,

$$\boxed{\frac{d\ln n!}{dn} \approx \ln n.} \tag{7}$$

Remark

More generally, the derivative of $\ln n!$ can be defined in terms of any small integral increment m by the relation

$$\frac{d\ln n!}{dn} = \frac{\ln (n+m)! - \ln n!}{m}.$$

Hence $$\frac{d\ln n!}{dn} = \frac{1}{m}\ln\left[\frac{(n+m)!}{n!}\right] = \frac{1}{m}\ln [(n+m)(n+m-1)\cdots(n+1)].$$

Since $m \ll n$, we obtain then

$$\frac{d\ln n!}{dn} \approx \frac{1}{m}\ln [n^m] = \ln n$$

which agrees with (7).

M.4 *Value of* ln n! *for large n*

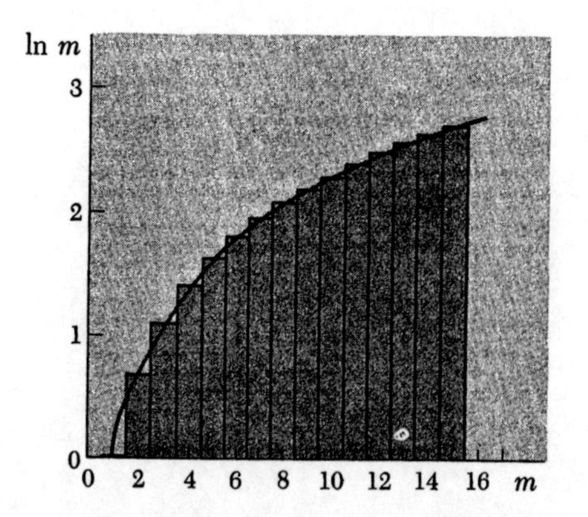

Fig. M.1 Behavior of ln m as a function of m.

Since the calculation of $n!$ becomes very laborious if n is large, we should like to find a simple approximation useful for computing $n!$ in this case. By its definition

$$n! = 1 \times 2 \times 3 \times \cdots \times (n-1) \times n.$$

Hence
$$\ln n! = \ln 1 + \ln 2 + \cdots + \ln n = \sum_{m=1}^{n} \ln m. \tag{8}$$

If n is large, all terms in the sum (8) (except the first few which are also the smallest) correspond to values of m large enough so that $\ln m$ varies only slightly when m is increased by unity. The sum (8) (given by the area under the rectangles of Fig. M.1) can then be approximated with little error by an integral (giving the area under the continuous curve of Fig. M.1). With this approximation (8) becomes

$$\ln n! \approx \int_1^n \ln x \, dx = \Big[x \ln x - x\Big]_1^n. \tag{9}$$

Hence,

if $n \gg 1$,
$$\boxed{\ln n! \approx n \ln n - n} \tag{10}$$

since the contribution of the lower limit in (9) is then negligible.

A better approximation, which is good within less than one per cent error in $n!$ even if n is as small as 10, is given by *Stirling's formula*

$$\ln n! = n \ln n - n + \tfrac{1}{2} \ln (2\pi n). \tag{11}$$

When n is quite large, $n \gg \ln n$ so that Stirling's formula reduces to the simpler form (10).

Note also that (10) implies the result

$$\frac{d \ln n!}{dn} = \ln n + n\left(\frac{1}{n}\right) - 1 = \ln n$$

which agrees with (7).

M.5 *The Inequality* $\ln x \le x - 1$

We wish to compare $\ln x$ with x itself for positive values of x. Consider the difference function

$$f(x) \equiv x - \ln x. \tag{12}$$

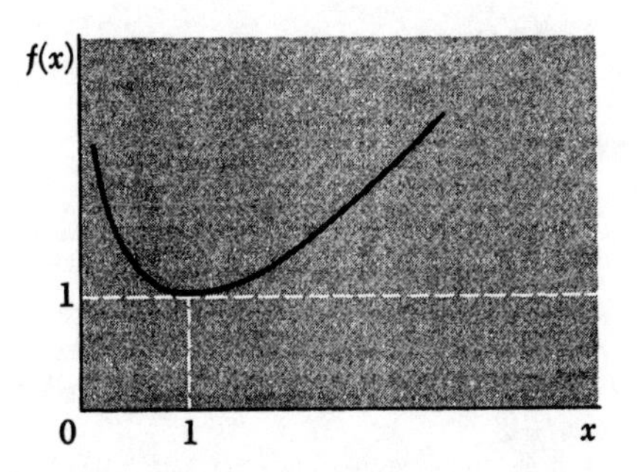

Fig. M.2 The function $f(x) \equiv x - \ln x$ as a function of x.

$$\left.\begin{array}{lll} \text{As } x \to 0, & \ln x \to -\infty; & \text{hence } f(x) \to \infty. \\ \text{As } x \to \infty, & \ln x \ll x; & \text{hence } f(x) \to \infty. \end{array}\right\} \tag{13}$$

To investigate the behavior of $f(x)$ between these limits, we note that

$$\frac{df}{dx} = 1 - \frac{1}{x} = 0 \qquad \text{for } x = 1. \tag{14}$$

Since $f(x)$ is a continuous function of x satisfying (13) and having a single extremum at $x = 1$, it follows that $f(x)$ must have the appearance shown in Fig. M.2 with a minimum at $x = 1$. Hence

$$f(x) \ge f(1) \qquad (= \text{sign if } x = 1),$$

or by (12),

$$x - \ln x \ge 1.$$

Thus

$$\ln x \le x - 1 \qquad (= \text{sign if } x = 1). \tag{15}$$

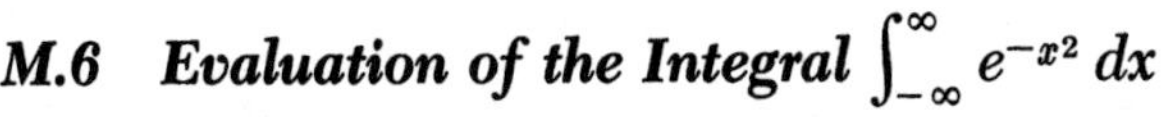

M.6 *Evaluation of the Integral* $\int_{-\infty}^{\infty} e^{-x^2}\,dx$

The *indefinite* integral $\int e^{-x^2}\,dx$ cannot be evaluated in terms of elementary functions. Let I denote the desired *definite* integral

$$I \equiv \int_{-\infty}^{\infty} e^{-x^2}\,dx. \tag{16}$$

This integral can be evaluated by exploiting the properties of the exponential function. Thus we can equally well write (16) in terms of a different variable of integration, i.e.,

$$I \equiv \int_{-\infty}^{\infty} e^{-y^2}\,dy. \tag{17}$$

Multiplication of (16) and (17) then yields

$$I^2 = \int_{-\infty}^{\infty} e^{-x^2}\,dx \int_{-\infty}^{\infty} e^{-y^2}\,dy$$

$$= \int_{-\infty}^{\infty}\int_{-\infty}^{\infty} e^{-x^2}e^{-y^2}\,dx\,dy$$

or

$$I^2 = \int_{-\infty}^{\infty}\int_{-\infty}^{\infty} e^{-(x^2+y^2)}\,dx\,dy. \tag{18}$$

This double integral extends thus over the entire xy plane.

Let us express the integration over this plane in terms of polar coordinates r and φ. Then one has simply $x^2 + y^2 = r^2$, and the element of area in these coordinates is given by $(r\,dr\,d\varphi)$. In order to cover the entire plane, the variables φ and r must range over the values $0 < \varphi < 2\pi$ and $0 < r < \infty$. Hence (18) becomes

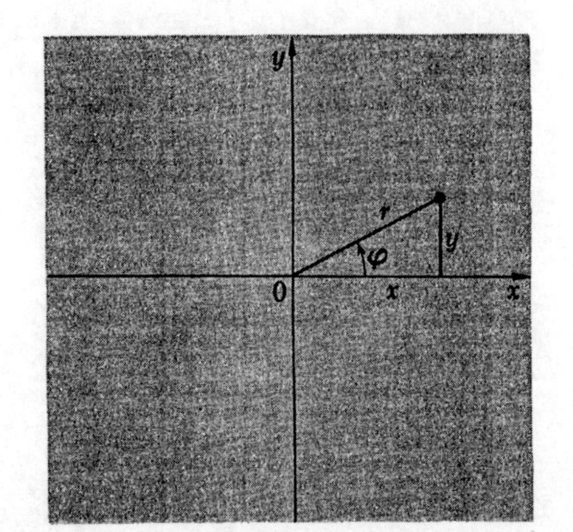

Fig. M.3 The polar coordinates r and φ used for evaluating the integral (18).

$$I^2 = \int_0^{\infty}\int_0^{2\pi} e^{-r^2}r\,dr\,d\varphi = 2\pi\int_0^{\infty} e^{-r^2}r\,dr \tag{19}$$

since the integration over φ is immediate. But the factor r in this integrand makes the evaluation of this last integral trivial. Thus

$$I^2 = 2\pi\int_0^{\infty}(-\tfrac{1}{2})\,d(e^{-r^2}) = -\pi\Big[e^{-r^2}\Big]_0^{\infty} = -\pi(0-1) = \pi$$

or

$$I = \sqrt{\pi}.$$

Thus

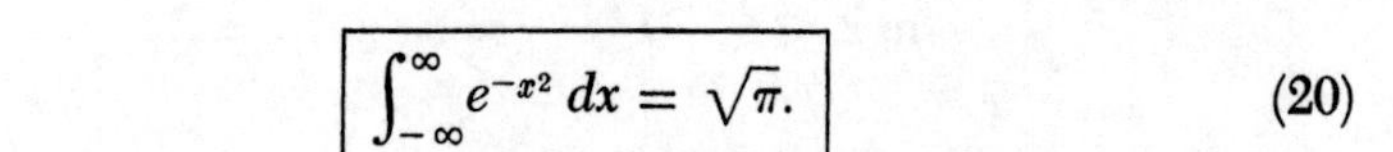

$$\boxed{\int_{-\infty}^{\infty} e^{-x^2}\,dx = \sqrt{\pi}.} \tag{20}$$

Since e^{-x^2} assumes the same value for x and $-x$, it also follows that

$$\int_{-\infty}^{\infty} e^{-x^2}\,dx = 2\int_0^{\infty} e^{-x^2}\,dx.$$

Hence

$$\int_0^{\infty} e^{-x^2}\,dx = \tfrac{1}{2}\sqrt{\pi}. \tag{21}$$

M.7 Evaluation of Integrals of the Form $\int_0^\infty e^{-\alpha x^2} x^n \, dx$

Denote the desired integral by

$$I_n \equiv \int_0^\infty e^{-\alpha x^2} x^n \, dx. \tag{22}$$

Putting $x = \alpha^{-1/2} y$, the integral becomes for $n = 0$

$$I_0 = \alpha^{-1/2} \int_0^\infty e^{-y^2} \, dy = \frac{\sqrt{\pi}}{2} \alpha^{-1/2} \tag{23}$$

where we have used the result (21). Similarly,

$$I_1 = \alpha^{-1} \int_0^\infty e^{-y^2} y \, dy = \alpha^{-1} \Big[-\tfrac{1}{2} e^{-y^2} \Big]_0^\infty = \tfrac{1}{2} \alpha^{-1}. \tag{24}$$

All other integrals where n is any integer such that $n \geq 2$ can then be calculated in terms of I_0 or I_1 by successive integrations by parts. Indeed,

$$\begin{aligned} \int_0^\infty e^{-\alpha x^2} x^n \, dx &= -\frac{1}{2\alpha} \int_0^\infty d(e^{-\alpha x^2}) x^{n-1} \\ &= -\frac{1}{2\alpha} \Big[e^{-\alpha x^2} x^{n-1} \Big]_0^\infty + \frac{n-1}{2\alpha} \int_0^\infty e^{-\alpha x^2} x^{n-2} \, dx. \end{aligned}$$

Since the integrated term vanishes at both limits, we thus obtain

$$\boxed{I_n = \left(\frac{n-1}{2\alpha} \right) I_{n-2}.} \tag{25}$$

For example,

$$I_2 = \frac{I_0}{2\alpha} = \frac{\sqrt{\pi}}{4} \alpha^{-3/2}. \tag{26}$$

Supplementary Problems

Supplementary Problems

P.1 *Simple application of the Gaussian approximation*

A coin is tossed 400 times. Find the probability of obtaining 215 heads.

P.2 *Gaussian probability density*

Consider an ideal system of N spins $\frac{1}{2}$, where each spin has a magnetic moment μ_0 which points up or down with probabilities p and q, respectively. Using the relation (2.74) and the Gaussian approximation (A.13) valid for large N, write down the Gaussian approximation for the probability $\mathcal{P}(M)\,dM$ that the total magnetic moment of the system has a value between M and $M + dM$.

P.3 *Accuracy of the Gaussian approximation*

To examine the extent of validity of the Gaussian approximation (A.13), evaluate the expression (A.7) up to terms involving y^3.

(*a*) Show that (A.9) can then be written in the form

$$\tilde{P}(n) = \tilde{P}e^{-(1/2)z^2} \exp\left[-\frac{p-q}{6(Npq)^{1/2}}z^3\right] \tag{i}$$

where

$$z \equiv \frac{y}{\sqrt{Npq}} \equiv \frac{n - Np}{\sqrt{Npq}}. \tag{ii}$$

(*b*) The first exponential factor makes the probability P negligibly small whenever $|z| \gg 1$. Hence it follows that P has appreciable magnitude only where $|z| \lesssim 1$, and in this domain the argument of the second exponential in (i) is much smaller than unity if $\sqrt{Npq} \gg 1$. Hence this exponential can be expanded in a power series. Show, therefore, that P can be written in the form

$$P(n) = \frac{1}{\sqrt{2\pi Npq}}e^{-(1/2)z^2}\left[1 - \frac{p-q}{6(Npq)^{1/2}}z^3 + \cdots\right]. \tag{iii}$$

(*c*) Show that the fractional error committed by using the simple Gaussian approximation is only of the order of $(Npq)^{-1/2}$ which becomes negligibly small to the extent that N is large enough so that $Npq \gg 1$. Show further that in the symmetric case when $p = q$, the correction term in (iii) vanishes and the fractional error is only of the order of $(Npq)^{-1}$.

P.4 *Properties of the Poisson distribution*

Consider the Poisson distribution (A.23).

(*a*) Verify that this distribution is properly normalized in the sense that $\sum_n P(n) = 1$.

(*b*) Calculate the dispersion of n and show that it is equal to λ.

P.5 ***Occurrence of misprints***

Assume that the typographical errors committed by a typesetter occur completely at random. Suppose that a book of 600 pages contains 600 such misprints. Use the Poisson distribution to calculate the probability

(*a*) that a page contains no misprints;

(*b*) that a page contains at least three misprints.

P.6 ***Radioactive decay***

Consider the alpha particles emitted by a radioactive source during some time interval t. One can imagine this time interval to be subdivided into many small intervals of length Δt. Since the alpha particles are emitted at random times, the probability of a radioactive disintegration occurring during any such time Δt is completely independent of what disintegrations occur at other times. Furthermore, Δt can be imagined to be chosen small enough so that the probability of more than one disintegration occurring during a time Δt is negligibly small. This means that there is some probability p of one disintegration occurring during a time Δt (with $p \ll 1$, since Δt was chosen small enough) and probability $(1 - p)$ of no disintegration occurring during this time. Each such time interval Δt can then be regarded as an independent trial, there being a total of $N = (t/\Delta t)$ such trials during a time t.

(*a*) Show that the probability $P(n)$ that n disintegrations occur during a time t is given by a Poisson distribution.

(*b*) Suppose that the strength of the radioactive source is such that the mean number of disintegrations per minute is 24. What is the probability of obtaining n counts in a time interval of 10 seconds? Obtain approximate numerical values for all integral values of n from 0 to 8.

P.7 ***Molecular collisions in a gas***

Imagine time to be subdivided into many small intervals of duration Δt. There is then a very small probability p that a molecule in a gas suffers a collision during any one such interval.

(*a*) Apply the Poisson distribution to show that the probability P_N of a molecule surviving N such consecutive time intervals without suffering a collision is simply $P_N = e^{-Np}$.

(*b*) Putting $p = w\,\Delta t$ (where w is the probability per unit time of suffering a collision) and expressing N in terms of the elapsed time t, show that the probability $P(t)$ that a molecule survives a time t without suffering a collision is given by $P(t) = e^{-wt}$. Compare this result with the one obtained in Prob. 8.12 by different reasoning.

P.8 *Thickness fluctuations of a thin film*

A metal is evaporated in vacuum from a hot filament. The resultant metal atoms are incident upon a quartz plate some distance away and form there a thin metallic film. This quartz plate is maintained at a low temperature so that any incident metal atom sticks at its place of impact without further migration. The metal atoms can be assumed to impinge with equal likelihood upon any element of area of the plate.

If one considers an element of quartz area of size b^2 (where b is of the order of a metal-atom diameter), show that the number of metal atoms piled up on this area should be approximately distributed according to a Poisson distribution. Suppose that one evaporates enough metal to form a film of mean thickness corresponding to 6 atomic layers. What fraction of the quartz area then is not covered by metal at all? What fraction is covered by metal layers 3 atoms thick and 6 atoms thick, respectively?

P.9 *Accuracy of the Poisson distribution*

To investigate the extent of validity of the Poisson distribution, carry the approximations made in Sec. A.2 to next higher order.

(*a*) By using the explicit expression for $N!/(N-n)!$ and expanding its logarithm, show that

$$\frac{N}{(N-n)!} \approx N^n \exp\left[-\frac{n(n-1)}{2N}\right].$$

(*b*) By expanding $\ln(1-p)$ to terms involving p^2, find an improved approximation for $(1-p)^{N-n}$.

(*c*) Hence show that the binominal distribution can be approximated by

$$P(n) \approx \frac{\lambda^n}{n!} e^{-\lambda} \exp\left[\frac{n-(n-\lambda)^2}{2N}\right].$$

(*d*) Use this result to show that the Poisson distribution is valid to the extent that $\lambda \ll N^{1/2}$ and $n \ll N^{1/2}$, the fractional error committed being smaller than, or of the order of, $(\lambda^2 + n^2)/N$.

P.10 ***Energy fluctuations of systems in thermal contact***

Consider two macroscopic systems A and A' in thermal equilibrium at the absolute temperature T. Let C and C' denote their respective heat capacities (when their external parameters are kept constant).

(*a*) Use the results (A.32) and (A.37) to show that the standard deviation of the energy E of system A is equal to

$$\Delta E = kT\left[\frac{CC'}{k(C + C')}\right]^{1/2}.$$

(*b*) What is ΔE when $C' \gg C$?

(*c*) Suppose that A and A' are both monatomic ideal gases containing N and N' molecules respectively. Calculate the relative magnitude of energy fluctuations $\Delta E/\bar{E}$, where $\bar{E}$ is the mean energy of system A.

(*d*) Examine the expression of part (*c*) in the limiting cases when $N' \gg N$ and when $N' \ll N$. Does your result agree with what you would expect for ΔE in the limit as $N' \to 0$?

Mathematical Symbols

$=$	is equal to
$\equiv$	is (by definition) identically equal to
$\approx$	is approximately equal to
$\sim$	is of the order of
$\propto$	is proportional to
$\neq$	is not equal to
$\not\approx$	is not close to
$>$	is greater than
$\gg$	is much greater than
$\ggg$	is very very much greater than
$\geq$	is greater than or equal to
$\gtrsim$	is greater than or approximately equal to
$<$	is smaller than
$\ll$	is much smaller than
$\lll$	is very very much smaller than
$\leq$	is smaller than or equal to
$\lesssim$	is smaller than or approximately equal to
$\exp u$	e^u
$\ln u$	natural logarithm of u (to the base e)

Greek Alphabet

A	α	alpha
B	β	beta
Γ	γ	gamma
Δ	δ, ∂	delta
E	ϵ	epsilon
Z	ζ	zeta
H	η	eta
Θ	θ, ϑ	theta
I	ι	iota
K	κ, $\varkappa$	kappa
Λ	λ	lambda
M	μ	mu
N	ν	nu
Ξ	ξ	xi
O	o	omicron
Π	π	pi
P	ρ	rho
Σ	σ	sigma
T	τ	tau
Υ	υ	upsilon
Φ	ϕ, φ	phi
X	χ	chi
Ψ	ψ	psi
Ω	ω	omega

Numerical Constants

Physical Constants

Quantity	*Value*	*Error*
Elementary charge	$e = 4.80298 \times 10^{-10}$ esu	± 7
	$= 1.60210 \times 10^{-19}$ coulomb	± 2
Speed of light in vacuum	$c = 2.997925 \times 10^{10}$ cm sec^{-1}	± 1
Planck's constant,	$h = 6.62559 \times 10^{-27}$ erg sec	± 16
$\hbar \equiv h/2\pi$	$\hbar = 1.054494 \times 10^{-27}$ erg sec	± 25
Electron rest mass	$m_e = 9.10908 \times 10^{-28}$ gm	± 13
Proton rest mass	$m_p = 1.67252 \times 10^{-24}$ gm	± 3
Bohr magneton, $e\hbar/2m_ec$	$\mu_B = 9.2732 \times 10^{-21}$ erg gauss^{-1}	± 2
Nuclear magneton, $e\hbar/2m_pc$	$\mu_N = 5.05050 \times 10^{-24}$ erg gauss^{-1}	± 13
Avogadro's number	$N_a = 6.02252 \times 10^{23}$ mole^{-1}	± 9
Boltzmann's constant	$k = 1.38054 \times 10^{-16}$ erg deg^{-1}	± 6
Gas constant	$R = 8.31434 \times 10^{7}$ ergs deg^{-1} mole^{-1}	± 35
	$= 1.98717$ calories deg $^{-1}$ mole^{-1}	± 8

Conversion Factors

Quantity	*Value*	*Error*
Triple point of water	$\equiv 273.16$°K	by definition
Celsius temperature	X°C $\equiv (273.15 + X)$°K	by definition
1 atmosphere $\equiv$ 760 mm Hg	$\equiv 1.01325 \times 10^{6}$ dynes cm^{-2}	by definition
1 joule	$\equiv 10^{7}$ ergs	by definition
1 thermochemical calorie	$\equiv 4.184$ joules	by definition
1 electron volt (ev)	$= 1.60210 \times 10^{-12}$ erg	± 2
1 ev per particle	$= 23.061$ kilocalories/mole	± 1
	$= 11604.9$°K	± 5

SOURCE: The values are those given by E. R. Cohen and J. W. M. DuMond, *Rev. Mod. Phys.* **37**, 589–591 (October 1965).

Each estimated error is one standard deviation applied to the final digits of the preceding column.

The mole is defined in accordance with the modern convention where ^{12}C is assigned the atomic weight 12.

The abbreviation "deg" refers to "degree Kelvin" (also denoted by °K).

Answers to Problems

Chapter 1

1.1 $\frac{1}{32}$, $\frac{5}{32}$, $\frac{10}{32}$, $\frac{10}{32}$, $\frac{5}{32}$, $\frac{1}{32}$.
1.2 Irreversible.
1.5 (*a*) Energy unchanged; (*b*) $p_f/p_i = V_i/V_f$.
1.6 2.1×10^{23} molecules $\sec^{-1}$ cm^{-2}.
1.7 Approximately 45 days.
1.8 6×10^{-10} sec.
1.10 (*a*) $\bar{v}_1/\bar{v}_2 = (m_2/m_1)^{1/2}$.
1.11 $\bar{p} = \frac{2}{3}(n_1 + n_2)\,\bar{\epsilon}$.
1.12 (*c*) Pressure $= \bar{p}_1 = \bar{p}_2$.
1.13 (*c*) 2 atmospheres.
1.14 (*a*) 3.4×10^4 cm/sec; (*b*) 2.3×10^{-8} cm;
(*c*) $(2.9 \times 10^4)x$ dynes/cm^2; (*d*) 2×10^{-9} cm.

Chapter 2

2.1 $5/54 \approx 0.092$.

2.2 $63/256 \approx 0.25$.

2.3 (*a*) $(5/6)^5 \approx 0.4$; (*b*) $1 - (5/6)^5 \approx 0.6$; (*c*) $\frac{1}{3}(5/6)^4 \approx 0.16$.

2.4 (*c*) $(5/6)^N$; (*d*) $(5/6)^{N-1}(1/6)$.

2.5 (*a*) $N![n!n'!]^{-1}p^n q^{n'}$.

2.6 (*a*) $N![(\frac{1}{2}N)!]^{-2}(\frac{1}{2})^N$; (*b*) 0.

2.7 (*b*) 0; (*c*) $(t/\tau)^{1/2}l$.

2.9 (*a*) $\bar{\mu} = (2p - 1)\mu_0$, $\overline{\mu^2} = \mu_0^2$.

2.12 (*a*) $1/\bar{c}$; (*b*) $\overline{(1/c)}$.

2.13 (*c*) $\overline{M} = 0$, $\overline{(\Delta M)^2} = 2Np\mu_0^2$.

2.15 (*a*) $N(V/V_0)$; (*b*) $N^{-(1/2)}[(V_0/V) - 1]^{1/2}$.

2.16 (*a*) $(t/\Delta t)pe$; (*b*) $(t/\Delta t)pe^2$; (*d*) 4×10^{-12} amp.

2.17 $(N^2v^2/R)p^2[1 + (1 - p)/Np]$.

2.18 0.82 cm.

2.19 (*a*) 0; (*b*) $N^{1/2}l$.

2.20 $(A^2 - x^2)^{-1/2}(dx/\pi)$ for $-A \le x \le A$; 0 otherwise.

Chapter 3

3.1 (*a*) $P(-3\mu_0) = \frac{1}{7}$, $P(\mu_0) = \frac{6}{7}$, $P(M) = 0$ otherwise; (*b*) $(\frac{3}{7})\mu_0$;
(*c*) same as in (*a*) and (*b*).

3.2 $\frac{1}{3}$.

3.3 (*a*) $N![n!(N-n)!]^{-1}$; (*c*) n'/n.

3.4 $(n'/n)^2$.

3.5 (*a*) $N![n!(N-n)!]^{-1}$; (*b*) $(E_r - E_0)/2\mu_0 B$; (*d*) $(n'/n)^{\Delta n}$;
(*e*) $\beta = \ln (n/n')/2\mu_0 B$.

3.6 (*a*) $(\pi^2\hbar^2/2m)(n_x^2/L_x^2)(2/L_x)$; (*b*) $\bar{F} = \frac{2}{3}(\bar{E}/L)$.

3.7 (*a*) 1.9×10^{29}; (*b*) 4.5×10^{18}.

3.9 (*a*) $N![n!(N-n)!]^{-1}(\delta E/2\mu_0 B)$.

Chapter 4

4.1 (*a*) No; (*b*) no.

4.2 0.025 electron volt.

4.3 (*a*) 4 percent; (*b*) 5×10^{43}.

4.4 1.1×10^{-2}.

4.5 1.5×10^{-5}.

4.6 Power $\propto T^{-1}$.

4.7 (*a*) $N_{3/2}/N_{1/2} \approx 0.5$; (*b*) $N_{3/2}/N_{1/2} \approx 1$.

4.8 (*a*) For $T \to 0$, $\bar{E} \to N\epsilon_1$; for $T \to \infty$, $\bar{E} \to \frac{1}{2}N(\epsilon_1 + \epsilon_2)$; change when $kT \sim (\epsilon_2 - \epsilon_1)$.
(*b*) $N[\epsilon_1 + \epsilon_2 e^{-\beta(\epsilon_2 - \epsilon_1)}][1 + e^{-\beta(\epsilon_2-\epsilon_1)}]^{-1}$.

4.9 $Na \tanh (Wa/kT)$.

4.10 $\frac{1}{2}Nea \tanh (ea\mathcal{E}/2kT)$.

4.12 (*a*) Separation between levels increases; (*b*) increases; (*c*) positive; (*d*) increases; (*e*) increases.

4.13 (*a*) Separation between levels increases; (*b*) decreases; (*c*) negative; (*d*) increases; (*e*) increases.

4.14 $(N_1 + N_2)kT/V$.

4.21 (*a*) $e^{\beta\mu_0 B} + e^{-\beta\mu_0 B}$; (*b*) $-\mu_0 B \tanh (\beta\mu_0 B)$.

4.22 (*a*) $e^{-\beta\hbar\omega/2}[1 - e^{-\beta\hbar\omega}]^{-1}$; (*b*) $\hbar\omega[\frac{1}{2} + (e^{\beta\hbar\omega} - 1)^{-1}]$; (*d*) $\frac{1}{2}\hbar\omega$; (*e*) kT.

4.23 (*a*) $2A/\beta\hbar^2$; (*b*) kT.

4.25 (*a*) $e^{-\beta n\epsilon}$; (*b*) $N![n!(N - n)!]^{-1}$ in each case.

4.26 (*a*) $C(M/2\pi\beta)^{3/2}(V/\hbar^3)$;
(*b*) $C[(M/2\pi\beta)^{3/2}(V/\hbar^3)][(m/2\pi\beta)^{3/2}(V/\hbar^3)]e^{-\beta\mu}$;
(*c*) $(mkT/2\pi\hbar^2)^{3/2}Ve^{-u/kT}$; (*d*) $(mkT/2\pi\hbar^2)^{3/4}(V/N)^{1/2}e^{-u/2kT}$;
(*e*) undissociated; (*f*) dissociated.

4.27 (*a*) $(\bar{n}/N)^2 = (m/2\pi)^{3/2}\hbar^{-3}(kT)^{5/2}\,\bar{p}^{-1}e^{-u/kT}$; (*b*) 0.4 percent.

4.28 (*a*) $\frac{3}{2}NkT$.

4.29 (*a*) $-N\mu_0 B \tanh (\mu_0 B/kT)$; (*b*) $N\mu_0 \tanh (\mu_0 B/kT)$.

Chapter 5

5.1 (*a*) 3°K; (*b*) 4×10^{-3} °K.
5.2 (*a*) 0.62 gauss; (*b*) 2×10^{-8} °K.
5.3 $\nu RT \ln (V_2/V_1)$.
5.4 3.6×10^{10} ergs.
5.5 $a \to c \to b$: $W = 7 \times 10^9$ ergs, $Q = -2.9 \times 10^{10}$ ergs;
$a \to d \to b$: $W = 2.1 \times 10^{11}$ ergs, $Q = 1.8 \times 10^{11}$ ergs;
$a \to b$: $W = 1.4 \times 10^{11}$ ergs, $Q = 1.1 \times 10^{11}$ ergs.
5.8 (*a*) Mg/A; (*b*) $T = \frac{3}{5}T_0 + (2MgV_0/5\nu RA)$,
$V = \frac{2}{5}V_0 + (3\nu RAT_0/5Mg)$.
5.9 (*b*) 14.8 joules/cm.
5.10 (*a*) 9.92×10^3 joules; (*b*) 1.35×10^3 joules.
5.11 (*b*) $\bar{E} = N[\epsilon_1 + \epsilon_2 e^{-(\epsilon_2-\epsilon_1)/kT}][1 + e^{-(\epsilon_2-\epsilon_1)/kT}]^{-1}$,
$C = (N/kT^2)(\epsilon_2 - \epsilon_1)^2 e^{-(\epsilon_2-\epsilon_1)/kT}[1 + e^{-(\epsilon_2-\epsilon_1)/kT}]^{-2}$.
5.12 (*a*) $-N\mu_0 B$, 0; (*b*) 0, 0; (*c*) $-N\mu_0 B \tanh (\mu_0 B/kT)$;
(*d*) $Nk(\mu_0 B/kT)^2[\cosh (\mu_0 B/kT)]^{-2}$.
5.13 (*a*) $2N\epsilon(e^{\epsilon/kT} + 2)^{-1}$;
(*b*) $(2N\epsilon^2/kT^2)e^{\epsilon/kT}(e^{\epsilon/kT} + 2)^{-2}$, $(2N\epsilon^2/9kT^2)$ for large T.
5.14 (*a*) $(C_A T_A + C_B T_B)/(C_A + C_B)$; (*b*) $\Delta S = C_A \ln (T/T_A) + C_B \ln (T/T_B)$.
5.15 (*a*) 1.27×10^3 joules/deg, -1.12×10^3 joules/deg,
1.5×10^2 joules/deg; (*b*) 1.1×10^2 joules/deg.
5.16 (*a*) 21.8 joules/deg; (*b*) $10^{6.8\times10^{24}}$.
5.17 (*a*) 12.6°C; (*b*) 12.8 joules/deg; (*c*) 9.4×10^3 joules.
5.18 (*a*) 0; (*b*) 0; (*c*) 0; (*d*) independent.
5.19 (*a*) $S_n = S_s$; (*b*) $C_n = \frac{1}{3}C_s$.
5.20 (*a*) $Nk(\hbar\omega/kT)^2 e^{\hbar\omega/kT}(e^{\hbar\omega/kT} - 1)^{-2}$; (*c*) Nk.
5.21 (*a*) $\frac{5}{2}kT$; (*b*) $\frac{5}{2}R = 20.8$ joule deg^{-1} mole^{-1}.
5.22 (*a*) $(\partial^2 \ln Z/\partial\beta^2) + (\partial \ln Z/\partial\beta)^2$; (*d*) $(\frac{3}{2}N)^{-1/2}$.

Chapter 6

6.2 (*a*) $e^{-\beta[p^2/2m+mgz]}\,d^3\mathbf{r}\,d^3\mathbf{p}$; (*b*) $e^{-(1/2)\beta mv^2}\,d^3\mathbf{v}$; (*c*) $e^{-\beta mgz}\,dz$.
6.3 $n \propto e^{-\beta mgz}$.
6.4 (*a*) $V[\ln(R/r_0)]^{-1}r^{-1}$; (*b*) $n \propto (r/R)^{-\beta eV/\ln(R/r_0)}$; (*c*) $kT \gg e^2n^{1/3}$.
6.5 (*a*) $\omega^2 r(m-\rho v)$; (*b*) $e^{-(1/2)\beta\omega^2r^2(m-\rho v)}\,dr$.
6.6 (*a*) $e^{\beta\mu_0(B_2-B_1)}$; (*b*) $\cosh(\beta\mu_0B_2)/\cosh(\beta\mu_0B_1)$;
(*c*) $1+(\mu_0/kT)^2(B_2{}^2-B_1{}^2)$; (*d*) 1.00015.
6.7 $\frac{1}{2}kT$, no.
6.8 (*a*) $2^{-(1/2)}$; (*b*) unchanged.
6.9 Greater.
6.10 $4V(\ln 2)/A\bar{v}$.
6.11 Approximately 4 seconds.
6.12 (*a*) $(c_2/c_1)(m_1/m_2)^{1/2}$.
6.13 $2^{(1-\sqrt{\mu_{\text{He}}/\mu_{\text{Ne}}})}$.
6.14 (*a*) 0; (*b*) kT/m; (*c*) 0; (*d*) 0; (*e*) $(kT/m)(1+b^2)$.
6.15 (*a*) ν_0; (*b*) $\nu_0{}^2(kT/mc^2)$.
6.16 R.
6.17 $\rho \propto T$.
6.18 (*a*) Mg/α; (*b*) kT/α; (*c*) $(\alpha kT)^{1/2}/g$.
6.19 (*a*) $\frac{1}{2}kT$; (*b*) $\frac{1}{4}kT$; (*c*) $\frac{3}{4}kT$; (*d*) $\frac{3}{4}R$.
6.20 R.
6.21 (*a*) $\bar{E} = 3N\hbar\omega[\frac{1}{2}+(e^{\hbar\omega/kT}-1)^{-1}]$;
(*b*) $3R(\hbar\omega/kT)^2e^{\hbar\omega/kT}(e^{\hbar\omega/kT}-1)^{-2}$; (*f*) $3R(\Theta/T)^2e^{-\Theta/T}$; (*h*) $\hbar\omega/k$.

Chapter 7

7.2 (*a*) 6.21 atmospheres; (*b*) 832°K.
7.3 (*a*) $(c_V/R)(\bar{p}_f V_f - \bar{p}_i V_i)$; (*b*) $\nu c_V(T_f - T_i)$.
7.4 (*b*) $\frac{5}{2}R = 20.8$ joules deg^{-1} mole^{-1}; (*c*) $\frac{5}{3}$.
7.5 (*a*) $\frac{5}{2}R$; (*b*) 1300 joules; (*c*) 1500 joules; (*d*) 23.6 joules/deg.
7.6 $\frac{3}{2}R \ln [(T/T_0)(V/V_0)^{2/3}]$ where T and V are given by answer to Prob. 5.8.
7.8 (*a*) 195°K;
(*b*) 3.12×10^4 joules/mole for sublimation,
2.55×10^4 joules/mole for vaporization;
(*c*) 5.7×10^3 joules/mole.
7.9 Zero.
7.10 $[(L/RT) - \frac{1}{2}]T^{-1}$.
7.11 (*a*) $(RT_r/L)(\mathcal{Q}/\mathcal{V})$.

(*b*) $$\left[\frac{1}{T_0} - \frac{R}{L}\ln\left(\frac{RT_r}{Lp_0}\frac{\mathcal{Q}}{\mathcal{V}}\right)\right]^{-1}$$

(*c*) 1.4°K.
7.15 (*a*) $T_i/(T_i - T_o)$; (*b*) 11.9.
7.16 (*a*) $C(T_1 + T_2 - 2T_f)$; (*b*) $T_f \geq (T_1T_2)^{1/2}$; (*c*) $C(T_1^{1/2} - T_2^{1/2})^2$.
7.17 (*a*) $\nu RT \ln (V_c/V_b)$; (*b*) $\nu RT \ln (V_a/V_d)$; (*c*) $V_a/V_b = V_d/V_c$;
(*d*) $q'/q = T'/T$.
7.18 $1 - (V_1/V_2)^{\gamma - 1}$.

Chapter 8

8.1 (*a*) 2; (*b*) 2; (*c*) unaffected.
8.2 (*a*) τ; (*b*) τ; (*c*) unaffected.
8.3 (*a*) $\frac{1}{2}t_c$; (*b*) $\frac{1}{2}t_c$; (*c*) at_c, $\frac{1}{2}at_c = a\tau$; (*d*) $\frac{1}{2}at_c^2$, $\frac{1}{2}at_c = a\tau$.
8.4 Decreases, unchanged.
8.5 (*a*) $2\pi\eta R^3 L\omega/\delta$; (*b*) Approximately 25 dyne cm.
8.6 1.4×10^{-4} gm cm^{-1} sec^{-1}.
8.7 (*a*) $\sigma \propto V^{-4/(s-1)}$; (*b*) $\eta \propto T^{(s+3)/2(s-1)}$.
8.8 (*a*) 1.4 watts/cm; (*b*) 4×10^{-3} mm of Hg.
8.9 (*a*) $(\eta_1/\eta_2)(\mu_2/\mu_1)^{1/2}$; (*b*) $(\eta_2/\eta_1)(\mu_1/\mu_2)$; (*c*) $(\eta_2/\eta_1)(\mu_1/\mu_2)$; (*e*) $d_1 \approx 1.9 \times 10^{-8}$ cm, $d_2 \approx 3.1 \times 10^{-8}$ cm.
8.10 Approximately 10 hours.
8.11 Approximately $\frac{3}{2}$ (ln 2)$(M/m)(n\bar{v}L^2)^{-1}$.
8.13 (*c*) $2\tau^2$.
8.14 $\dfrac{\partial T}{\partial t} = \dfrac{\kappa}{\rho c}\dfrac{\partial^2 T}{\partial z^2}$.
8.15 $\Delta T = (I^2R/2\pi b\kappa) \ln (b/a)$.
8.16 (*a*) $(\pi/8)(\rho a^4/\eta L)(p_1 - p_2)$; (*b*) $(\pi/16)(\mu a^4/\eta RTL)(p_1^2 - p_2^2)$.

Supplementary Problems

P.1 0.013.

P.2 $(2\mu_0)^{-1}(2\pi Npq)^{-1/2} \exp \{-[M - N(p - q)\mu_0]^2/8Npq\mu_0^2\}$.

P.5 (*a*) 0.37; (*b*) 0.08.

P.8 0.0025, 0.090, 0.162.

P.10 (*b*) $T(kC)^{1/2}$; (*c*) $[\frac{2}{3}N'/N(N + N')]^{1/2}$;
(*d*) $(\frac{2}{3}N)^{1/2}$ if $N \ll N'$, $N^{-1}(2N'/3)^{1/2}$ if $N \gg N'$.

Index

Entries followed by the letter *n* refer to a footnote, those followed by the letter *d* to a list of definitions.

D

E

F

J

K

L

M

N

O

P

T

U

V

W

Z